Lecture Notes in Mathematics

A collection of informal reports and seminars
Edited by A. Dold, Heidelberg and B. Eckmann, Zürich

280

Conference on the Theory of Ordinary and Partial Differential Equations

Held in Dundee/Scotland, March 28-31, 1972

Edited by W. N. Everitt and B. D. Sleeman
University of Dundee, Dundee/Scotland

Springer-Verlag
Berlin · Heidelberg · New York 1972

AMS Subject Classifications (1970): 34 A xx, 34 B xx, 34 C 10, 34 D 10, 34 D 15, 34 D 20, 34 H 05, 34 J 99, 35 A 20, 35 A 35, 35 A 40, 35 D 10, 35 F xx, 35 B 20, 35 G xx, 35 B 25, 35 J xx, 35 K 05, 35 P 05, 35 Q 10, 35 Q 15, 46 E 20, 47 A 10, 47 A 55, 47 F 05, 47 E 05, 49 A 20, 49 A 10, 81 A 09, 81 A 10, 93 D 05.

ISBN 3-540-05962-8 Springer-Verlag Berlin · Heidelberg · New York
ISBN 0-387-05962-8 Springer-Verlag New York · Heidelberg · Berlin

Offsetdruck: Julius Beltz, Hemsbach/Bergstr.

PREFACE

These Proceedings form a record of the lectures delivered at the Conference on the Theory of Ordinary and Partial Differential Equations held at the University of Dundee, Scotland during the four days 28 to 31 March 1972.

The Conference was attended by 140 mathematicians from the following countries Belgium, Canada, Denmark, France, Germany, Ireland, Italy, The Netherlands, Poland, Sweden, Switzerland, the United Kingdom and the United States of America.

Invited lectures were delivered by B. L. J. Braaksma (Groningen), L. Collatz (Hamburg), R. Conti (Florence), G. Fichera (Rome), C. Goulaouic (Paris), E. M. de Jager (Amsterdam), K. Jörgens (Munich), H. W. Knobloch (Würzburg), C. Lobry (Bordeaux), E. Mohr (Berlin), C. Olech (Warsaw), Å. Pleijel (Uppsala), G. Stampacchia (Pisa), F. Stummel (Frankfurt-am-Main), W. Walter (Karlsruhe), W. Wendland (Darmstadt).

Contributed lectures formed an important part of the work of the Conference and a list of these will be found in the contents of these Proceedings.

The Conference was organised by the following committee: W. N. Everitt (Chairman), J. S. Bradley, B. D. Sleeman and I. M. Michael. Dr. Sleeman and Dr. Michael acted as Organising Secretaries for the Conference.

The Committee takes this opportunity to thank all mathematicians who took part in the work of the Conference for their contribution. The Committee thanks the University of Dundee for providing facilities which made it possible to hold the Conference in Dundee, and to many members of the University for freely giving of their help and advice, and to colleagues in the Department of Mathematics.

In particular the Committee wish to thank Mrs. Norah Thompson, Secretary in the Department of Mathematics, for considerable assistance over several weeks which made possible the preparation of the papers for the Conference and many of the manuscripts of these Proceedings.

Some of the expenses involved in organising the conference were met by the balance of a NATO (Scientific Affairs Division) grant for the conference on differential equations held at the University of Edinburgh in 1967.

W.N. Everitt B.D. Sleeman
Editors

CONTENTS

Invited lectures

Contributed lectures

Lectures whose proceedings do not appear here

E.A.Catchpole:	**An initial value problem for an integro-differential equation**
C.Comstock:	**Dirichlet conditions for the wave equation**
F.Cooper:	**A maximum principle for equations with non-local potentials**
E.Hille:	**On a class of non-linear second-order differential equations**
R.Kress:	**Boundary value problems for inhomogeneous harmonic tensor fields**
H.Lemei:	**Integral transforms related to a class of even order linear differential equations**
A.H.M.Levelt:	**Algebraic invariants for singular points of linear ordinary differential equations with analytic coefficients**
J.H.Miller:	**New algorithms for checking the stability of differential equations**
E.Mohr:	**Das spectrum von eigenwertproblemen elliptischer differentialgleichungen**
C.Olech:	**On global asymptotic stability on the plane**
L.A.Peletier:	**Downstream asymptotic behaviour of solutions of the Prandtl boundary layer equations**
F.Ursell:	**On the exterior problems of acoustics**

LIST OF SPEAKERS

Invited Speakers

B. L. J. Braaksma: Mathematical Institute, PO Box 800, Groningen, The Netherlands.

L. Collatz: Institut für Angewandte Mathematik, Universität Hamburg, 2 Hamburg 13, Rothenbaumchaussee 41, West Germany.

R. Conti: Universita Degli Studi, Instituto Matematico, Viale Morgagni 67/A, Firenze, Italy.

G. Fichera: Ordinario Nell'Universita di Roma, Via Pietro Mascagni 7, Roma, Italy.

C. Goulaouic: Departement de Mathematiques, Faculte de Sciences, Universite de Paris, 91 - Orsay, France.

E. M. de Jager: Mathematisch Instituut, Universiteit Van Amsterdam, Roetersstraat 15, Amsterdam, The Netherlands.

K. Jörgens: Mathematisches Institut, Universität München, 8 München 13, Schellingstrasse 2-8, West Germany.

H. W. Knobloch: Mathematisches Institut der Universität Würzburg, D-87 Würzburg, Klinikstrasse 6, West Germany.

C. Lobry: Universite de Bordeaux 1, 251 Cours de la Liberation, 33 Talence, France.

E. Mohr:	Technische Universität Berlin, Institut für Matematik, 1 Berlin 12, Strasse des 17 Juni 135, West Germany.
C. Olech:	Mathematical Institute of the Polish Academy of Sciences, Warszawa/Poland, Sniadeckich 8.
Å. V. C. Pleijel:	Mathematiska Institutionen, Sysslomansgatan 8, 75223 Uppsala, Sweden.
G. Stampacchia:	Sculo Normale Superiore, Piazza dei Cavalieri, 56100 Pisa, Italy.
F. Stummel:	Johann Wolfgang Goethe-Universität, Mathematisches Seminar, 6 Frankfurt am Main, Robert-Mayer-Strasse 10, West Germany.
W. Walter:	Mathematisches Institut I, Universität Karlsruhe, 75 Karlsruhe, West Germany.
W. Wendland:	Mathematisches Institut, Der Technischen Hochschule Darmstadt, 61 Darmstadt, Hochschulstrasse 1 West Germany.

Other Speakers (contributed lectures)

F. M. Arscott: Department of Mathematics, University of Reading, Whiteknights, Reading, Berkshire, England.

M. Authier: Département de Mathématiques, Académie de Reims, Moulin de la Housse, Reims 51, Boîte Postale 347, France.

J. Barros-Neto: School of Mathematics, Institute for Advanced Study, Princeton, New Jersey 08540, USA.

I. E. C. Bennewitz: Department of Mathematics, Uppsala University, Sysslomansgatan 8, S-75223, Sweden.

J. Bona: Department of Mathematics, University of Essex, Wivenhoe Park, Colchester, Essex, England.

J. S. Bradley: Department of Mathematics, University of Tennessee, Knoxville, Tennessee 37916, USA.

D. W. Bresters: Institute of Applied Mathematics, University of Amsterdam, Roetersstraat 15, Amsterdam, The Netherlands.

E. A. Catchpole: Department of Mathematics, University of Essex, Wivenhoe Park, Colchester, Essex, England.

C. Comstock: Department of Mathematics, Naval Postgraduate School, Monterey, California, 93940, USA.

C. C. Conley: IBM T J Watson Research Center, PO Box 218, Yorktown Heights, New York 10598, USA.

F. H. Cooper: School of Mathematical and Physical Sciences, University of Sussex, Falmer, Brighton, Sussex, England.

J. M. Cushing: IBM T J Watson Research Center, PO Box 218, Yorktown Heights, New York 10598, USA.

M. S. P. Eastham: Department of Mathematics, Chelsea College of Science and Technology, Manresa Road, London SW3, England.

M. R. Essén: Department of Mathematics, Royal Institute of Technology, Stockholm 70, Sweden.

W. N. Everitt: Department of Mathematics, University of Dundee, Dundee, Scotland.

W. Förster: Department of Mathematics, University of Dundee, Dundee, Scotland.

M. Giertz: Department of Mathematics, Royal Institute of Technology, Stockholm 70, Sweden.

P. P. N. de Groen: Wiskundig Seminarium, Vrije Universiteit, de Boelelaan 1081, Amsterdam-Buitenveldert, The Netherlands.

H. P. Heinig: Mathematical Institute, University of Edinburgh, Chambers Street, Edinburgh, Scotland.

E. Hille: Department of Mathematics and Statistics, University of New Mexico, Albuguergue, New Mexico 87106, USA.

H. B. Keller: Department of Applied Mathematics, California Institute of Technology, Pasadena, California 91109, USA.

K. Kreith: Department of Mathematics, University of California, Davis, California 95616, USA.

R. Kress: Institut für Numerische und Angewandte Mathematik, D 34 Göttingen, Bürgerstrasse 32, West Germany.

H. Lemei: Afdeling Wiskunde, T H Delft, Julianalaan 132, Delft, The Netherlands.

A. H. M. Levelt: Mathematisch Institut, Toernooiveld, Nijmegen, The Netherlands.

N. G. Lloyd: Department of Pure Mathematics, University of Cambridge, 16 Mill Lane, Cambridge, England.

D. L. Lovelady: Department of Mathematics, University of South Carolina, Columbia, South Carolina 29208, USA.

J. B. McLeod: Wadham College, Oxford, England.

J. J. H. Miller: School of Mathematics, Trinity College, Dublin 2, Ireland.

R. E. O'Malley: Courant Institute of Mathematical Sciences, New York University, New York NY 10012, USA.

L. A. Peletier: School of Mathematical and Physical Sciences, University of Sussex, Falmer, Brighton Sussex, England.

C. J. S. Petrie:	Department of Engineering Mathematics, The University, Newcastle upon Tyne, NE1 7RU England.
J. W. de Roever:	Mathematisch Centrum, 2 Boerhaavestraat 49, Amsterdam (oost) The Netherlands.
B. V. Schmitt:	Departement de Mathematique, Institut de Recherche Mathematique, Avancée de Strasbourg, Rue - René Descartes, 67 - Strasbourg, France.
K. Schmitt:	Department of Mathematics, University of Utah, Salt Lake City, Utah, USA.
C. A. Stuart:	School of Mathematical and Physcial Sciences, University of Sussex, Falmer, Brighton, Sussex, England.
B. D. Sleeman:	Department of Mathematics, University of Dundee, Dundee, Scotland.
K. Unsworth:	Department of Mathematics, Chelsea College of Science and Technology, Manresa Road, London SW3, England.
F. Ursell:	Department of Mathematics, University of Manchester, Manchester M13 9PL, England.
P. Waltman:	Department of Mathematics, University of Iowa, Iowa City, Iowa 5 2240, USA.
J. R. L. Webb:	School of Mathematical and Physical Sciences, University of Sussex, Falmer, Brighton, Sussex, England.
D. H. Wood:	New London Laboratory, Naval Underwater Systems Center, New London, Connecticut 06320, USA.

Recessive Solutions of Linear Differential Equations with Polynomial Coefficients

B.L.J. Braaksma

0. Introduction

We consider the differential equation

$$(0.1)\qquad \frac{d^n y}{dx^n} + \sum_{j=1}^{n} a_j(x) \frac{d^{n-j} y}{dx^{n-j}} = 0,$$

where $a_j(x)$ is a polynomial of degree m_j $(j = 1, \ldots, n)$. We assume that

$$(0.2)\qquad j^{-1} m_j < n^{-1} m_n \ , \quad j = 1, \ldots, n-1,$$

and that the leading coefficient of $a_n(x)$ is $(-1)^{n+1}$.

The purpose of this paper is to construct and investigate a solution of (0.1) which behaves as

$$\exp\left(-\frac{n}{m_n+n}\ x^{\frac{m_n+n}{n}}\right)$$

as $x \to \infty$ on $|\arg x| < \frac{n+1}{m_n+n}\pi$ and which is an entire function of the coefficients of the polynomials $a_j(x)$. This solution is recessive on the sector $|\arg x| < \frac{1}{2}\frac{n}{m_n+n}\pi$.

Let

$$(0.3)\qquad a_n(x) = (-1)^{n+1} \ \{P(x) + \lambda Q(x)\},$$

where $P(x)$ and $Q(x)$ are polynomials of degree $p = m_n$ and q respectively, $p > q$, and with leading coefficients equal to one. We show that the solution mentioned above is an entire function of λ of order $\frac{p+n}{n(p-q)}$.

The case $n = 2$, $q = 0$ has been considered by P.F. Hsieh and Y. Sibuya [1] and Y. Sibuya [2].

In section 1 we give some preliminary estimates for eigenvalues associated with (0.1). In section 2 we transform (0.1) to system (2.13) where D is diagonal and B satisfies (2.16). Then the solution is constructed by means of an integral equation in section 3. In sections 4 and 5 we deduce estimates as $\lambda \to \infty$ analogously to Sibuya [2].

1. Estimates for eigenvalues

A system equivalent to (0.1) is

$$\frac{d\mathfrak{y}}{dx} = A(x)\mathfrak{y}, \tag{1.1}$$

where

$$A(x) = \begin{pmatrix} 0 & 1 & & & \bigcirc \\ 0 & 0 & \ddots & & \\ & & & \ddots & \\ & & & & 1 \\ -a_n(x) & -a_{n-1}(x) & & & -a_1(x) \end{pmatrix}, \quad \mathfrak{y} = \begin{pmatrix} y \\ y' \\ \vdots \\ y^{(n-1)} \end{pmatrix} \tag{1.2}$$

In this section we derive some estimates for the eigenvalues of A(x), so for the roots of

$$z^n + \sum_{j=1}^{n} a_j(x) z^{n-j} = 0. \tag{1.3}$$

We first consider the function $s(x, \lambda)$ defined by

$$s(x, \lambda) = - \{-a_n(x)\}^{\frac{1}{n}} = \{P(x) + \lambda Q(x)\}^{\frac{1}{n}}, \tag{1.4}$$

where we choose the branch which tends to $+\infty$ as $x \to +\infty$.
This function is an analytic function of $x^{\frac{1}{n}}$ for sufficiently large $|x|$, with a pole of order p at ∞. Some estimates are given by

<u>Lemma 1</u>. Let ε be a constant such that $0 < \varepsilon < \pi$.

Then

$$s(x, \lambda) = x^{\frac{p}{n}} (1 + o(1)), \text{ as } x \to \infty, \ \lambda x^{q-p} \to 0, \tag{1.5}$$

$$s(x, \lambda) = \lambda^{\frac{1}{n}} O(1), \text{ as } \lambda \to \infty \tag{1.6}$$

uniformly on any interval $\alpha \leq x \leq \beta$. Furthermore,

$$s^{-1}(x, \lambda) = O(x^{-\frac{p}{n}}), \tag{1.7}$$

$$\frac{s'(x, \lambda)}{s(x, \lambda)} = O\left(\frac{1}{x}\right), \tag{1.8}$$

$$\left\{\frac{s'(x, \lambda)}{s(x, \lambda)}\right\}' = O\left(\frac{1}{x^2}\right), \tag{1.9}$$

in the following cases :

i) as $x \to \infty$ uniformly for λ on any compact set in the complex plane,

ii) as $x \to \infty$, $x > 0$ uniformly for $|\arg \lambda| \leq \pi - \varepsilon$,

iii) as $x \to \infty$, $\arg x = \frac{\pi}{2(p+n)}$ uniformly for $\pi - \varepsilon \leq \arg \lambda \leq \pi$.

Finally, there exists a positive constant x_0 such that for every $x_1 \geq x_0$:

$$(1.10) \quad s^{-1}(x, \lambda) = O(\lambda^{-\frac{1}{n}}), \ \frac{s'(x, \lambda)}{s(x, \lambda)} = O(1), \ \left\{\frac{s'(x, \lambda)}{s(x, \lambda)}\right\}' = O(1)$$

as $\lambda \to \infty$ uniformly on $x_0 \leq |x| \leq x_1$.

All estimates hold uniformly for the variable coefficients of $P(x)$ and $Q(x)$ on compact sets. The function $s(x, \lambda)$ is analytic in these coefficients and in λ for $|\arg \lambda - (p-q) \arg x| \leq \pi - \varepsilon$, if $|x|$ is sufficiently large.

Proof : The proofs of (1.5), (1.6) and (1.10) are straight-forward. The same applies to the proof of (1.7), (1.8) and (1.9) in case i). Therefore we only prove (1.7) - (1.9) in the cases ii) and iii).

If x is sufficiently large positive, then $|\arg \frac{Q(x)}{P(x)}| \leq \frac{1}{2}\varepsilon$.

Hence, if $|\arg \lambda| \leq \pi - \varepsilon$, then for these values of x :

$$(1.11) \quad |\arg \lambda \frac{Q(x)}{P(x)}| \leq \pi - \frac{1}{2}\varepsilon, \ |1 + \lambda \frac{Q(x)}{P(x)}| \geq \sin \frac{1}{2}\varepsilon.$$

In the same way we see that for sufficiently large $|x|$ and $\arg x = \frac{\pi}{2(p+n)}$, $\pi - \varepsilon \leq \arg \lambda \leq \pi$, then

$$0 \leq \arg \lambda \frac{Q(x)}{P(x)} \leq \frac{1}{2(p+n)} \{\pi(p + q + 2n) + \varepsilon\} \leq \pi - \frac{\pi - \varepsilon}{2(p+n)},$$

$$(1.12) \quad |1 + \lambda \frac{Q(x)}{P(x)}| \geq \sin \frac{\pi - \varepsilon}{2(p+n)}.$$

From this and (1.4) we deduce (1.7) in the cases ii) and iii).

To prove (1.8) we write

$$\frac{s'(x,\lambda)}{s(x,\lambda)} = \frac{1}{n}\left\{\frac{P'(x)}{P(x)} + \lambda \frac{Q(x)}{P(x)} \frac{Q'(x)}{Q(x)}\right\}\left\{1 + \lambda \frac{Q(x)}{P(x)}\right\}^{-1} =$$

$$= \frac{1}{n}\left\{\frac{P(x)}{\lambda Q(x)} \frac{P'(x)}{P(x)} + \frac{Q'(x)}{Q(x)}\right\}\left\{1 + \frac{P(x)}{\lambda Q(x)}\right\}^{-1}.$$

We use the first part and second part of this formula if $|\lambda Q(x) / P(x)| \leq 2$ and ≥ 2 respectively. Taking (1.11) and (1.12) into account we obtain (1.8) in the cases ii) and iii). In the same way we may prove (1.9).

We denote the eigenvalues of $A(x)$ by $\mu_1(x,\lambda), \ldots, \mu_n(x,\lambda)$ such that for fixed λ these functions are analytic for sufficiently large $|x|$ and that

(1.13) $$\mu_j(x,\lambda) \sim - x^{\frac{p}{n}} \omega^{j-1} , x \to \infty, \text{ where } \omega = e^{2\pi i/n} .$$

Estimates for these eigenvalues are given by

<u>Lemma 2</u>. Let

(1.14) $$\mu_j(x,\lambda) = - \omega^{j-1} s(x,\lambda) g_j(x,\lambda), \quad j = 1, \ldots, n.$$

Then $g_j(x,\lambda)$ is an analytic function of $x^{\frac{1}{n}}$ in the neighborhood of $x^{\frac{1}{n}} = \infty$ and

(1.15) $$g_j(x,\lambda) = 1 + O(x^{-\frac{1}{n}}), \; g'_j(x,\lambda) = O(x^{-1-\frac{1}{n}}), \quad j = 1, \ldots, n$$

in the cases i), ii) and iii) of lemma 1. Furthermore, there exists a positive constant x_0 such that for every $x_1 \geq x_0$:

(1.16) $$g_j(x,\lambda)=1+O(\lambda^{-\frac{1}{n}}) , \quad g'_j(x,\lambda) = O(\lambda^{-\frac{1}{n}}) , \quad j = 1, \ldots, n,$$

as $\lambda \to \infty$ uniformly on $x_0 \leq |x| \leq x_1$.

The estimates hold uniformly for the variable coefficients of the polynomials $a_1(x), \ldots, a_{n-1}(x)$, $P(x)$ and $Q(x)$ on compact sets.

The functions $g_j(x,\lambda)$ and $\mu_j(x,\lambda)$ ($j = 1, \ldots, n$) are analytic in λ and in these coefficients for $|\arg \lambda-(p - q) \arg x| \leq \pi - \varepsilon$, if $|x|$ is sufficiently large.

<u>Proof</u> : From (0.2), (1.7) and (1.10) we deduce

$$\frac{a_j(x)}{\{s(x,\lambda)\}^j} = O(x^{-\frac{1}{n}}) , \qquad j = 1, \ldots, n - 1,$$

in the cases i), ii) and iii) and

$$\frac{a_j(x)}{\{s(x,\lambda)\}^j} = O(\lambda^{-\frac{j}{n}}) , \qquad j = 1, \ldots, n - 1$$

as $\lambda \to \infty$ uniformly on $x_0 \leq |x| \leq x_1$.

The assertions now easily follow from the relations(cf.(1.3))

$$g_j^n = 1 + \sum_{h=1}^{n-1} \frac{a_h(x)}{\{-\omega^{j-1}s(x,\lambda)\}^h} g_j^{n-h} ,$$

$$\{ng_j^{n-1} - \sum_{h=1}^{n-1} (n - h) \frac{a_h(x)}{\{-\omega^{j-1}s(x,\lambda)\}^h} g_j^{n-h-1} \} g_j' =$$

$$= - \sum_{h=1}^{n-1} \left\{ \frac{a'_h(x)}{\{-\omega^{j-1}s(x,\lambda)\}^h} - h \frac{a_h(x)}{\{-\omega^{j-1}s(x,\lambda)\}^h} \frac{s'(x,\lambda)}{s(x,\lambda)} \right\} g_j^{n-h} ,$$

and lemma 1.

2. Transformation of the differential equation

Let $R(x,\lambda)$ be the matrix of eigenvectors of $A(x)$ defined by

(2.1) $R_{jh}(x,\lambda) = \{\mu_h(x,\lambda)\}^{j-1}$, $j,h = 1,\dots,n.$

Substituting

(2.2) $\hat{y} = R(x,\lambda)u$

in (1.1) we obtain

(2.3) $\frac{du}{dx} = A_1(x,\lambda)u,$

where

(2.4) $\begin{cases} A_1(x,\lambda) = D(x,\lambda) - R^{-1}(x,\lambda)R'(x,\lambda), \\ D(x,\lambda) = \text{diag }\{\mu_1(x,\lambda),\dots,\mu_n(x,\lambda)\}. \end{cases}$

Let $R_0(x,\lambda)$ be defined by

$[R_0(x,\lambda)]_{jh} = \{-\omega^{h-1}s(x,\lambda)\}^{j-1}$, $j,h = 1,\dots n.$

Then

$[R_0^{-1}(x,\lambda)]_{jh} = \frac{1}{n}\,[R_0(x,\lambda)]_{hj}^{-1}$, $j,h = 1,\dots,n.$

Defining $R_1(x,\lambda)$by

$R_1(x,\lambda) = R_0^{-1}(x,\lambda)\,R(x,\lambda)$

we see that

(2.5) $[R_1(x,\lambda)]_{jh} = \frac{1}{n}\sum_{k=0}^{n-1}\{\omega^{h-j}g_h(x,\lambda)\}^k$, $j,h = 1,\dots,n.$

Lemma 2 implies

(2.6) $R_1(x,\lambda) = I + O(x^{-\frac{1}{n}})$, $R_1'(x,\lambda) = O(x^{-1-\frac{1}{n}})$

as $x \to \infty$ in the cases i), ii) and iii) of lemma 1, and

(2.7) $R_1(x,\lambda) = I + O(\lambda^{-\frac{1}{n}})$, $R_1'(x,\lambda) = O(\lambda^{-\frac{1}{n}})$

as $\lambda \to \infty$ uniformly on $x_0 \leq |x| \leq x_1$. Moreover, we easily see that

(2.8) $R_1^{-1}(x,\lambda) = I + O(x^{-\frac{1}{n}})$

as $x \to \infty$ in the cases i), ii) and iii), and

(2.9) $R_1^{-1}(x,\lambda) = I + O(\lambda^{-\frac{1}{n}})$

as $\lambda \to \infty$ uniformly on $x_0 \leq |x| \leq x_1$. In (2.4) we may use the following relation

(2.10) $R^{-1}(x,\lambda)\, R'(x,\lambda) = \frac{s'(x,\lambda)}{s(x,\lambda)}\{\frac{n-1}{2} I + R_1^{-1}(x,\lambda)\, C\, R_1(x,\lambda)\} + R_1^{-1}(x,\lambda) R_1'(x,\lambda)$,

where $C_{jj} = 0$, $C_{jh} = \frac{1}{\omega^{h-j} - 1}$, if $j \neq h$, $j,h = 1,\dots,n$. Hence

$A_1(x,\lambda) = D(x,\lambda) + O(x^{-1}) + O(x^{-\frac{1}{n}})$

as $x \to \infty$ for fixed λ.

However, we need a transformation of (1.1) to an equation of the same type such that the new factor $A(x)$ equals $D(x) + O(x^{-1-\varepsilon})$, $\varepsilon > 0$. This may be accomplished by the substitution

(2.11) $u = \{s(x,\lambda)\}^{-\frac{1}{2}(n-1)}\{I + W(x,\lambda)\}v$,

where $W(x,\lambda)$ is the matrix defined by

(2.12) $W_{jj} = 0$, $W_{jh}(x,\lambda) = \{(\mu_j(x,\lambda) - \mu_h(x,\lambda))(\omega^{h-j} - 1)\}^{-1} \frac{s'(x,\lambda)}{s(x,\lambda)}$,

for $j \neq h$; $j,h = 1,\dots,n$. Then

$D(x,\lambda)W(x,\lambda) - W(x,\lambda)D(x,\lambda) = \frac{s'(x,\lambda)}{s(x,\lambda)} C$.

From this, (2,4) and (2.10) we may deduce that the substitution (2.11) transforms (2.3) into

(2.13) $\frac{dv}{dx} = \{D(x,\lambda) + B(x,\lambda)\}v$,

where

(2.14) $B = (I + W)^{-1}[\frac{s'}{s}\{(I - R_1^{-1})CR_1 + C(I - R_1) - R_1^{-1}CR_1W\} - R_1^{-1}R_1'(I + W) - W']$.

Applying lemmas 1 and 2, (2.5) - (2.9) and (2.12) we infer to

Lemma 3. The transformation

(2.15) $\hat{y} = \{s(x,\lambda)\}^{-\frac{1}{2}(n-1)}R(x,\lambda)\{I + W(x,\lambda)\}v$

where R and W are defined by (2.2) and (2.12), transforms (1.1) into (2.13), where D and B are given by (2.4) and (2.14). The matrix $B(x,\lambda)$ is an analytic function of $x^{\frac{1}{n}}$ in a neighborhood of $x^{\frac{1}{n}} = \infty$. Furthermore,

(2.16) $B(x,\lambda) = O(x^{-1-\frac{1}{n}})$

in the cases i), ii) and iii) of lemma 1, and

(2.17) $B(x,\lambda) = O(\lambda^{-\frac{1}{n}})$

as $\lambda \to \infty$ uniformly on $x_0 \leq |x| \leq x_1$, if x_0 is chosen sufficiently large. These estimates hold uniformly for the variable coefficients of the polynomials $a_1(x),\dots,a_{n-1}(x)$, $P(x)$ and $Q(x)$ on compact sets. The matrix

$B(x,\lambda)$ is an analytic function of these coefficients and λ for

(2.18) $|\arg \lambda - (p - q) \arg x| \leq \pi - \varepsilon$,

if $|x|$ is sufficiently large.

3. The integral equation and its solution

A fundamental matrix of the differential equation $\frac{dv}{dx} = D(x,\lambda)v$ is

$V(x,\lambda) = \text{diag}\,\{\exp \int^x \mu_1(\tau,\lambda)d\tau, \ \ldots, \ \exp \int^x \mu_n(\tau,\lambda)d\tau\}$.

Hence, the solutions of (2.13) are the solutions of

(3. 1) $v(x,\lambda) = V(x,\lambda)c + \int^x \text{diag}\,\{\exp \int_t^x \mu_1(\tau,\lambda)d\tau, \ \ldots, \ \exp \int_t^x \mu_n(\tau,\lambda)d\tau\}\ B(t,\lambda)v(t,\lambda)dt$

where c is an arbitrary n-vector independent of x.

Expanding $\mu_1(x,\lambda)$ in a Laurent series near $x^{\frac{1}{n}} = \infty$ we obtain

(3. 2) $\mu_1(x,\lambda) = \ x^{\frac{p}{n}} \sum_{j=0}^{\infty} \alpha_j(\lambda)x^{-\frac{j}{n}}$, $\alpha_0(\lambda) = -1$,

which converges for sufficiently large $|x|$. Now we choose c such that

(3. 3) $V(x,\lambda)c = \{\exp \chi(x,\lambda)\}\ e_1$

where e_1 denotes the first unit-vector and

(3. 4) $\chi(x,\lambda) = \int_\infty^x \{\mu_1(\tau,\lambda) - \sum_{j=0}^{n+p} \alpha_j(\lambda)\tau^{\frac{p-j}{n}}\}d\tau + \sum_{j=0}^{n+p-1} \frac{n\alpha_j(\lambda)}{n+p-j}\, x^{\frac{n+p-j}{n}} + \alpha_{n+p}(\lambda) \log x$

If

(3. 5) $w(x,\lambda) = \{\exp - \chi(x,\lambda)\}\ v(x,\lambda)$,

we obtain for the kth component $w_k(x,\lambda)$ of $w(x,\lambda)$ the following integral equation

(3. 6) $w_k(x,\lambda) = \delta_{k1} + \sum_{j=1}^{n} \int_{d_k}^{x} \{\exp \int_t^x (\mu_k(\tau,\lambda) - \mu_1(\tau,\lambda))d\tau\}\ B_{kj}(t,\lambda)\ w_j(t,\lambda)dt$,

where d_k will be chosen appropriately . This equation may be solved by iterations.

From the lemmas 1 and 2 we deduce

(3. 7) $f_k(x,t,\lambda) = \int_t^x \{\mu_k(\tau,\lambda) - \mu_1(\tau,\lambda)\}d\tau \ = -\frac{2in}{p+n} (\sin \frac{k-1}{n}\pi)e^{\pi i(k-1)/n}$

$\left(x^{\frac{p+n}{n}} - t^{\frac{p+n}{n}}\right)\ (1 + o(1)\ )$

in the following cases :

i) as $x \to \infty$, $t \to \infty$ uniformly for λ on a compact set,

and

ii) as $x \to \infty$, $\lambda x^{q-p} \to 0$ and

$t \to \infty$, $\lambda t^{q-p} \to 0$.

The function $f_k(x,t,\lambda)$ is analytic in λ and the coefficients of the polynomials $a_1(x), \ldots, a_{n-1}(x)$, $P(x)$ and $Q(x)$, and (3.7) holds uniformly for these coefficients on a compact set.

We now assume that λ and these coefficients belong to a compact set S, and that

$$|\arg x| \leq \frac{n+1}{n+p}\pi - \varepsilon. \tag{3.8}$$

Then there exists a positive constant K such that $\mu_j(x,\lambda)$ $(j=1, \ldots, n)$ and $B(x,\lambda)$ are analytic functions of $x^{\frac{1}{n}}$ for $|x| \geq K$.

If $k \geq 2$, we put in (3.6)

$$x^{\frac{p+n}{n}} = \xi, \quad t^{\frac{p+n}{n}} = \xi(1+\sigma). \tag{3.9}$$

Then

$$|\arg \xi| \leq \pi(1+\frac{1}{n}) - \varepsilon_1, \text{ where } \varepsilon_1 = (n+p)\frac{\varepsilon}{n}. \tag{3.10}$$

Let ϕ be an arbitrary real number such that

$$|\phi| \leq \frac{1}{n}\pi - \frac{1}{2}\varepsilon_1 \text{ and } |\phi - \arg \xi| \leq \pi - \frac{1}{2}\varepsilon_1. \tag{3.11}$$

Such numbers ϕ exist for all values of $\arg \xi$ satisfying (3.10) since $-\frac{1}{n}\pi + \frac{1}{2}\varepsilon_1 \leq \arg \xi + \pi - \frac{1}{2}\varepsilon_1$ and $\arg \xi - \pi + \frac{1}{2}\varepsilon_1 \leq \frac{1}{n}\pi - \frac{1}{2}\varepsilon_1$.

In (3.6) we choose the path of integration such that its image in the σ-plane is the half line L : $\arg \sigma = \phi - \arg \xi$ with starting point $\sigma = \infty$.

Then

$$|\arg \sigma| \leq \pi - \frac{1}{2}\varepsilon_1,$$

$$\arg \xi \leq \arg\{\xi(1+\sigma)\} \leq \phi \leq \frac{1}{n}\pi - \frac{1}{2}\varepsilon_1 \quad \text{if } \arg \xi \leq \phi,$$

$$-\frac{1}{n}\pi + \frac{1}{2}\varepsilon_1 \leq \phi \leq \arg\{\xi(1+\sigma)\} \leq \arg \xi, \text{ if } \arg \xi \geq \phi.$$

Hence, if σ on L then

$$|\arg t| \leq \pi\frac{n+1}{n+p} - \varepsilon \quad \text{and} \quad |t| \geq |x| (\sin \tfrac{1}{2}\varepsilon_1)^{\frac{n}{n+p}}. \tag{3.12}$$

From this, (3.7) and (3.9) we deduce that there exist positive constants K_0 and η such that the function $f_k(x,t,\lambda)$ and $B(t,\lambda)$ are analytic in ξ and σ, and

(3.13) $\operatorname{Re} f_k(x,t,\lambda) \le -\eta\, |\xi\sigma|, \quad k = 2, \ldots, n$

if $|x| \ge K_0$ and $\arg\sigma = \phi - \arg\xi$ for all values of ϕ satisfying (3.11).
Now suppose $r \ge 0$ and $\Psi(x)$ is an analytic function of x on (3.8) for $|x| \ge K_1 \ge K_0$ and

(3.14) $|\psi(x)| \le K_2 |x^{-\frac{r}{n}}|$,

for these values of x. Moreover, let $\psi(x)$ be an analytic function of λ and the coefficients of the polynomials $a_1(x), \ldots, a_{n-1}(x)$, $P(x)$ and $Q(x)$, if these numbers belong to a compact set S and suppose (3.14) holds uniformly on S. Then we deduce from lemma 3, (3.12), (3.13) and (3.14) that there exists a constant K independent of x, the parameters λ etc., K_2 and r such that

(3.15) $\left| \int_{d_k}^{x} \{\exp \int_t^x (\mu_k(\tau,\lambda) - \mu_1(\tau,\lambda))d\tau\} B_{kj}(t,\lambda)\, \psi(t)dt \right| \le KK_2 |x^{-\frac{r+1}{n}}|$

for $|x| \ge K_1$ on (3.8), whereas the integral in (3.15) is analytic in x, λ and the parameters for $|x| \ge K_1$, (3.8), λ and the parameters on S.
It is now easily seen that the usual method of iterations leads to a solution $w(x,\lambda)$ of (3.6) which is analytic in x, λ and the coefficients of the polynomials for sufficiently large $|x|$ on (3.8), and that

(3.16) $w(x,\lambda) = e_1 + O(x^{-\frac{1}{n}})$

as $x \to \infty$ on (3.8) uniformly for λ and these coefficients on compact sets.
If we expand a finite number of the successive approximations asymptotically in powers of $x^{-\frac{1}{n}}$ as $x \to \infty$ we may derive

(3.17) $w(x,\lambda) \sim e_1 + \sum_{j=1}^{\infty} \beta_j(\lambda) x^{-\frac{j}{n}}$

as $x \to \infty$ on (3.8) uniformly for λ and the coefficients on compact sets.
Taking into account the substitutions (1.2), (2.15),(3.5) with (3.4) we obtain

Theorem 1. The differential equation (0.1) has a solution $\tilde{y}(x,\lambda)$ which is an entire function of x, λ and the coefficients of the polynomials $a_1(x), \ldots, a_{n-1}(x)$, P(x) and Q(x) and which admits the asymptotic representation

$$(3.18) \quad \tilde{y}(x,\lambda) \sim x^{\alpha_{n+p}(\lambda) - \frac{1}{2n}(n-1)p} \left\{ \exp \sum_{j=0}^{n+p-1} \frac{n\alpha_j(\lambda)}{n+p-j} x^{\frac{n+p-j}{n}} \right\} \sum_{\nu=0}^{\infty} A_\nu x^{-\frac{\nu}{n}}$$

as $x \to \infty$ uniformly on (3.8) for any positive constant $\varepsilon < \frac{n+1}{n+p}\pi$ and for λ and the coefficients of the polynomials on compact sets.

Here $\alpha_0(\lambda), \ldots, \alpha_{n+p}(\lambda)$ are defined by (3.2) with $\alpha_0 = -1$. The coefficients A_ν are functions of λ, $A_0 = 1$.

Remark. The solution (3.18) is recessive on $|\arg x| < \frac{n}{n+p} \frac{\pi}{2}$ and therefore it is uniquely determined. The existence of recessive solutions and the expansion (3.18) on (3.8) may also be deduced from the results of W.J. Trjitzinsky [3 , section 8] (cf. also W. Wasow [4 , section 15]).

4. Behavior of the solution as $\lambda \to \infty$ on $|\arg \lambda| \leq \pi - \varepsilon$

We first consider this behavior for $x \geq x_0$, where x_0 will be a suitably chosen positive number. From lemma 3 we deduce

$$(4.1) \quad \int_{x_0}^{\infty} ||B(t,\lambda)||dt = o(1),$$

as $\lambda \to \infty$, $|\arg \lambda| \leq \pi - \varepsilon$, if x_0 is a sufficiently large constant. In view of (1.4) and lemma 2 we may choose x_0 so large that

$$|\arg s(x,\lambda)| \leq \frac{1}{n}(\pi - \frac{1}{2}\varepsilon), \quad |g_j(x,\lambda) - 1| \leq \frac{1}{4n}\varepsilon \sin\frac{\pi}{n},$$

if $x \geq x_0$, $|\arg \lambda| \leq \pi - \varepsilon$. Now (3.7) and (1.14) imply that Re $f_k(x,t,\lambda) \leq 0$ if $t \geq x \geq x_0$, $|\arg \lambda| \leq \pi - \varepsilon$. Combining this with (3.6) and (4.1) we deduce

$$(4.2) \quad w(x,\lambda) = e_1 + o(1),$$

as $\lambda \to \infty$ $|\arg \lambda| \leq \pi - \varepsilon$ uniformly for $x \geq x_0$.

Next we consider the function $\chi(x,\lambda)$ in (3.4). Let

$$(4.3) \quad t = \lambda^{-\frac{1}{p-q}} x.$$

Then we infer from (1.4) and lemma 2 that $s(x,\lambda)$ and $g_1(x,\lambda)$ are analytic functions of $t^{1/n}$ for $|t| \geq K$ ($K > 0$ independent of λ).

Furthermore,

$$s(x,\lambda) = \lambda^{\frac{p}{n(p-q)}} t^{\frac{p}{n}} (1 + t^{q-p} + \lambda^{-\frac{1}{p-q}} \sum_{j=1}^{p} p_j(\lambda) t^{-j})^{1/n},$$

$$g_1(x,\lambda) = 1 + \lambda^{-\frac{1}{n(p-q)}} \sum_{j=1}^{\infty} q_j(\lambda) t^{-\frac{j}{n}},$$

hence

$$(4.4)\quad \mu_1(x,\lambda) = -\lambda^{\frac{p}{n(p-q)}} \{(t^p + t^q)^{\frac{1}{n}} + \lambda^{-\frac{1}{n(p-q)}} \sum_{j=1}^{\infty} r_j(\lambda) t^{\frac{p-j}{n}}\},$$

where $p_j(\lambda)$, $q_j(\lambda)$ and $r_j(\lambda)$ are bounded for $|\arg \lambda| \leq \pi - \varepsilon$.

Now (3.2), (4.3) and (4.4) imply

$$(4.5)\quad \begin{cases} \lambda^{-\nu} \alpha_{\nu n(p-q)}(\lambda) = -\binom{\frac{1}{n}}{\nu} + O(\lambda^{-\frac{1}{n(p-q)}}), \text{ if } \nu = 1,2,\ldots, \\ \lambda^{-\frac{j}{n(p-q)}} \alpha_j(\lambda) = O(\lambda^{-\frac{1}{n(p-q)}}), \text{ if } j \neq \nu n(p-q),\ \nu = 0,1,2,\ldots. \end{cases}$$

Let

$$(4.6)\quad \kappa = \frac{n+p}{n(p-q)}.$$

Using the lemmas 1 and 2, and (4.1) we obtain

$$(4.7)\quad \int_{\infty}^{x} \{\mu_1(\tau,\lambda) - \sum_{j=0}^{n+p} \alpha_j(\lambda) \tau^{\frac{p-j}{n}}\} d\tau =$$

$$\lambda^{\kappa} \int_{x\lambda^{-\frac{1}{p-q}}}^{\infty} \{(t^p + t^q)^{\frac{1}{n}} - \sum_{\nu=0}^{[\kappa]} \binom{\frac{1}{n}}{\nu} t^{(p-q)(\kappa-\nu)-1}\} dt + \lambda^{\kappa - \frac{1}{n(p-q)}} O(1),$$

as $\lambda \to \infty$, $|\arg \lambda| \leq \pi - \varepsilon$ uniformly for $x \geq x_0$.

From (2.15), (3.4), (3.5), (4.2), (4.5) and (4.7) we deduce

Theorem 1^a. The solution $\tilde{y}(x,\lambda)$ of theorem 1 satisfies

$$(4.8)\quad \frac{d^k}{dx^k} \tilde{y}(x,\lambda) = \exp [\lambda^{\kappa} \{\delta \log \lambda (1 + O(\lambda^{-\frac{1}{n(p-q)}})) + A(1 + O(\lambda^{-\frac{1}{n(p-q)}}))\}]$$

$k=0,\dots,n-1$, as $\lambda \to \infty$, $\lambda x^{q-p} \to \infty$, $|\arg \lambda| \leq \pi - \varepsilon$, $x \geq x_0$. Here

(4. 9) $\delta = \frac{-1}{p-q}\binom{\frac{1}{n}}{\kappa}$, if $\kappa = 1,2,\dots$ and $\delta = 0$ otherwise,

(4.10) $$A = \int_1^\infty \{(t^p + t^q)^{\frac{1}{n}} - \sum_{\nu=0}^{[\kappa]} \binom{\frac{1}{n}}{\nu} t^{(p-q)(\kappa-\nu)-1}\}\, dt + \int_0^1 (t^p + t^q)^{\frac{1}{n}}\, dt$$

$$- \sum_{0 \leq \nu < \kappa} \frac{1}{(p-q)(\kappa-\nu)} \binom{\frac{1}{n}}{\nu}.$$

Moreover,

(4.11) $$\frac{d^k}{dx^k} \tilde{y}(x,\lambda) = \exp\Big[- \frac{1}{p-q} \{ \sum_{\substack{\nu=0 \\ \nu\neq\kappa}}^{\infty} \frac{1}{\kappa-\nu} \binom{\frac{1}{n}}{\nu} (\lambda x^{q-p})^{\nu} + O(x^{-\frac{1}{n}}) \} x^{\frac{p+n}{n}}$$

$$+ \delta\lambda^{\kappa} \{1 + O(\lambda^{-\frac{1}{n(p-q)}})\} \log x \Big],$$

$k=0,1,\dots,n-1$, as $\lambda \to \infty, \lambda x^{q-p} \to 0$, $|\arg \lambda| \leq \pi - \varepsilon$, $x > 0$.

Next we deduce from this a corresponding result for $\hat{y}(x,\lambda)$. In (1.1) we substitute

(4.12) $\hat{y} = \text{diag}\ \{1, \lambda^{\frac{1}{n}}, \dots, \lambda^{\frac{n-1}{n}}\}z.$

The transformed equation reads

(4.13) $\frac{dz}{dx} = \lambda^{\frac{1}{n}} \tilde{A}(x,\lambda)z,$

where $\tilde{A}(x,\lambda)$ arises from the matrix $A(x)$ by replacing $a_j(x)$ by $\lambda^{-\frac{j}{n}} a_j(x)$, $j = 1, \dots, n-1$ and $-a_n(x)$ by $(-1)^n\{\lambda^{-1} P(x) + Q(x)\}$.

We now solve the integral equation

$$z(x,\lambda) = z(x_0,\lambda) + \lambda^{\frac{1}{n}} \int_{x_0}^{x} \tilde{A}(t,\lambda)\, z(t,\lambda)\, dt,$$

where $z(x_0,\lambda)$ is given by (4.9) and (4.12), by means of successive approximations. Since $\tilde{A}(t,\lambda) = O(1)$ as $\lambda \to \infty$ uniformly on any compact set of the t-plane, the successive approximations can be easily estimated, and we find

(4.14) $z(x,\lambda) = z(x_0,\lambda) \exp O(\lambda^{\frac{1}{n}}).$

Combining (4.8), (4.12), (4.14) and (1.2) we obtain

Theorem 2. The recessive solution $\tilde{y}(x,\lambda)$ of theorem 1 possesses the following asymptotic behavior as $\lambda \to \infty$ on $|\arg \lambda| \le \pi - \varepsilon$, when x is fixed :

$$\frac{d^k}{dx^k}\tilde{y}(x,\lambda) = \exp\{\lambda^{\kappa}(\delta \log \lambda + A + o(1))\}, \quad k = 0, 1, \ldots, n-1, \tag{4.15}$$

where κ, δ and A are defined in (4.6), (4.9) and (4.10).

5. Behavior of the solution as $\lambda \to \infty$ on $\pi - \varepsilon \le |\arg \lambda| \le \pi$

It is sufficient to consider the case $\pi - \varepsilon \le \arg \lambda \le \pi$, because the case $-\pi \le \arg \lambda \le -\pi + \varepsilon$ may be treated analogously. Let $\psi = \frac{\pi}{2(p+n)}$ and $\arg x = \psi$. First we consider Re $f_k(x, t, \lambda)$. Using lemmas 1 and 2 and (3.7) in case ii) we see that there exists a positive constant K such that Re $f_k(x,t,\lambda) \le 0$, $k = 2, \ldots, n$ if

$$|x| \ge K\,|\lambda|^{\frac{1}{p-q}}, \quad |\lambda| \ge K \text{ and } \arg(t-x) = \arg x = \psi.$$

Let η be a positive constant. Then according to lemma 3 we may choose K so large that moreover

$$\left|\int_{x_0}^{\infty \exp i\psi} ||B(t,\lambda)||\,dt\right| \le \tfrac{1}{2}\eta, \quad x_0 = K\,|\lambda|^{\frac{1}{p-q}}\,e^{i\psi}$$

if $|\lambda| \ge K$ and $\pi - \varepsilon \le \arg \lambda \le \pi$. Using these estimates in the successive approximations to the solution of (3.6) we see that (4.2) holds as $\lambda \to \infty$ on $\pi - \varepsilon \le \arg \lambda \le \pi$. In an analogous manner as in section 4 we may derive

$$\hat{y}_k(x_0,\lambda) = \exp\{\lambda^{\kappa}(\delta \log \lambda + O(1))\}$$

as $\lambda \to \infty$ on $\pi - \varepsilon \le \arg \lambda \le \pi$.

Now we substitute in (1.1)

$$x = \lambda^{\frac{1}{p-q}}\,t, \quad \hat{y} = \operatorname{diag}\{1, \lambda^{\alpha}, \ldots, \lambda^{(n-1)\alpha}\}\,z, \quad \alpha = \frac{p}{n(p-q)},$$

and we get

$$\frac{dz}{dt} = \lambda^{\kappa}\,\hat{A}(t,\lambda)z,$$

where $\hat{A}(t,\lambda) = O(1)$ as $\lambda \to \infty$ uniformly for t on compact sets.

In the same way as in section 4 we then deduce

$\hat{y}(x,\lambda) = \exp\{\lambda^{\kappa}(\delta \log \lambda + O(1))\}$

as $\lambda \to \infty$ on $\pi - \varepsilon \leq \arg \lambda \leq \pi$. Hence

$\tilde{y}^{(k)}(x,\lambda) = \exp\{\lambda^{\kappa}(\delta \log \lambda + O(1))\}, \quad k = 0, \ldots, n-1,$

as $\lambda \to \infty$ on $\pi - \varepsilon \leq |\arg \lambda| \leq \pi$. Together with theorem 2 this yields

Theorem 3. The solution $\tilde{y}(x,\lambda)$ of theorem 1 is an entire function of λ of order $\kappa = \dfrac{n+p}{n(p-q)}$.

6. Recessive solutions of related differential equations

In (0.1) we replace x by $x + x_0$. Then we obtain an equation of the same type

$$y^{(n)} + \sum_{j=1}^{n} a_j^*(x) y^{(n-j)} = 0, \tag{6.1}$$

$$\text{where } a_j^*(x) = a_j(x + x_0),\ a_n^*(x) = P(x + x_0) + \lambda Q(x + x_0). \tag{6.2}$$

Now $\tilde{y}^*(x,\lambda)$ and $\tilde{y}(x + x_0,\lambda)$ are recessive solutions of (6.1). We have

Theorem 4. The recessive solutions $\tilde{y}(x,\lambda)$ and $\tilde{y}^*(x,\lambda)$ of (0.1) and (6.1) are related by

$$\tilde{y}(x + x_0, \lambda) = \{\exp \sum_{h=1}^{[\frac{p}{n}]+1} \frac{1}{h} \alpha_{p+n-nh}(\lambda)\, x_0^h\}\, \tilde{y}^*(x,\lambda), \tag{6.3}$$

where the $\alpha_j(\lambda)$ are determined by (3.2).

Proof. From the uniqueness of recessive solutions apart from a factor we deduce $\tilde{y}(x + x_0, \lambda) = c\tilde{y}^*(x,\lambda)$.

Formula (3.2) yields

$$\mu_1^*(x) = \mu_1(x + x_0) = \sum_{j=0}^{\infty} \tilde{\alpha}_j(\lambda) x^{\frac{p-j}{n}},\quad \tilde{\alpha}_j = \sum_{h=0}^{[\frac{j}{n}]} \binom{\frac{p-j}{n} + h}{h} \alpha_{j-nh}\, x_0^h .$$

Hence $\alpha_{p+n} = \tilde{\alpha}_{p+n}$ and

$$\sum_{j=0}^{n+p-1} \{\alpha_j (x + x_0)^{\frac{p-j}{n} + 1} - \tilde{\alpha}_j x^{\frac{p-j}{n} + 1}\} \frac{n}{p-j+n} =$$

$$= \sum_{h=1}^{[\frac{p}{n}]+1} \frac{1}{h} \alpha_{p+n-nh} x_0^h + O(x^{-\frac{1}{n}}) ,$$

as $x \to \infty$. This and theorem 1 imply (6.3).

REFERENCES

[1] P.F. Hsieh and Y. Sibuya, On the asymptotic integration of second order linear ordinary differential equations with polynomial coefficients, J. Math. Anal. Appl. 16 (1966), 84-103.

[2] Y. Sibuya, Subdominant solutions of linear differential equations with polynomial coefficients, Michigan Math. J. 14 (1967),53-63.

[3] W.J. Trjitzinsky, Analytic theory of linear differential equations, Acta Math. 62 (1934), 167-226.

[4] W. Wasow, Asymptotic expansions for ordinary differential equations, Interscience, New York, 1965.

Approximation by Functions of Fewer Variables

L. Collatz

Summary: There are different possibilities to approximate a continuous function of n independent real variables by functions of fewer veriables and their combinations. Here we consider several special types which occur in the applications: The sum-type, product-sum-type, parametric type and combined type. For these types applications, especially to partial differential equations are given. For the product-sum-type approximation an inclusion theorem for the minimal distance is given which in special cases allows us to prove the optimality of a given approximation.

1. General formulation

The representation of a function $f(x_1,\dots,x_n)$ or briefly $f(x)$, by an expression, which is composed of functions of fewer variables, is one of the famous problems posed by David Hilbert; here we don't ask for the representation but for the approximation by an expression of the form

$$(1.1)\quad \phi(u_1(x_{11},x_{12},\dots,x_{1m_1}),\ u_2(x_{21},x_{22},\dots,x_{2m_2}),\dots,\dots,\ u_q(x_{q1},\dots,x_{qm_q}))$$

where every x_{jk} is identical with a certain x_ν, and

$$(1.2)\qquad m_\mu < n \quad \text{for} \quad \mu=1,\dots,q\ .$$

The approximation should hold in a given domain B of the n-dimensional point space R^n.

2. Special cases

Frequently occur following special cases:

A) Sum-type:

(2.1) $$\phi = \sum_{\nu=1}^{n} u_\nu(x_\nu)$$

or more general

(2.2) $$\phi = \sum_{\nu=1}^{q} u_\nu(x_{\nu 1}, x_{\nu 2}, \ldots, x_{\nu m_\nu})$$

B) Product-sum-type:

(2.3) $$\phi = \sum_{\nu=1}^{q} u_\nu(x_1, \ldots, x_m)\; v_\nu(x_{m+1}, \ldots, x_n) \quad \text{with} \quad m < n$$

C) Parametric-type:

$$\phi = W(x,a) = \{w(x,a)\}$$

Here w is a given function of the coordinates $x_1, \ldots, x_n$ and finitely many parameters $a_1, \ldots, a_p$, which have to be determined to solve the optimization problem:

(2.4) $$-\delta_1 \leq w(x,a) - f(x) \leq \delta_2 \quad \text{for all } x \in B$$

$$\delta_1 \geq 0, \; \delta_2 \geq 0, \quad \delta_1 + \delta_2 = \text{Min.}$$

This is the problem of Tschebyscheff-Approximation which was considered very often. Here are included as special cases

1. $\delta_1 = \delta_2$, classical Tschebyscheff-Approximation
2. $\delta_1 = 0$, onesided Tschebyscheff-Approximation
3. δ_1, δ_2 free, unsymmetric Tschebyscheff-Approximation

These cases which have many applications in differential- and integral equations, see for instance Collatz [69] [71], will not be considered here.

D) Combined type:

$$(2.5)\quad \phi = \sum_{\nu=1}^{n} u_\nu(x_\nu)\, w_\nu(x_1,\ldots,x_{\nu-1},x_{\nu+1},\ldots,x_n,a_1,\ldots,a_{p_\nu})$$

Here the functions w_ν are given functions of their arguments and the a_{jk} are parameters to be determinated as in C)

Of course, many other special cases could be considered, but in the applications to be mentioned below these cases occur.

Applications

3. Computer

Perhaps one of the most important application is the approximation of a given function f(x) by a term ϕ in (1.1), which can be calculated easily on a computer; (2.1) and (2.3) are common forms.

4. Hammerstein-Equation

Another very important application is the approximation of a kernel K(x,t) a "degenerated" kernel K*(x,t) where $x = \{x_1,\ldots,x_m\}$, $t = \{t_1,\ldots,t_m\}$ are points in a given domain B of the n-dimensional point space R^m. K*(x,t) has the form of a product-sum (2.3)

$$(4.1)\qquad K^*(x,t) = \sum_{\nu=1}^{q} u_\nu(x_1,\ldots,x_m)\, v_\nu(t_1,\ldots,t_m)\ .$$

Consider the wellknown nonlinear Hammerstein-Integral Equation

$$(4.2)\qquad u(x) = \int_B K(x,t)\, \varphi(u(t))\, dt$$

with a given real-valued function $\varphi(u)$.

The corresponding integral equation

$$(4.3)\qquad u^*(x) = \int_B K^*(x,t)\ \varphi(u^*(t))\ dt$$

can be solved explicitly by solving a finite system of nonlinear equations for unknown coefficients c_ν in the expression

$$(4.4)\qquad u^*(x) = \sum_{\nu=1}^{q} c_\nu\, u_\nu(x)$$

In the linear case $\varphi(z) = \gamma_1 + \gamma_2 z$ with given constants γ_1, γ_2 Kantorowitsch-Kryloff [56] gave an error estimation

$$(4.5)\qquad |\varepsilon(x)| = |u^*(x) - u(x)| \leq \psi = [\underset{\substack{x\in B\\ t\in B}}{\mathrm{Max}} |K(x,t) - K^*(x,t)|\,]\ \psi_1$$

This bound suggests the approximation of the function K(x,t) by K*(x,t) in the Tschebyscheff-sense (Tschebyscheff-Approximation, shortly T.A.).

5. Dirichlet-Problem

Let B be a domain as in No.4 with boundary ∂B. We consider the Laplace-Equation for a function u(x)

$$(5.1)\qquad \Delta u = \sum_{j=1}^{n} \frac{\partial^2 u}{\partial x_j^{\,2}} = f(x_1,\ldots,x_n) \quad \text{for } x \in B$$

with the boundary condition

$$(5.2)\qquad u = \psi(x) \quad \text{for} \quad x \in \partial B$$

We approximate f by a sum as in (2.1)

$$(5.3)\qquad f(x) \approx \varphi(x) = \sum_{\nu=1}^{n} u_\nu(x_\nu)\ .$$

We choose functions $h_\nu(x_\nu)$ with

(5.4) $$\frac{d^2h_\nu}{dx_\nu^2} = u_\nu(x_\nu)$$

and arbitrary harmonic functions $z_\sigma(x)$ with $\Delta z_\sigma = 0$ ($\sigma=1,\ldots,s$). Then

(5.5) $$z(x) = \sum_{\sigma=1}^{s} \alpha_\sigma z_\sigma(x) + \sum_{\nu=1}^{n} h_\nu(x_\nu)$$

satisfies the equation

(5.6) $$\Delta z = \varphi$$

and one has to determine the coefficients α_σ in such a way that the boundary condition (5.2) is satisfied as good as possible. This is a T.A. along the boundary ∂B.

6. Combined type

One can improve the method of No.5 by approximating the given function $f(x)$ not by the expression (5.3) but by

(6.1) $$f(x) \approx \hat{\varphi}(x) = \sum_{\nu=1}^{n} u_\nu(x_\nu)\lambda_\nu(x)$$

where λ_ν is linear in x but independent of x_ν:

(6.2) $$\lambda_\nu(x) = \lambda_{\nu 0} + \sum_{\mu=1}^{n} \lambda_{\nu\mu} x_\mu \quad \text{with } \lambda_{\nu\nu} = 0 \quad (\nu=1,2,\ldots,n)$$

Then

(6.3) $$\hat{h} = \sum_{\nu=1}^{m} h_\nu(x_\nu)\lambda_\nu(x)$$

solves

(6.4) $$\Delta \hat{h} = \hat{\varphi} .$$

Introducing the errors

(6.5) $$\begin{cases} \varepsilon = \varphi - f & \text{for (5.3)} \\ \hat{\varepsilon} = \hat{\varphi} - f & \text{for (6.1)} \end{cases}$$

there seems not to be considered hitherto, which improvement can be reached by using $\hat{\varphi}$ instead of φ.

7. Variable coefficients

We consider the Tschaplygin-Differential-Equation (with x,y instead of x_1, x_2)

$$(7.1) \qquad L\,u = K(y)\,\frac{\partial^2 u}{\partial x^2} + \frac{\partial^2 u}{\partial y^2} = f(x,y)$$

with given functions K,f. The equation reduces for K(y)=1,-1,y to the Laplace-, Wave-, Tricomi-Equation, which is considered for transonic flow. We approximate f by the combined type

$$(7.2) \qquad f(x,y) \approx \varphi(x,y) = (a_1+a_2 y)\,K(y)\,u_1(x) + (a_3+a_4 x)\,u_2(y)$$

Choosing functions h_ν as in (5.4)

$$(7.3) \qquad h_1''(x) = u_1(x), \quad h_2''(y) = u_2(y)\ ,$$

one has a special solution h(x,y) of $L\,h = \varphi$ in the form

$$(7.4) \qquad h(x,y) = (a_1+a_2 y)\,h_1(x) + (a_3+a_4 x)\,h_2(y).$$

Further procedure as in No.5.

8. More general type of differential equations

The combined type occurs also for the equation

$$(8.1) \qquad L\,u = \sum_{j=1}^{n} K_j(x)\,L_j u = f(x)$$

with

$$(8.2) \qquad L_j(u) = \sum_{m=1}^{m_j} b_{jm}(x_j)\,\frac{\partial^m u}{\partial x_j^{\,m}}$$

with given functions b_{jm}, K_j, f; here $K_j(x)$ should be independent of the variable x_j. We choose functions $A_j(x)$ and $h_j(x_j)$ with

(8.3) $$\frac{\partial A_j}{\partial x_j} = 0; \quad L_k A_j = 0 \quad \text{for all } j,k; \quad L_j h_j = u_j(x_j);$$

we approximate f by the combined type

(8.4) $$f(x) \approx \varphi(x) = \sum_{j=1}^{n} A_j(x)\, K_j(x)\, u_j(x_j)$$

then

(8.5) $$h(x) = \sum_{j=1}^{n} A_j(x)\, h_j(x_j)$$

solves the equation $L\,h = \varphi$. Further procedure as in No.5.

Examples for equation (8.1) are (7.1) or

$$L\,u = k(x,y)\frac{\partial u}{\partial t} - \frac{\partial^2 u}{\partial x^2} - \frac{\partial^2 u}{\partial y^2} \quad \text{or} \quad L\,u = \frac{\partial^2 u}{\partial t^2} + \frac{\partial^4 u}{\partial x^4}$$

and others.

9. Inverse problems

One has observed a distribution of temperature $f(x_1,\ldots,x_n)$ for $n=2$ or $n=3$ in a certain domain B of R^2 or R^3.
Let us suppose, that this domain B is heated uniformly and f(x) satisfies approximately a differential equation of the form

(9.1) $$-\Delta u = -\sum_{j=1}^{n} \frac{\partial^2 u}{\partial x_j^2} = q = \text{const. in B}$$

with the solution

(9.2) $$u(x) = -\frac{q}{2n} \sum_{j=1}^{n} x_j^{\,2} + w(x_\nu)$$

where w solves the homogeneous equation $\Delta w = 0$. One wishes to approximate the observed distribution $f(x)$ by a function $u(x)$ of the form (9.2) and to determine the unknown constant q (combined-type-problem).

10. T.- and L_2-Approximation

Let (8.1) $L\,u = f(x)$ be the equation to be solved and h a function with $L\,h = \varphi$. We wish to estimate the error $\varepsilon = h-u$ from the defect $\eta = \varphi - f$. Suppose for simplicity, that we have given boundary values and that $\varepsilon = 0$ on ∂B. Suppose furthermore that there exists a Green's function $G(x,s)$ with

$$(10.1) \qquad \varepsilon(x) = \int_B G(x,t)\eta(t)\,dt \ .$$

For the estimation

$$(10.2) \qquad |\varepsilon(x)| \le \|\eta\|_\infty \cdot \int_B |G(x,t)|\,dt \quad \text{with } \|\eta\|_\infty = \max_{x \in B} |\eta(x)|$$

it is useful to have $\|\eta\|_\infty$ as small as possible and to approximate f by φ in the T.A.-sense.

By using Schwarz's inequality we get from (10.1)

$$(10.3) \qquad |\varepsilon(x)| \le \|\eta\|_2\, K(x) \text{ with } K(x)^2 \le \int_B [G(x,t)]^2 dt$$

$$\|\eta\|_2^2 = \int_B \eta^2(t)\,dt$$

and one approximates f by φ in the L_2-sense. The values of $K(x)$ for instance for $L\,u = -\Delta u$ exist for $n=2$ or 3, but not for $n > 3$.

11. Sum-type T.-Approximation, n=2

The T,-approximation of $f(x,y)$ by $\phi = u(x)+v(y)$ in a convex domain B of the two-dimensional x-y-plane is considered by many authors, Diliberto-Strauss [51], Golomb [59], Ofmann [65], Rivlin-Sibner [65], Aumann [68], Sprecher [67,68] and others.

Let ρ_0 be the minimal distance

$$(11.1) \qquad \rho_0 = \inf_{u,v} \| f-u-v \|_\infty \quad .$$

Let us introduce the difference $\delta_R \varphi$ for a function φ defined in B, with respect to a rectangle R in B with corners $P_1P_2P_3P_4$, fig. 1

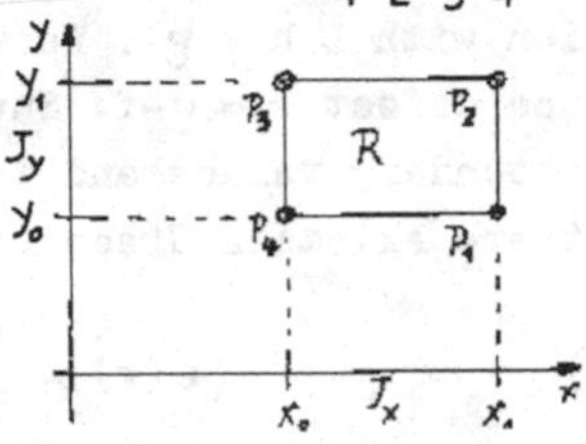

fig.1

$$(11.2) \qquad \delta_R \varphi = \varphi(P_1)-\varphi(P_2)+\varphi(P_3)-\varphi(P_4)$$

Then there are two possibilities:

Case I: $\delta_R f = 0$ for every $R \subset B$; then $\rho_0 = 0$ and f(x,y) can be represented in the form $f(x,y) = u(x)+v(y)$.

Case II: There exists a rectangle $R \subset B$ with $\delta_R f \neq 0$; then f(x,y) cannot be represented in the form u(x)+v(y) and we have the lower bound for the minimal distance

$$(11.3) \qquad \frac{1}{4}|\delta_R f| \leq \rho_0$$

12. Sum-type L_2-Approximation, q=2

We wish to approximate a function $f(x_1,\dots,x_k, y_1,\dots,y_m)$ by $\phi = u(x_1,\dots,x_k) + v(y_1,\dots,y_m)$ as in (2.2) in the L_2-norm (10.3) in a domain $B = B_x \times B_y$

$$(12.1) \qquad \| f-u-v \|_2^2 = \text{Min}$$

Here B_x, B_y are measurable domains in the x-, resp. y-space.

If there exists a solution u_0, v_0, the calculus of variations

$$u = u_o + \varepsilon u_1, \qquad v = v_o + \eta v_1$$

gives immediately the necessary conditions for u_o, v_o:

$$(12.2) \qquad u_o = \frac{\int_{B_y} f(x,y)dy}{\int_{B_y} dy} + \gamma_1, \quad v_o = \frac{\int_{B_x} f(x,y)dx}{\int_{B_x} dx} + \gamma_2 ,$$

γ_1 and γ_2 are constants.

13. T- and L_2-sum-type-approximations

Let the function

$$f(x,y,z) = \frac{1}{1+x+yz}$$

be approximated by

$$w = u(x,y) + v(z)$$

in the domain B: $0 \leq x \leq 1, \quad 0 \leq y \leq 1, \quad 0 \leq z \leq 1$.

The set M of solutions in T-sense is a continuum, one of the solutions being

$$w = \frac{1}{2}\left[\frac{1}{1+x} + \frac{1}{1+x+y}\right] + \frac{1}{8} - \frac{z}{4} .$$

The L_2-approximation w_L is uniquely determined by (12.2).

$$w_L = \frac{1}{y}\,\ell n\left(\frac{1+x+y}{1+x}\right) + \frac{1}{z}\left[(2+z)\ell n(2+z) - (1+z)\ell n(1+z) - 2\ell n 2\right] + \gamma_1$$

which doesn't belong to the continuum M.

14. Product-sum-type

For approximating functions of type (2.3) the L_2-Approximation is leading to eigenvalue problems (Golomb [59]) but the T-Approximation seems not to be considered even in the case q=2: Let us look at the simplest case that a function f(x,y) of two real variables x,y should be approximated by

$$(14.1) \qquad f(x,y) \approx u_1(x)\, v_1(y) + u_2(x)\, v_2(y) .$$

A corresponding finite problem is the approximation of a $n \times n$ matrix A by a sum

$$B = \sum_{\nu=1}^{q} u_\nu v_\nu^T$$

with vectors u_ν, v_ν; for practical applications with filters see Treitel-Shanks [71].

Let be for instance $n=10^3$; for A is needed a memory of 10^6 places on a computer. If one can reach a good approximation of B to A with q=20, one needs 40 000 places.

(14.1) is a nonlinear approximation problem; but if one can guess the function $v_1(y)$ and $u_2(x)$, for instance by $\varphi(y)$ and $\psi(x)$, one can get similar results as in No.11.

15. Alternating-property for a combined type approximation

Let us write instead of (14.1)

$$(15.1) \qquad f(x,y) \approx w(x,y) = \varphi(y)\, u(x) + \psi(x)\, v(y) .$$

Here let $\psi(x)$ and $\varphi(y)$ be given positive functions in intervals $J_x : x_0 \leq x \leq x_1$, resp. $J_y : y_0 \leq y \leq y_1$ and f(x,y) given in the rectangle $R = J_x \times J_y$. All functions considered here are supposed to be continuous. The class of all functions w is called W. For brevity let us write $\varphi(y_0) = \varphi_0$ and analogously $\varphi_1, \psi_0, \psi_1$, and $f(P_j) = f_j$ (j=1,2,3,4) where P_j are the corners of R, fig. 1.

We introduce the following class of functions, the linear manifold

(15.2) $$W^* = \{a_1\psi(x) + a_2\varphi(y) + a_3\psi(x)\,\varphi(y)\}$$

with a_j as free real constants. Then the value σ_R in (11.2) corresponds here to the value σ (except a factor 4) in such a way, that by adding σ to the values f_1, f_3 and subtracting σ from f_2, f_4 one gets values, which are assumed by a function of W*:

Lemma (Alternating property): There exists in the case $\varphi_1 \neq \varphi_0$, $\psi_1 \neq \psi_0$ a constant σ, for which the values

(15.3) $$f_j^* = f_j - (-1)^j \sigma \quad (j=1,2,3,4)$$

are assumed by a function

(15.4) $$w^* = a_1\psi(x)+a_2\varphi(y)+a_3\psi(x)\varphi(y) \in W^*$$

with

(15.5) $$w^*(P_j) = f_j^* \qquad (j=1,2,3,4)$$

Proof. The equations

(15.6) $$a_1\psi(P_j)+a_2\varphi(P_j)+a_3\psi(P_j)\varphi(P_j)+(-1)^j\sigma = f_j \quad (j=1,2,3,4)$$

for the unknowns a_1, a_2, a_3, σ have an uniquely determined solution, because the determinant D doesn't vanish:

(15.7) $$D = -(\varphi_0+\varphi_1)(\psi_0+\psi_1)(\varphi_1-\varphi_0)(\psi_1-\psi_0) \neq 0 .$$

One gets

$$(15.8) \qquad \delta = \frac{-\psi_0\varphi_1 f_1+\psi_0\varphi_0 f_2-\psi_1\varphi_0 f_3+\psi_1\varphi_1 f_4}{(\varphi_0+\varphi_1)(\psi_0+\psi_1)}$$

16. H-set-Property

Lemma: Introducing the sets: $M_1 = \{P_1,P_3\}$, $M_2 = \{P_2,P_4\}$, $M = M_1 \cup M_2$ one gets as M an H-set for the linear manifold W* in (15.2).

H-sets are defined by the property: There exists no pair w_1^*, w_2^* in W* with

$$(16.1) \qquad w_1^* - w_2^* \quad \begin{cases} > 0 \text{ in } M_1 \\ < 0 \text{ in } M_2 \end{cases}.$$

With respect to the linearity of W* this is equivalent to: There exists no function $w^* \in W^*$ with

$$(16.2) \qquad w^* > 0 \text{ in } M_1, \quad w^* < 0 \text{ in } M_2 .$$

Proof (indirect). Suppose w* as in (15.4) satisfies (16.2). Then we have the following inequalities (1)(2)(3)(4) [in the schema for instance the first row means $a_1\psi_1+a_2\varphi_0+a_3\psi_1\varphi_0 > 0$, compare schema Collatz [66] p.425]:

		a_1	a_2	a_3	
Point P_1	(1)	ψ_1	φ_0	$\psi_1\varphi_0$	>0
" P_2	(2)	$-\psi_1$	$-\varphi_1$	$-\psi_1\varphi_1$	>0
" P_3	(3)	ψ_0	φ_1	$\psi_0\varphi_0$	>0
" P_4	(4)	$-\psi_0$	$-\varphi_0$	$-\psi_0\varphi_0$	>0
(1) φ_1+(2)φ_0	(5)	$-\varphi_d\psi_1$	0	0	>0
(3) φ_0+(4)φ_1	(6)	$-\varphi_d\psi_0$	0	0	>0
(5) ψ_0+(6)ψ_1	(7)	0	0	0	>0

Here we have got the inequalities (5)(6)(7) by linear combinations of the earlier inequalities (1),...(4) with positive coefficients $\varphi_0, \varphi_1, \psi_0, \psi_1$ as indicated. This gives in (7) the contradiction $0 > 0$.

17. Lower bound for the minimal distance

Theorem: For the T-approximation of a function f(x,y) by functions w(x,y) of the form (15.1) with given functions $\varphi(y)$, $\psi(x)$ [with $\varphi_0 \neq \varphi_1$, $\psi_0 \neq \psi_1$] and arbitrary functions u(x),v(y) in a rectangle R one has the lower bound for the minimal distance ρ_0

(17.1) $$|\delta| \leq \rho_0 \ .$$

Here all functions are considered as continuous, ρ_0 is defined by

(17.2) $$\rho_0 = \inf_w \|w-f\|_\infty$$

and δ by (15.8).

Proof (indirect). Suppose $\rho_0 < |\delta|$, then there exists a function $\tilde{w} = \varphi\tilde{u}+\psi\tilde{v} \in W$ with $\rho_0 < \|\tilde{w}-f\| < |\delta|$, and especially we have

(17.3) $$|\tilde{w}(P_j)-f_j| < |\delta| \quad \text{for } j=1,2,3,4 \ .$$

Now we consider another approximation problem (P*):
Approximate the given function f(x,y) only in the 4 points P_j (j=1,2,3,4) by functions of the class W* of (15.4). Let the minimal distance for this problem be ρ_0^*. For this problem the function w* of (15.4) is a minimal solution, because it has the alternating property, and it holds $\rho_0^* = |\delta|$. This means: There exists no better approximation $\hat{w} \in W^*$ for this problem; for no element $\hat{w} \in W^*$ holds $|\hat{w}(P_j)-f_j| < |\delta|$ for j=1,2,3,4.
This contradicts (17.3).

Remarks:

I. Let B be any convex domain in the x-y-plane and R any rectangle in B. Then still holds (17.1) with (15.8).

II. The theorem is an inclusion theorem for the minimal distance ρ_0, because it is trivial, that $\rho_0 \leq \|w-f\|_\infty$ for any $w \in W$. If one can find a $w \in W$ with $\|w-f\|_\infty = |\delta|$ then w is a minimal solution.

18. Numerical example

Let us consider the function

$$(18.1) \qquad f(x,y) = \frac{3x^2-x^3}{2} \cdot \frac{3y^2-y^3}{2} \quad \text{in the domain } B: \begin{cases} 0 \leq x \leq 1 \\ 0 \leq y \leq 1 \end{cases}$$

as a simple example in which an exact best T-approximation can be given.

A) If we take as approximating function

$$\tilde{w}(x,y) = u(x)+v(y)$$

we have the lower bound for the minimal distance ρ_0 from (11.3)

$$\frac{1}{4}|\delta_R f| = \frac{1}{4}\,|0-1+0-0| = \frac{1}{4} \leq \tilde{\rho}_0$$

B) For comparison we take the class of approximating functions

$$w(x,y) = (1+y)\,u(x) + (1+x)\,v(y)$$

Here (17.1) gives the lower bound for the minimal distance ρ_0

$$\delta = \frac{1}{9} \leq \rho_0$$

For the function

$$(18.2) \qquad w = \frac{1}{18}\left[(1+y)(-1+x+4x^2) + (1+x)(-1+y+4y^2)\right]$$

we have $|w-f| \leq \frac{1}{9}$ in B and therefore the function w given in (18.2) is a best T-approximation.

References

G. Aumann [68] Approximation von Funktionen, in R. Sauer - I. Szabó, Math. Hilfsmittel des Ingenieurs, Springer 1968, Teil III, 320-351, speziell S.348-351.

L. Collatz [66] Functional Analysis and Numerical Mathematics, Academic Press 1966, 472 p.

L. Collatz [69] Nichtlineare Approximationen bei Randwertaufgaben, V. IKM, Internationaler Kongreß für die Anwendungen der Mathematik, Weimar 1969, 169-182.

L. Collatz [71] Some applications of Functional Analysis to Analysis, particularly to Nonlinear Integral Equations, Proc. Symp. Nonlin. Funct. Anal. Edited by Rall, Academic Press 1971, 1-43.

S.P. Diliberto - E.G. Strauss [51] On the approximation of functions of several variables by the sum of functions of fewer variables, Pacific Journal of Math. 1 (1951) 195-210.

U. Golomb [59] Approximation by Functions of fewer variables, Part B, in R.E. Langer: On numerical Approximation (Madison 1959), 311-327.

L.W. Kantorowitsch - W.I. Kryloff [56] Näherungsmethoden der höheren Analysis, Berlin 1956, 611 p.

J.P. Ofmann [65] Best approximation of functions of two variables by functions of the form $\varphi(x)+\psi(y)$, Americ. Math. Soc. Translations, Ser.2, 44 (1965), 12-29 (translated from Russian).

T.J. Rivlin - R.J. Sibner [65] The degree of approximation of certain functions of two variables by a sum of functions of one variable, Amer. Math. Monthly 72 (1965), 1101-1103.

D.A. Sprecher [67] On best approximations in several variables, Crelles Journ. Reine Angew. Math. 229 (1967), 117-130.

D.A. Sprecher [68] On best approximations of functions of two variables, Duke Math. Journal 35 (1968) 391-397.

S. Treitel - J.L. Shanks [71] The Design of Multistage Separable Planar Filters, IEEE Transactions on Geoscience Electronics 9 (1971), 10-27.

Affine Control Processes

R. Conti

0. Introduction

This is an expository paper surveying some essential features of controllability theory for affine control processes. Our analysis is based on a scheme proposed by S.K. Mitter at the 1967 Conference on Mathematical Theory of Control at the University of Southern California, L.A. (see S.K. Mitter [38]; also, M.C. Delfour [8], M.C. Delfour - S.K. Mitter [9]). The generality and flexibility of Mitter's scheme are brought into evidence by many examples of "concrete" control systems described by a variety of functional equations (ordinary differential, partial differential, delay, etc.). However, we limited our illustration to a differential equation of evolutive type (Sec. 9).

The whole exposition centers around relation (2.1) (and its consequences (5.3), (6.1), (6.5), (7.2)) whose proof depends on the following "strict separation theorem for convex sets": "If A and B are disjoint convex closed subsets of a locally convex topological vector space $\mathfrak{X}$ and one of them is compact, then they are strictly separated by a closed hyperplane, i.e., there exist numbers $\alpha, \beta \in \mathbb{R}$ and a continuous linear functional $x' \in \mathfrak{X}'$, such that

$$\langle a, x'\rangle \leq \alpha < \beta \leq \langle b, x'\rangle, \quad a \in A, \ b \in B."$$

From this follows that the convex closure $\overline{\text{co}}\, A$ of a subset A of a Hausdorff l.c.t. vector space is characterized by

$$(0.1) \qquad \overline{\text{co}}\, A = \{x \in \mathfrak{X} : \langle x, x'\rangle \leq \sup_{a \in A} \langle a, x'\rangle, \quad x' \in \mathfrak{X}'\}.$$

The essential role played in control theory by separation and support properties of convex sets was the motivation recently of an excellent survey paper of V. Klee [26].

1. Affine control processes. Reachable sets. Bang-bang controls

Given two real vector spaces $\mathcal{U}$, $\mathcal{X}$, vectors $u \in \mathcal{U}$ are thought of as controls, and vectors $x \in \mathcal{X}$ as states of a control system. Then, given a fixed $x_o \in \mathcal{X}$ and a linear map $\mathcal{T} : \mathcal{U} \rightarrow \mathcal{X}$, the affine map

$$u \rightarrow x(u) = \mathcal{T} u + x_o$$

is, by definition, an affine control process.

Given a set $U \subset \mathcal{U}$ of admissible controls, the set $R \subset \mathcal{X}$ defined by

$$R = \{x \in \mathcal{X} : x = \mathcal{T} u + x_o, \ u \in U\} = \\ = \mathcal{T} U + \{x_o\}$$

is the reachable set, relative to x_o, $\mathcal{T}$, U.

The theory of affine control processes can be identified with the study of the properties of R. This obviously requires some assumptions, among them the fundamental hypothesis

H) $\mathcal{T}$ U is a convex subset of $\mathcal{X}$,

which will be tacitly assumed from now on.

It should be noted that H) can be verified without U being convex. In general $\mathcal{T}$ is not one-to-one, so that it makes sense and is of interest to look for proper subsets $V \subset U$ such that $\mathcal{T} V = \mathcal{T} U$.

For instance, the validity of $\mathcal{T} \ddot{U} = \mathcal{T} U$, where $\ddot{U}$ denotes the set of extreme points of U, is known as the "Bang-Bang principle" and the $u \in \ddot{U}$ are called Bang-Bang controls.

2. Characterization of the reachable set

When $\mathcal{X}$ is a Hausdorff l.c.t. space, $\mathcal{X}'$ its topological dual, then, according to (0.1) the closure $\bar{R}$ of R is characterized by

(2.1) $$\bar{R} = \{x \in \mathcal{X} : \langle x,x' \rangle \leq \langle x_o,x' \rangle + \sup_{u \in U} \langle \mathcal{T}u, x' \rangle, \ x' \in \mathcal{X}'\}$$

that is, by means of an infinite set of inequalities. When both $\mathcal{U}$ and $\mathcal{X}$

are Hausdorff l.c.t. vector spaces and $\mathcal{T}$ is continuous, it is of interest to find under which conditions on U the set $\mathcal{T}$U, i.e., the reachable set R, is closed, in which case (2.1) actually characterizes R itself.

A situation commonly met is the following. Both $\mathcal{U}$ and $\mathcal{X}$ are normed dual spaces, that is $\mathcal{U} = \mathcal{V}'$, $\mathcal{X} = \mathcal{Y}'$ for some pair of normed vector spaces $\mathcal{V}$, $\mathcal{Y}$, and $\mathcal{T}$ is the adjoint $\mathcal{T} = \mathcal{S}'$ of a continuous linear map $\mathcal{S} : \mathcal{Y} \to \mathcal{V}$. This is equivalent to saying that $\mathcal{T}$ is weak* continuous. Then $\mathcal{T}$U is weakly* (compact, hence, weakly*) closed in $\mathcal{X}$ if U is a closed ball of $\mathcal{U}$, by Alaoglu's theorem.

If; further, $\mathcal{X}$ is reflexive, then $\mathcal{T}$U, being convex, is norm closed.

3. Complete controllability

When U = $\mathcal{U}$, $\mathcal{T}$U = $\mathcal{T}\mathcal{U}$ is a subspace of $\mathcal{X}$ and R is an affine subspace of $\mathcal{X}$.

If R = $\mathcal{X}$, i.e.,

(3.1) $$\mathcal{T}\mathcal{U} = \mathcal{X}$$

the control process is said to be <u>completely controllable in</u> $\mathcal{U}$.

More generally, if $\mathcal{X}$ is a Hausdorff l.c.t. vector space with a topology $T_{\mathcal{X}}$ we say that the control process is <u>completely controllable in</u> $\mathcal{U}$, <u>relative to</u> $T_{\mathcal{X}}$, if

(3.2) $$\overline{\mathcal{T}\mathcal{U}} = \mathcal{X} .$$

Since (3.2) is equivalent to $\bar{R} = \mathcal{X}$, it follows from (2.1) that (3.2) is also equivalent to

$$\sup_{u \in \mathcal{U}} \langle u, {}^t\mathcal{T} x' \rangle = +\infty, \quad x' \neq 0',$$

(${}^t\mathcal{T}$, the transpose of $\mathcal{T}$), which, in turn, is equivalent to

(3.3) $${}^t\mathcal{T} x' = 0' \Rightarrow x' = 0'.$$

4. Extremal and maximal controls. Normality. Synthesis problem

Let $\mathcal{X}$ be a Hausdorff l.c.t. vector space. An extremal control is any $u \in U$ such that $x(u) \in \partial R$, the boundary of R. If, further, the convex set R has a supporting hyperplane at $x(u)$, then u is a maximal control.

Such terminology is accounted for by the fact that, by (2.1), u_M will be maximal if and only if

$$<\mathcal{T} u_M, x'> \leq \sup_{u \in U} <\mathcal{T} u, x'>, \quad x' \in \mathcal{X}'$$

and

$$<\mathcal{T} u_M, x'_M> = \sup_{u \in U} <\mathcal{T} u, x'_M>$$

for some $x'_M \neq 0'$.

All the extremal controls are maximal in some cases, for instance when $\mathcal{X}$ is finite dimensional, or the interior int R of R is non-empty. (see V. Klee, [26], p.242-243). Clearly, int $R \neq \emptyset$ if int $U \neq \emptyset$ and $\mathcal{T}$ is an open mapping. This will happen, for instance, when $\mathcal{T}$ is continuous, $\mathcal{U}$ is a complete pseudo-metrizable topological vector space, $\mathcal{X}$ is a Hausdorff topological vector space and $\mathcal{T}$ is onto, that is, the control process is completely controllable in $\mathcal{U}$.

The equation

$$(4.1) \qquad <u, {}^t\mathcal{T} x'> = \sup_{v \in U} <v, {}^t\mathcal{T} x'>$$

plays an important role in control theory. The control process is said to be normal, relative to the set U, when (4.1) has just one solution $u = u(x')$ in U for each $x' \neq 0'$.

This implies that $\mathcal{T} U$ be strictly convex, not viceversa.

Also, normality relative to any set U containing at least two points implies complete controllability in $\mathcal{U}$, by virtue of (3.3).

The construction of $x' \to u(x')$ from (4.1) is known as a synthesis problem.

5. Controllability

Given a target set $W \subset \mathfrak{X}$, the control process is controllable in U, relative to W, if

$$(5.1) \qquad R \cap W \neq \emptyset .$$

More generally, if $\mathfrak{X}$ is a Hausdorff l.c.t. vector space with a topology $T_{\mathfrak{X}}$, the control process is controllable in U, relative to W and $T_{\mathfrak{X}}$, if

$$(5.2) \qquad \bar{R} \cap \bar{W} \neq \emptyset .$$

$\bar{R} \cap \bar{W} \neq \emptyset$ is equivalent to $0 \in -\bar{W} + \bar{R}$. Therefore, if both γU and W are convex sets and if $-\bar{W} + \overline{\gamma U}$ is closed, then from (1.1) follows that (5.2) is equivalent to

$$(5.3) \qquad \inf_{x \in W} \langle x, x' \rangle \leq \langle x_0, x' \rangle + \sup_{u \in U} \langle \gamma u, x' \rangle, \quad x' \in \mathfrak{X}' .$$

Therefore, (5.3) and (5.1) are equivalent if γU, W are both convex and $T_{\mathfrak{X}}$-closed and either γU or W is $T_{\mathfrak{X}}$-compact.

6. Multiple controls

Let the space $\mathcal{U}$ be the product $\mathcal{U}_1 \times \dots \times \mathcal{U}_k$ of k spaces $\mathcal{U}_1$, ..., $\mathcal{U}_k$ and let the mapping $\gamma : \mathcal{U} \to \mathfrak{X}$ be defined by

$$\gamma(u_1, \dots, u_k) = \gamma_1 u_1 + \dots + \gamma_k u_k$$

where $\gamma_s : \mathcal{U}_s \to \mathfrak{X}$, (s = 1, ..., k), are linear mappings.

If also the set $U \subset \mathcal{U}$ of admissable controls is the product $U_1 \times \dots \times U_k$ of k sets U_s, then $\gamma U = \gamma_1 U_1 + \dots + \gamma_k U_k$ and the reachable set R relative to x_0, γ, $U_1 \times \dots \times U_k$ is

$$R = \{x \in \mathfrak{X} : x = \sum_1^k {}_s \gamma_s u_s + x_0\} =$$

$$= \sum_1^k {}_s \gamma_s U_s + \{x_0\} .$$

If each set $\gamma_s U_s$, $s = 1, \ldots, k$, is convex, then γU is convex and viceversa.

If, further, $\mathcal{X}$ is a Hausdorff l.c.t. vector space, then, by virtue of (2.1), $\bar{R}$ is characterized by

(6.1) $$\bar{R} = \{x \in \mathcal{X} : <x, x'> \leqslant <x_o, x'> + \sum_1^k \sup_{u_s \in U_s} < \gamma_s u_s, x'>, \ x' \in \mathcal{X}'\}.$$

It should be noted that $\overline{\gamma_s U_s} = \gamma_s U_s$, $s = 1, \ldots, k$, does not imply $\overline{\gamma U} = \gamma U$, i.e., $\bar{R} = R$.

Equation (4.1) is now equivalent to the system of k equations

(6.2) $$< \gamma_s u_s, x'> = \sup_{v_s \in U_s} < \gamma_s v_s, x'>, \quad s = 1, \ldots, k$$

i.e., denoting by ${}^t\gamma_s$ the transpose of γ_s, to the system

(6.3) $$<u_s, {}^t\gamma_s x'> = \sup_{v_s \in U_s} <v_s, {}^t\gamma_s, x'>, \quad s = 1, \ldots, k.$$

Consequently, the notion of normality is "partialized" in the sense that the control process

(6.4) $$(u_1, \ldots, u_k) \rightarrow x(u_1, \ldots, u_k) = \sum_1 \gamma_s u_s + x_o$$

will be <u>normal relative to</u> U_s, for a given s, when the s.th equation in (6.2) (or (6.3)) has just one solution $u_s = u_s(x')$ for each $x' \neq 0'$. Normality relative to U will be equivalent to the normality relative to U_s for each $s = 1, \ldots, k$.

Also the notion of complete controllability is "partialized" and the control process (6.4) will be <u>completely controllable</u> in $\mathcal{U}_s$ for a given s, when $\gamma_s \mathcal{U}_s = \mathcal{X}$, <u>completely controllable in</u> $\mathcal{U}_s$, <u>relative to</u> $T_{\mathcal{X}}$, when $\overline{\gamma_s \mathcal{U}_s} = \mathcal{X}$. This will be equivalent to

$${}^t\gamma_s x' = 0' \Rightarrow x' = 0'.$$

When a target set W is given, the controllability inequalities (5.3) may be written

$$\text{(6.5)}\qquad \inf \langle x, x'\rangle \leq \langle x_o, x'\rangle + \sum_1^k {}_s \sup_{u_s \in U_s} \langle \gamma_s u_s, x'\rangle, \quad x' \in \mathcal{X}'$$

and they are equivalent to (5.2) provided that both $\sum_1^k {}_s \gamma_s U_s$ and W are convex sets and $-\bar{W} + \overline{\sum^k {}_s \gamma_s U_s}$ is closed.

7. Perturbed control processes

An extension of the scheme in Sec. 1 is the following.

Usually, the vector x_o is to be considered as a "perturbing" term in the control process

$$u \to x(u) = \gamma u + x_o.$$

More generally, one has to deal with control processes

$$\text{(7.1)}\qquad u \to x(u) = \{\gamma u\} + X_o$$

where the "perturbing" term X_o is a given subset of $\mathcal{X}$, not necessarily consisting of a single point. The perturbed control process (7.1) is then an "affine" mapping of $\mathcal{U}$ into $\mathbb{P}(\mathcal{X})$, the family of subsets of $\mathcal{X}$.

Given a target set $W \subset \mathcal{X}$ we say that (7.1) is strongly controllable in U (or else, controllable in U, against the perturbation X_o) if there exist $\bar{u} \in U$ such that

$$\{\gamma \bar{u}\} + X_o \subset W.$$

If one defines the modified target set as

$$W_o = \{x \in \mathcal{X} : \{x\} + X_o \subset W\}$$

it follows that (7.1) is strongly controllable in U if and only if

$$\gamma U \cap W_o \neq \emptyset.$$

Therefore, if both $\mathcal{T}U$ and W_o are convex sets and $-\bar{W}_o + \overline{\mathcal{T}U}$ is closed, then (7.1) is strongly controllable in U if and only if

(7.2) $$\inf_{x \in W_o} \langle x, x' \rangle \leq \sup_{u \in U} \langle \mathcal{T}u, x' \rangle, \quad x' \in \mathcal{X}' .$$

8. t-controllability

The scheme described in Sec. 1 can also be extended so as to cover "t-controllability" problems.

Given $\mathcal{U}$ and $\mathcal{X}$, let us denote by $\{x_t\}$ a given family of states $x_t \in \mathcal{X}$, and by $\{\mathcal{T}_t\}$ a given family of linear maps $\mathcal{T}_t : \mathcal{U} \to \mathcal{X}$, both depending on a real index $t \in]\alpha,\omega[$, called time. Then the family of affine maps

(8.1) $$(t,u) \to x(t,u) = \mathcal{T}_t u + x_t$$

is an affine t-control process.

It can happen that for each fixed $t \in]\alpha,\omega[$, the set

$$\mathcal{T}_t \quad + \{x_t\}$$

is an affine subspace properly contained in $\mathcal{X}$, not even dense in $\mathcal{X}$, when $\mathcal{X}$ is a Hausdorff l.c.t. vector space. Therefore, it makes sense to inquire whether the control process (8.1) is completely controllable in $\mathcal{U}$ at an unprescribed future or past time, which means

(8.2) $$\bigcup_t [\mathcal{T}_t \mathcal{U} + \{x_t\}] = \mathcal{X}$$

the union $\bigcup_t$ being taken over an interval $[\tau, \omega[$ or $]\alpha, \tau]$, respectively.

Replacing the set $U \subset \mathcal{U}$ of admissible controls by a family $\{U_t\}$, $t \in]\alpha,\omega[$, of such sets, we shall also have a family of reachable sets R_t relative to x_t, $\mathcal{T}_t$, U_t, namely

$$R_t = \{x \in \mathcal{X} : x = \mathcal{T}_t u + x_t, \; u \in U_t\} =$$
$$= \mathcal{T}_t U_t + \{x_t\}, \quad t \in]\alpha,\omega[.$$

The control process (8.1) is <u>completely controllable in $\{U_t\}$ at an unprescribed, future or past time</u> if

(8.3) $$\bigcup_t R_t = \mathcal{X}$$

with $t \in [\tau, \omega[$, or $t \in]\alpha, \tau]$, respectively.

When $\mathcal{X}$ is a topological vector space, (8.2), (8.3) can be replaced by the weaker requirements

$$\overline{\bigcup_t [\gamma_t + \{x_t\}]} = \mathcal{X}$$

$$\overline{\bigcup_t R_t} = \mathcal{X}$$

respectively.

Given a <u>moving target</u>, i.e., a family $\{W_t\}$, $t \in]\alpha,\omega[$, of target sets $W_t \subset \mathcal{X}$, the control process (8.1) is <u>controllable in</u> $\{U_t\}$ <u>relative to</u> $\{W_t\}$ <u>at an unprescribed, future or past time</u> if

$$(\bigcup_t R_t) \cap (\bigcup_t W_t) \neq \emptyset,$$

or, more in general, when $\mathcal{X}$ is a topological vector space, if

$$(\overline{\bigcup_t R_t}) \cap (\overline{\bigcup_t W_t}) \neq \emptyset .$$

9. Examples

Let us consider the function

(9.1) $$t \to \int_\tau^t G(t,s)B(s)u(s)\, ds + G(t,\tau)v + \int_\tau^t G(t,s)c(s)\, ds$$

defined for $t \geq \tau$, with range in a Banach space $\mathcal{B}$.

Under suitable assumptions (9.1) can be considered as the solution x of a differential equation

(9.2) $$\dot{x} - A(t)x = B(t)u(t) + c(t), \quad t \in]\alpha,\omega[$$

with initial datum

(9.3) $$x(\tau) = v.$$

When $\mathcal{B} = \mathcal{R}^n$, then $G(t,s) = Y(t)Y^{-1}(s)$, where Y is any non-singular n×n matrix solution of

$$\dot{y} - A(t)y = 0$$

and the integrals in (9.1) are in the sense of Cauchy-Riemann or Lebesgue, according to whether $t \to B(t)u(t)$, $t \to c(t)$ are continuous or measurable and Lebesgue (locally) integrable.

When $\mathcal{B}$ is infinite dimensional, then $A : t \to A(t)$ is a family of linear, possibly unbounded operators in $\mathcal{B}$, generating an evolution operator $G : (t,s) \to G(t,s)$. Correspondingly, the integrals in (9.1) are Bochner integrals.

Given a Banach space Ω, let $u(t) \in \Omega$ for $t \in]\alpha,\omega[$. Then, for a fixed $T > \tau$, let $\mathcal{U} = L^p(\tau, T, \Omega)$, $1 \leq p \leq +\infty$, and let γu be defined by

$$(9.4) \qquad \gamma u = \int_{\tau}^{T} G(T,s)B(s)u(s)\,ds$$

so that $\mathcal{X} = \mathcal{B}$. Let further

$$(9.5) \qquad x_0 = G(T,\tau)v + \int_{\tau}^{T} G(T,s)c(s)\,ds.$$

The finite dimensional case (<u>f.d.c.</u>), corresponding to finite dimensional Ω and $\mathcal{B}$, is by far the most investigated, but the attention and efforts of pure and applied research workers are more and more attracted by infinite dimensional cases (<u>i.d.c.</u>). This is also due to the fact that (9.2) with an unbounded A is apt to describe several "distributed parameter" control problems based on partial differential equations of evolutionary type, like heat equation, hyperbolic equations, etc. The literature, both theoretical and technical, is by now so extensive and rapidly increasing that it would be impossible to refer, even briefly, on particular problems. Therefore we have to confine our references in the i.d.c. only to papers of general interest.

To begin with we shall quote the results of <u>J.P. LaSalle</u> [32], <u>L.M. Sonneborn</u> - <u>F.S. Van Vleck</u> [47], <u>H. Halkin</u> [21], <u>N. Levinson</u> [34], <u>J. Schmets</u> [45] on the existence of bang-bang controls, in the f.d.c. In the i.d.c. bang-bang controls are to be replaced by "generalized" bang-bang controls and their existence gives a weakened form of the "Bang-bang principle" (see P.L. Falb [11], H.O. Fattorini [12], [13], [14]).

Conditions to insure the closedness of γU were given by P.L. Falb [11] in the i.d.c. The family of sets U such that γU is closed includes all convex, bounded closed sets of $L^p(0,T,\Omega)$, with Ω reflexive, if $1 < p < +\infty$. This is no longer true, even in the f.d.c., for $p = +\infty$, as was shown by H. Hermes [22]. The case $p = 1$ is completely anomalous since γU might not be closed even in the simple case of U being the unit ball of $L^1(0,T,E^2)$. This was the motivation for replacing the first integral in (9.1) by a Riemann-Stieltjes integral

$$\int_{\tau}^{t} G(t,s)B(s)\, d\omega(s)$$

and the space $L^1(\tau,T,E^m)$ by $VL_0(\tau,T,E^m)$, the dual space of $L^C(\tau,T,E^m)$ consisting of functions ω of bounded variation, as some Authors did (see : N.N. Krasovskii [27], [28]; L.W. Neustadt [39]).

When $\mathcal{X} = \mathcal{B}$ is the euclidean n-space there is no distinction between complete controllability and complete controllability relative to the norm topology since all the subspaces of $\mathcal{X}$ are finite dimensional, hence closed, and complete controllability has been characterized in terms of A and B (see: R.E. Kalman [24], R.E. Kalman - Y.C. Ho - K.S. Narendra [25], J.P. LaSalle [32], also H. Hermes - J.P. LaSalle [23]). The distinction arises in the i.d.c. to which are devoted papers by H.O. Fattorini [15], [16], [17], [18], L.M. Kuperman - Yu.M. Repin [30], A.B. Kurjanskii [31].

The solutions of some optimum control problems (like time optimum or minimum final distance problem) are maximal controls, according to the well-known Pontryagin's maximum principle. Examples of extremal controls which are not maximal cannot occur in the f.d.c., but they do occur in i.d.c. (see for instance Yu.A. Egorov [10]). Conditions insuring the existence of maximal controls in i.d.c. were given by A.V. Balakrishnan [2], A. Friedman [19], W.A. Porter [40].

The notion of normality with respect to particular sets U of L^∞ in the f.d.c. is due to J.P. LaSalle [32] and was used in connection with the uniqueness of optimum controls.

For the synthesis problem, i.e., the construction of $x' \to u(x')$ from (4.1) in the f.d.c. we refer, among others, to N.N. Krasovskii [27], [28], E. Kreindler [29], R. Conti [3].

Controllability relative to a target set W when W is a single point set was dealt with as a finite "moment problem" in the f.d.c. by N.N. Krasovskii [27] who obtained the inequalities (5.3) in this particular case (see also N.N. Krasovskii [28], W.T. Reid [43]). H.A. Antosiewicz [1] was the first to use the strict separation theorem for convex sets to obtain (5.3) with W a closed ball in E^n. (See also R. Conti [3]). Later on, Antosiewicz's result was extended by W. Miranker [37] to i.d.c. with W in a Hilbert space.

An example of multiple controls is obtained from (9.1) when the initial state v is allowed to vary within some convex set V. In this case, admissible controls are the pairs $(u,v) \in U \times V \subset L^p \times \mathcal{B}$ and we have

$$(9.6) \qquad \gamma(u,v) = \gamma_1 u + \gamma_2 v = \int_\tau^T G(T,s)B(s)u(s)\, ds + G(T,\tau)v$$

$$(9.7) \qquad x_o = \int_\tau^T G(T,s)c(s)\, ds .$$

Inequalities (6.5) were given in the f.d.c. by R. Gabasov - F.M. Kirillova [20] when U, V and W are closed balls. In i.d.c. they were given by R. Conti [4] who also gave conditions for the existence of maximal controls ([5], [6], [7]). The same questions, in Orlicz spaces instead of L^p, were studied by G. Pulvirenti - G. Santagati [41], [42].

A perturbed control process can be obtained by replacing the n-vector valued function $t \to c(t)$ in (9.1) by a set valued one, for instance by a function $t \to C(t)\sigma(t)$ where $C(t)$ is an $n \times k$ matrix and $t \to \sigma(t)$ is any selection of a function $t \to \Sigma(t)$ of t into the family of convex compact sets of E^k (See M.C. Delfour - S.K. Mitter [9]). Perhaps the connection of perturbed control processes with differential games would deserve further investigation.

A family $\mathcal{Y}_t$ of linear mappings is obtained from (9.1) by taking

$$\mathcal{Y}_t u = \int_{\tau}^{t} G(t,s)B(s)u(s)\, ds$$

and $t \in [\tau, \omega[$ or, $t \in]\alpha, \tau]$. Let further

$$x_t = G(t,\tau)v + \int_{\tau}^{t} G(t,s)c(s)\, ds.$$

This time, $\mathcal{X} = \mathcal{B}$ again, but $\mathcal{U} = L^p_{loc}(\tau, \omega, \Omega)$, or $\mathcal{U} = L^p_{loc}(\alpha, \tau, \Omega)$.

The influence of x_t on controllability problems at an unprescribed time can be decisive, so that there is a difference between the general case and the special case $v = 0$, $c(t) = 0$, i.e., $x_t = 0$.

Several important results about complete controllability at an unprescribed time in the i.d.c. were obtained by H.O. Fattorini [12], [16], [17], [14]; for other results see also K. Tsujioka [49], F.A. Sholokovitch [46].

The best known problem of controllability in $\{U_t\}$ relative to $\{W_t\}$ corresponds to $W_t = 0$, i.e., to the steering to the origin in a finite time by using bounded sets of controls. In this connection we shall quote J.P. LaSalle [32], R. Conti [3], P. Santoro [44], E.B. Lee - L. Markus [33].

Starting from (9.1) it is possible to define $\mathcal{Y}$ and x_o in a different way from (9.4) (9.5) or (9.6) (9.7).

For instance if we look for solutions of (9.1) satisfying a "periodicity" conditions (L. Markus [36]

(9.8) $$x(T) - x(0) = 0$$

one can define $\mathcal{Y}$ by

$$\gamma(u,v) = \gamma_1 u + \gamma_2 v = \int_0^T G(T,s)B(s)u(s)\,ds + [G(T,0) - I]v,$$

I, the identity operator in $\mathcal{B}$, and x_0 by (9.7). We thus have a controllability problem relative to the single-point target set $W = \{0\} \subset \mathcal{B}$. Controllability inequalities (6.5) then become

$$0 \leq \sup_{u \in U} \int_0^T \langle B(s)u(s), G'(T,s)x' \rangle\,ds + \sup_{v \in V} \langle v, [G(T,0) - I]'\,x' \rangle, \quad x' \in \mathcal{X}'$$

where G' and (G-I)' are adjoint operators.

In all the preceding examples the space $\mathcal{X}$ was identified with the space $\mathcal{B}$ of vector values x(t). More generally, let us denote by $C(\tau,T,\mathcal{B})$ the space of continuous functions of $t \in [0,T]$ into $\mathcal{B}$, so that $x \in C(\tau,T,\mathcal{B})$ for all the solutions of (9.2), and by L a linear operator from $C(\tau,T,\mathcal{B})$ into a Banach space $\mathcal{X}$, not necessarily $\mathcal{B}$. In this case x_0 shall be defined by

$$\gamma(u,v) = \gamma_1 u + \gamma_2 v = L\left(t \to \int_\tau^t G(t,s)B(s)u(s)\,ds\right) + L(t \to G(t,\tau)v)$$

$$x_0 = L\left(t \to \int_\tau^t G(t,s)c(s)\,ds\right).$$

There are a number of controllability problems which can be dealt with this way. For instance let $\mathcal{B} = E^n$, $\mathcal{X} = E^k$, and let L be represented by

$$L\,x = M\,x(\tau) + N\,x(T) + \int_\tau^T dF(s)\,x(s)$$

where M, N, F(t) are $n \times k$ matrices, $t \to F(t)$ is of bounded variation on $[\tau,T]$ and the integral is a Riemann-Stieltjes one. The target set W in this will be a given subset of E^k (C. Marchiò [35]).

Or else, let $\mathcal{X} = C(\tau,T,\mathcal{B})$, let L be identity operator in $C(\tau,T,\mathcal{B})$, and let W be defined by

$$\sup_{[\tau,T]} |x(t) - w_0(t)|_{\mathcal{B}} \leq \rho$$

or by

$$\int_{\tau}^{T} |x(t) - w_0(t)|^2_{\mathcal{B}}\, dt \leq \rho$$

with $w_0 \in C(\tau,T,\mathcal{B})$, $\rho > 0$. A problem of this kind, for a perturbed control process, is that considered by <u>M.C. Delfour</u> - <u>S.K. Mitter</u> [9].

In concluding this review we shall recall the very general controllability problems suggested (for nonlinear differential equations) by <u>I. Tarnove</u> [48] at the 1967 Los Angeles Conference. His results, in our opinion, deserve being developed for linear differential equations both in the f.d.c. and in i.d.c.

I wish to thank Jack K. Hale and George R. Sell for reading this manuscript and making useful comments and criticisms.

REFERENCES

[1] H. A. Antosiewicz, Linear control systems, Arch. Rat. Mech. Anal. 12 (1963), 313-324 ;

[2] A. V. Balakrishnan, Optimal control problems in Banach spaces, J.SIAM Control A, 3 (1965),152-180 ;

[3] R. Conti, Contributions to linear control theory, J. Diff. Eqs., 1 (1965), 427-444 ;

[4] R. Conti, Problems in linear control theory, Acta Fac. Rer. Nat. Univ. Comenianae, (Bratislava), Math. 17 (1967), 73-80 ;

[5] R. Conti, On some aspects of linear control theory, Math. Theory of Control, edited by A.V. Balakrishnan and Lucien W. Neustadt, Acad. Press 1967, 285-300 ;

[6] R. Conti, Time optimal solution of a linear evolution equation in Banach spaces, J. Opt. Theory & Appls., 2 (1968), 277-280 ;

[7] R. Conti, A convex programming problem in Banach spaces and applications to optimum control theory, J. Comp. Syst. Sci., 4 (1970), 38-48 ;

[8] M. C. Delfour, Generalized controllability for perturbed linear systems, SRC-68-5, Case Western Reserve Univ., June 1968 ;

[9] M. C. Delfour - S.K.Mitter, Reachability of perturbed systems and min sup properties, J. SIAM Control, 7 (1969), 521-533 ;

[10] Yu. V. Egorov, Necessary conditions for optimum controls in Banach spaces (Russian), Matem. Sbornik, 64 (1964), 79-101 ;

[11] P. L. Falb, Infinite dimensional control problems.I : On the closure of the set of attainable states for linear systems, J. Math. Anal. & Appls., 9 (1964), 12-22 ;

[12] H. O. Fattorini, Control in finite time of differential equations in Banach space, Comm. on pure & appl. Math., 19 (1966), 17-34 ;

[13] H.O. Fattorini , On Jordan operators and rigidity of linear control systems, Revista Un. Mat. Argentina, 23 (1967),67-75 ;

[14] H. O. Fattorini, A remark on the "Bang-Bang" principle for linear control systems in infinite dimensional space, J. SIAM Control, 6 (1968), 109-113 ;

[15] H. O. Fattorini, On complete controllability of linear systems, J. Diff. Eqs., 3(1967), 391-402 ;

[16] H. O. Fattorini, Some remarks on complete controllability, J. SIAM Control A, 4 (1966), 686-694 ;

[17] H. O. Fattorini, Controllability of higher order linear systems, Conference on the Math. ~~Control~~ Theory of Control, L.A. 1967, edited by A.V.Balakrishnan and Lucien W. Neustadt, Acad. Press 1967, 301-311 ;

[18] H. O. Fattorini, Boundary control systems, J. SIAM Control, 6 (1968), 349-385 ;

[19] A. Friedman, Optimal control in Banach spaces, J. Math. Anal. & Appls. 19 (1967), 35-55 ;

[20] R. Gabasov - F. M. Kirillova, The solution of some problems in the theory of optimal processes (Russian), Avtom. i Telem. 25 (1964), 1058-1066 ; engl. transl. in Autom. & Remote Control, 25, 945-955

[21] H. Halkin, A generalization of LaSalle's "Bang-Bang" principle, J. SIAM Control A, 2 (1965), 199-202 ;

[22] H. Hermes, On the closure and convexity of attainable sets in finite and infinite dimensions, J. SIAM Control, 5 (1967), 409-417 ;

[23] H. Hermes - J.P.LaSalle, Functional Analysis and Time Optimal Control, Academic Press,1969 ;

[24] R. E. Kalman, Contributions to the theory of optimal control, Symp. Intern. de Ecuaciones diferenciales ordinarias (1959, publ. 1961 ; 102-119 ;

[25] R. E. Kalman - Y.C. Ho - K. S. Narendra, Controllability of linear dynamical systems, Contribs. to Diff. Eqs., 1 (1963), 189-213 ;

[26] V. Klee, Separation and support properties of convex sets.A Survey., Control Theory and the Calculus of Variations, edited by A.V. Balakrishnan, U.C.L.A., July 1968, Academic Press,1969, 235-303 ;

[27] N. N. Krasovskii, On the theory of optimal regulation, (Russian), Prikl. Mat. i Meh., 23 (1959), 625-639 ; engl. transl. J. Appl. Math. Mech., 23, 899-919 ;

[28] N. N. Krasovskii, Control theory of motion. Linear systems. (Russian) Izd. Nauka, Moscow, 1968 ;

[29] E. Kreindler, Contributions to the theory of time optimal control, J. Franklin Inst., 275 (1963), 314-344 ;

[30] L. M. Kuperman - Yu. M. Repin, On controllability in infinite dimensional spaces (Russian), Dokl. Akad. Nauk SSSR, 200 (1971),767-769

[31] A. B. Kurjanskii, On controllability in Banach spaces (Russian), Diff. Uravnenia, 5 (1969), 1715-1718 ;

[32] J. P. LaSalle, The time optimal control problem, Contrs. to the Theory of Non-lin. Oscill., 5 (1960), 1-24 ;

[33] E. B. Lee - L. Markus , Foundations of Optimal Control Theory, J. Wiley & Sons, 1967 ;

[34] N. Levinson, Minimax, Liapunov and "Bang-Bang", J. Diff. Eqs., 2 (1966), 218-241 ;

[35] C. Marchiò, Unpublished paper ;

[36] L. Markus, Optimal control of limit cycles, Three Lectures on Control Theory, Warwick Control Theory Centre, July 1971, Report 1, 1-26 ;

[37] W. L. Miranker, Approximate controllability for distributed linear systems, J. Math. Anal. Appls., 10 (1965), 378-387 ;

[38] S. K. Mitter, Theory of inequalities and the controllability of linear systems, Math. Theory of Control, edited by A.V.Balakrishnan and Lucien W. Neustadt, Academic Press,1967, 203-212 ;

[39] L. W. Neustadt, Optimization, a moment problem, and nonlinear programming, J. SIAM Control A, 2 (1964), 33-53 ;

[40] W. A. Porter, On the optimal control of distributed systems, J. SIAM Control, 4 (1966), 466-472 ;

[41] G. Pulvirenti - G. Santagati, Controlli lineari negli spazi di Orlicz Ann. di Mat. pura ed appl., (4) 76 (1967), 165-202 ;

[42] G. Pulvirenti - G. Santagati, Sulla teoria dei controlli lineari negli spazi di Orlicz di funzioni a valori vettoriali, Ann. di Mat. pura ed appl., (4) 78 (1968), 279-322 ;

[43] W. T. Reid, Ordinary linear differential operators of minimum norm, Duke Math. J., 29 (1962), 591-606 ;

[44] P. Santoro, Condizioni di non controllabilità per i sistemi lineari, Boll. Un. Mat. Ital., (3) 19 (1964), 400-406;

[45] J. Schmets, Sur le principe du "Bang-Bang", Revue Roum. de Math. pures et appl., 15 (1970), 633-641 ;

[46] F. A. Sholokovitch, On controllability in Hilbert space (Russian), Diff. Urawnenia, 3 (1967), 479-484 ;

[47] L. M. Sonneborn - F. S. Van Vleck, The Bang-Bang principle for linear control systems, J. SIAM Control A, 2 (1965), 151-159 ;

[48] I. Tarnove, A controllability problem for nonlinear systems, Math. Theory of Control, edited by A.V. Balakrishnan and Lucien W. Neustadt, Academic Press, 1967, 170-179 ;

[49] K. Tsujioka, Remarks on controllability of second order evolution equations in Hilbert spaces, J. SIAM Control, 8 (1970), 90-99.

Abstract Unilateral Problems

G. Fichera

Boundary value problems for differential operators of elliptic type lead to the following problems of functional analysis. Let H be a Hilbert space, which we suppose for simplicity to be real. Let V be a closed linear subspace of H and $B(u,v)$ a bounded real bilinear form defined on $V \times V$. Let $F(v)$ be a bounded linear functional defined on V. The problem to be considered is the following

$$B(u,v) = F(v) \quad , \quad u \in V \, , \quad \forall v \in V. \tag{1}$$

Conditions to be satisfied by B and F for the existence and the uniqueness of a solution u of problem (1) are well known. In the case that $B(u,v)$ is symmetric, equations (1) are obtained by imposing upon the functional

$$\mathfrak{J}(v) = \frac{1}{2} B(v,v) - F(v) \tag{2}$$

the necessary condition for attaining a minimum on V.

In these last nine years, problems connected with "unilateral constraints" have been extensively considered. These problems consist in substituting for the classical equations, either in the domain or at the boundary, conditions expressed by several kinds of inequalities. These problems are mainly suggested by Mechanics and Physics, when the material structure under consideration is submitted to unilateral constraints, i.e. when the class V of the admissible displacements contains vectors which are irreversible. That means that $v \in V$ does not imply $-v \in V$.

The prototype of this kind of problems is the one proposed in 1933 and, more explicitly, in 1959 by Antonio Signorini [1],[2], and nowadays known as the Signorini problem. The analytical investigation of this problem, published in 1963 [3], started the theory of unilateral problems, which some authors, rather improperly, denote as variational inequalities.(1)

Let us suppose that $B(u,v)$ is symmetric and let V be a closed convex set of H. If we impose upon the functional $\mathfrak{J}(v)$ given by (2) the necessary condition for attaining a minimum in V we easily get

$$B(u, v-u) \geq F(v-u) \quad , \quad u \in V, \ \forall v \in V. \tag{3}$$

This is an abstract unilateral problem. Of course in investigating this problem we may also consider the more general case when B is not symmetric.

---

(1) The results contained in paper [3], concerning existence and uniqueness for the Signorini problem, are the first ones connected with the investigation of a unilateral problem. These results were announced in February 1963 (see [4]) and published in a paper [3] submitted for publication in September 1963. They were also communicated in two Symposia: the first held in Tbilisi (September 1963, [5]), the second in Turin (October 1963, [6]).

It must be remarked that the problem (3) has quite different degrees of difficulty according to the circumstance that $B(v,v)$ is coercive or is not. If $B(v,v)$ is coercive on $\|v\|^2$ i.e. if there exists a positive constant c such that

$$B(v,v) \geq c\,\|v\|^2 ,$$

then the proof of the existence and of the uniqueness of a solution u of the unilateral problem (3) is almost trivial in the case that $B(u,v)$ is symmetric. In fact in this case the problem is easily reduced to the existence of a unique projection of a point of a Hilbert space H into a closed convex subset of H. The nonsymmetric case is easily reduced to the symmetric one by a simple argument founded on the use of a suitable contraction mapping.(2)

Much more difficult is the case when $B(v,v)$ is not coercive on $\|v\|^2$, but satisfies only a semi-coerciveness condition. This is just the case of the Signorini problem.

Let us consider the specific case concerning the Signorini problem. Let A be a bounded domain of the cartesian space X^r with a piece-wise smooth boundary ∂A and satisfying a restriced cone-hypothesis. Such kind of domains, which I have been considering since 25 years, are denoted in my papers as "properly regular domains".

Let $v(x)$ be a smooth r-vector valued functions defined on A. The space H is the one obtained by functional completion through the norm

$$\|v\|^2 = \int_A |v|^2 dx + \frac{1}{4}\int_A \left(v_{i/k} + v_{k/i}\right)\left(v_{i/k} + v_{k/i}\right) dx \,. \tag{4}$$

---

(2) See [9].

We denote by $\varepsilon_{ik}(v)$ the strain components

$$\varepsilon_{ik}(v) = \frac{1}{2}(v_{i/k} + v_{k/i})$$

and consider the symmetric bilinear form

$$B(u,v) = \int_A a_{ik,jh}(x)\, \varepsilon_{ik}(u)\, \varepsilon_{jh}(v)\, dx . \tag{5}$$

The real valued functions $a_{ik,jh}$ are supposed smooth in $\bar{A}$ - for instance belongingto $C^{\infty}(\bar{A})$ - and satisfying the conditions

$$a_{ik,jh}(x) \equiv a_{jh,ik}(x) \quad , \quad a_{ik,jh}(x) \equiv a_{ki,jh}(x) .$$

Moreover the quadratic form

$$a_{ik,jh}(x)\, \varepsilon_{ik}\, \varepsilon_{jh}$$

in the variables ε_{ik} (with $\varepsilon_{ik} = \varepsilon_{ki}$) is strictly positive.

Let Σ be a subset of ∂A (eventually coinciding with ∂A) consisting of a finite set of non-overlapping regular open $(r-1)$-dimensional surfaces.

Let $\mathcal{U}_{\Sigma}$ be the subset of H formed by all the functions v of H satisfying on Σ the unilateral condition

$$v_i \nu_i \geq 0 . \tag{6}$$

This condition must be understood in the sense of the H_1 space, which is isomorphic to our space H, because of the hypotheses we have assumed on A.

If A is conceived as the natural configuration of an elastic body and Σ as a frictionless surface, the condition (6) means that the unstrained body A rests on Σ and in any of its admissible motions it is not permitted to go below Σ . Let f be a given r-vector valued function, belonging to $L^2(A)$, and let φ be a given r-vector valued function, belonging to $L^2(\Sigma^*)$ $\{\Sigma^* = \partial A - \Sigma\}$; f and φ determine, respectively, the body forces acting on A and the surface forces acting on Σ^*. The equilibrium configuration of A is the one which minimizes in the class $\mathcal{U}_\Sigma$ the <u>energy functional</u> (2), where $B(u,v)$ is given by (5) and

$$F(v) = \int_A f v \, dx - \int_{\Sigma^*} \varphi v \, d\sigma .$$

The necessary and sufficient condition for the existence of a minimum of the functional $\mathcal{J}(v)$ in the set $V \equiv \mathcal{U}_\Sigma$ leads to the unilateral problem (3).

As it is remarked above, the main difficulty in handling this problem consists in the fact that the bilinear form (5) is not coercive on the norm $\|v\|$ given by (4).

Let R be the kernel of the quadratic form $B(v,v)$:

$$R \equiv \{ \varrho \, ; \, B(\varrho,\varrho) = 0 \}$$

and let R_Σ be the intersection of R with $\mathcal{U}_\Sigma$

$$R_\Sigma = R \cap \mathcal{U}_\Sigma .$$

It is easily seen that the condition

(7) $$F(\varrho) \leq 0 \qquad \varrho \in R_\Sigma$$

is necessary for the existence of a solution of problem (3). In [3] it is shown that if (7) is satisfied in the <u>strong sense</u> (i.e. $F(\varrho) = 0$, $\varrho \in R_\Sigma$ imply $-\varrho \in R_\Sigma$) then there exists a solution u of the unilateral problem (3) and every other solution is given by $u+\rho$, where ρ is any vector of R such that $F(\rho)=0$, $u+\rho \in \mathcal{U}_\Sigma$.

The fact that (7) is satisfied in the strong sense means that it must be $F(\rho) < 0$ for every $\rho \in R_\Sigma$ which is irreversible, i.e. such that $-\rho \notin R_\Sigma$.

In [3] it is also shown that this strong condition, not only is sufficient, but, under suitable assumptions for Σ (for instance when Σ is planar) is also necessary for the existence of a solution.

We refer the reader to paper [3] for the details of the proofs. However we wish to review here some of the most significant aspects of the analytical situation in the Signorini problem.

1) Set

$$p_0(v) = \left(\int_A |v|^2 dx\right)^{1/2}, \quad p_1(v) = \int_A \varepsilon_{ik}(v)\, \varepsilon_{ik}(v)\, dx,$$

then

$$\|v\|^2 = [p_0(v)]^2 + [p_1(v)]^2. \tag{8}$$

The space H is a pre-Hilbert space (i.e. non necessarily complete Hilbert space) with respect to the norm $p_0(v)$; $p_1(v)$ is a semi-norm in H (see [3] p.109).

2) The space

$$R \equiv \{v;\ p_1(v) = 0\}$$

is finite dimensional (see [3], pp.112-113).

3) There exists a positive constant c_0 such that

$$\inf_{\rho \in R} p_0(v-\rho) \le c_0\, p_1(v)$$

(see [3] , p.115).

4) There exists a positive constant c such that

$$B(v,v) \geq c\left[p_1(v)\right]^2$$

(see [3] , p.117).

5) The condition (7') is satisfied in the strong sense. (see [3] ,p.117 and pp. 131-135).

Let us now suppose that H is an abstract Hilbert space and the norm of H may be decomposed as indicated by (8). Suppose, moreover, that H is a pre-Hilbert space with respect to p_0 and p_1 is a semi--norm in H .

Let us assume that, given the bounded bilinear form (not necessarily symmetric) $B(u,v)$, conditions 2),3),4) are satisfied. Assume that also 5) is satisfied, in the sense that if $R \cap V \neq \{0\}$ then the strong condition

$$(7') \qquad F(\rho) < 0 \qquad \text{for } \forall \rho \in R \cap V, \ \rho \neq 0,$$

holds.

It has been shown (see [9]) that under these assumptions the results of [3] still hold so that they may be presented as very particular case of a general abstract theorem.

Unfortunately imitation of the results of [3] in the general abstract setting, which has been considered, leads to unsatisfactory results. In fact it must be pointed out that results of [3] <u>relate to a particular convex set</u>: the set $\mathcal{U}_\Sigma$, which has the peculiarity of being a <u>cone</u> (i.e. $v \in \mathcal{U}_\Sigma$ implies $tv \in \mathcal{U}_\Sigma$ for $t>0$). Under this circumstance the conditions stated in [3] are the best obtainible ones, but one has to expect that, for a general convex set, the conditions considered for a cone are not appropriate, and more sophisticated conditions, which consider the geometrical nature of V, must be discovered.

In fact, applying the above mentioned general abstract theorem to very simple unilateral problems, not related to a cone, one gets unacceptable answers.

Let us consider, as an example, the two-dimensional space.

Condition 1) is satisfied assuming $(v \equiv v_1, v_2)$:

$$\|v\|^2 = v_1^2 + v_2^2 ,$$

$$[p_0(v)]^2 = \frac{3}{4}(v_1^2 + v_2^2) - \frac{1}{2} v_1 v_2 , \qquad [p_1(v)]^2 = \frac{1}{4}(v_1 + v_2)^2 .$$

The space R is determined by the condition $v_1 + v_2 = 0$ and is one--dimensional. Thus condition 2) is satisfied. We have

$$\inf_{\rho \in R} [p_0(v-\rho)]^2 = \frac{7}{8}(v_1 + v_2)^2 = \frac{7}{2}[p_1(v)]^2$$

which proves that condition 3) holds.

Let us assume

$$B(u,v) = u_1 v_1 + u_1 v_2 + u_2 v_1 + u_2 v_2 ,$$

then

$$B(v,v) = (v_1 + v_2)^2 = 4[p_1(v)]^2 ,$$

which proves 4).

Assume as V the closed convex set defined by the condition

$$0 \leq v_2 \leq 1$$

and consider the unilateral problem (3), which in this particular case becomes

$$(9) \quad u_1(v_1 - u_1) + u_1(v_2 - u_2) + u_2(v_1 - u_1) + u_2(v_2 - u_2) \geq F_1(v_1 - u_1) + F_2(v_2 - u_2),$$

F_1 and F_2 are given real numbers.

According to the condition (7') this problem has a solution if

$$-F_1 \rho + F_2 \rho < 0$$

for every ρ such that $0 \leq \rho \leq 1$. This implies that a solution exists if $F_2 - F_1 < 0$. However the problem has a solution for every choice of F_1, F_2. In fact from (9), supposing $v_1 > 0$ (supposing $v_1 < 0$) and making $v_1 \to +\infty$ ($v_1 \to -\infty$), we get

$$u_1 + u_2 \geq F_1 \qquad (u_1 + u_2 \leq F_1).$$

Hence $u_1 + u_2 = F_1$. From (9) we obtain

$$(F_1 - F_2)(v_2 - u_2) \geq 0,$$

which implies

$$u_2 \begin{cases} = 1 & \text{if} \quad F_1 - F_2 < 0, \\ \in [0,1] & \text{if} \quad F_1 - F_2 = 0, \\ = 0 & \text{if} \quad F_1 - F_2 > 0. \end{cases}$$

It is trivial to verify that the values of u_1 and u_2, which we have found, satisfy (9).

I want to refer here on results, concerning unilateral problems in a Hilbert space for semi-coercive bilinear forms, which are contained in a monograph of mine published, quite recently, on Handbuch der Physik [10]. From these results it is clearly seen how the geometric structure of the convex set V enters into the picture and how the particular case of a cone is misleading if taken as a pattern for the general case.

We shall denote by (u,v) the scalar product in the Hilbert space H . Let $B(u,v)$ be a bounded bilinear form defined on $H \times H$. Let us first suppose that B be symmetric. Let T be the bounded linear operator defined by the condition $B(u,v) = (Tu,v)$. Let $N(T)$ be the kernel of the operator T . Let us assume that the quadratic form $B(v,v)$ is non-negative. The linear subspace $N(T)$ of H is also the kernel of the quadratic form $B(v,v)$, i.e.

$$N(T) \equiv \{v \,;\, B(v,v) = 0\}.$$

Let Q be the orthogonal projector of H onto the kernel of $B(v,v)$ and set $P = I - Q$ (I = identity operator). We shall assume the following hypotheses.

(I) Semi-coerciveness hypothesis

(10) $$B(v,v) \geq c \|Pv\|^2 \,, \quad \forall\, v \in H .$$

c is a positive constant independent on v .

(II) The kernel of the quadratic form $B(v,v)$ is finite-dimensional.

Let U be an arbitrary subset of H containing some $u \neq 0$. Let us consider for any $u \in U$, $u \neq 0$, the set of non-negative numbers t such that $t\|u\|^{-1}u \in U$. We shall denote by $p(u, U)$ the supremum of this set, i.e.

$$p(u, U) = \sup\left\{t\ ;\ u \in U,\ u \neq 0,\ t\|u\|^{-1}u \in U\right\}.$$

If u_0 is a point of H we shall denote by $U(u_0)$ the set of all the points u of H such that $u + u_0 \in U$. If T is a mapping from H into H and U a subset of H, by $T[U]$ we shall indicate the image of U under the mapping T.

Let $N(F)$ be the kernel of the linear bounded functional F, i.e.

$$N(F) = \left\{v\ ;\ F(v) = 0\right\}.$$

Set

$$L = N(F) \cap N(T) \quad , \quad N(T) = L \oplus L_1 .$$

Let $\tilde{Q}$ be the orthogonal projector of H onto L. Set $\tilde{P} = I - \tilde{Q}$. Let Q_1 be the orthogonal projector of H onto L_1.

The following theorem holds:

THEOREM 1 - The unilateral problem (3) has a solution if there exists $u_0 \in V$ such that

(i) $F(\varrho) < 0$ for $\varrho \in N(T) \cap V(u_0)$, $p\left\{Q_1\varrho, Q_1[N(T) \cap V(u_0)]\right\} = +\infty$.

(ii) The set $\tilde{P}[V(u_0)]$ is closed.

The following theorem proves that condition (i) of Theorem 1 is necessary.

THEOREM 2. ***If problem (3) has a solution, for any*** $u_0 \in V$ ***and any*** ϱ ***such that***

$$\varrho \in N(T) \cap V(u_0), \quad p\{Q_1 \varrho, Q_1 [N(T) \cap V(u_0)]\} = +\infty,$$

one must have $F(\varrho) < 0$.

In [10] an example is given which proves that if condition (ii) does not hold, Theorem 1 could fail to be true.

In [10] it is also shown that the vector Pu is uniquely determined by (3) and, moreover, if u is a solution of (3), every other solution is given by $u+\varrho$, where ϱ is any vector of L such that $u+\varrho \in V$.

The analysis in the case when B is not symmetric is more complicated.

Let us mantain for $B(u,v)$, for V and for F the hypotheses already stated, but we no longer assume $B(u,v)$ to be symmetric. Q is the orthogonal projector of H onto the kernel of $B(v,v)$ and set, as before, $P = I - Q$. (10) is supposed to hold and hypothesis (II) to be satisfied.

If we set $B(u,v) = (Tu,v)$ and denote by T^* the adjoint operator of T, the kernel of $B(v,v)$ coincides with the kernel $N(T+T^*)$ of the operator $T+T^*$ and moreover

$$N(T) \equiv N(T^*) \subset N(T+T^*).$$

Let us denote by $K(T)$ the orthogonal complement of $N(T)$ with respect to $N(T+T^*)$, i.e.

$$N(T+T^*) = N(T) \oplus K(T).$$

We define $N(F)$, L, L_1, $\tilde{Q}$, $\tilde{P}$, Q_1 as before. The following theorem generalizes Theorem 1 to a non symmetric form:

THEOREM 3. The unilateral problem (3) for the (not necessarily symmetric) bilinear form $B(u,v)$ has a solution u if there exists a $u_0 \in V$ such that the following conditions are satisfied:

(i) $F(\rho)<0$ for $\rho \in N(T)\cap V(u_0)$, $p\{Q_1\rho, Q_1[N(T)\cap V(u_0)]\} = +\infty$.

(ii) The set $\tilde{P}[V(u_0)]$ is closed.

(iii) Let Q_0 be the orthogonal projector of H onto $K(T)$. For every ρ satisfying the conditions

$$Q_0\rho \neq 0, \quad \rho \in N(T+T^*)\cap V(u_0),$$

$$p\left\{(Q_0+Q_1)\rho, (Q_0+Q_1)\left[N(T+T^*)\cap V(u_0)\right]\right\} = +\infty,$$

there exists a vector $v_\rho \in V$ such that

$$F(\rho) + B(\rho, v_\rho) < 0.$$

As in the symmetric case it is possible to state a theorem which proves the necessity of condition (i). Analogous results we have for condition (ii) and (iii).

We refer for the proofs of the theorems stated in this paper to the monograph [10].

References

[1] A.SIGNORINI - Sopra alcune questioni di elastostatica - Atti Soc. Ital. per il Progresso delle Scienze,1933.

[2] A.SIGNORINI - Questioni di elasticità non linearizzata o semi-linearizzata - Rend.di Matem.e delle sue appl. 18, 1959.

[3] G.FICHERA - Problemi elastostatici con vincoli unilaterali: il problema di Signorini con ambigue condizioni al contorno - Mem.Acc.Naz.Lincei,s.VIII,7,fasc.5, 1964.

[4] G.FICHERA - Sul problema elastostatico di Signorini con ambigue condizioni al contorno - Rend.Acc.Naz.Lincei,s.VIII,34,1963.

[5] G.FICHERA - The Signorini elastostatics problem with ambiguous boundary conditions - Proc.of the Internat. Symp."Applications of the theory of functions in continuum mechanics",vol.1, Tbilisi, sept.1963.

[6] G.FICHERA - Un teorema generale di semicontinuità per gli integrali multipli e sue applicazioni alla fisica-matematica - Atti del Convegno Lagrangiano,Acc.Sci.Torino,1963.

[7] G.STAMPACCHIA - Variational inequalities - in "Theory and applications of monotone operators" edited by A.Ghizzetti, Proc. of a NATO Advanced Study Institute, Venice 1968, Publ.Oderisi,Gubbio,1969.

[8] G.STAMPACCHIA - Formes bilinéaires coercitives sur les ensembles convexes - Compt.Rend.258, 1964.

[9] J.L.LIONS - G.STAMPACCHIA - Variational inequalities - Comm.Pure Appl.Math.20, 1967.

[10] G.FICHERA - Boundary Value Problems of Elasticity with Unilateral Constraints - Handbuch der Physik, Bd.VIa/2, Springer Verlag,1972.

Regularity for Some Degenerate Differential Operators

C. Goulaouic (*)

We give some results of existence and regularity (C^∞, Gevrey, analytic...) of solutions for degenerate linear operators and for associated boundary value problems, which may be considered as generalizations of the elliptic cases.

First, we consider the most simple example, the Legendre operator; then we will see some "natural generalisations" and some applications.

Notations: If Ω is an open subset of $\mathbb{R}^n$, we denote by $\mathcal{A}(\Omega)$ the space of analytic functions on Ω; more generally we denote by $G_s(\Omega)$ the space of Gevrey of order s ($s \geq 1$):

$$G_s(\Omega) = \Big\{ u \in C^\infty(\Omega);\ \text{for every compact } K \subset \Omega,\ \text{there exists } L > 0 \text{ such that: } \|D^\alpha u\|_{L^2(K)} \leq L^{|\alpha|+1} (\alpha!)^s \text{ for } \alpha \in \mathbb{N}^n \Big\}.$$

If F is a (closed) subset of $\mathbb{R}^n$, we denote by $C^\infty(F)$, $\mathcal{A}(F)$, $G_s(F)$ the spaces of restrictions to F of functions respectively C^∞, analytic, Gevrey of order s, in some neighborhood of F.

I. An example: Legendre's operator

We note:

$$\text{(I)} \qquad \mathcal{L} = -\frac{d}{dx}(1-x^2)\frac{d}{dx} \quad \text{on }]-1,+1[= I,$$

and we have:

Proposition I: $\mathcal{L}+I$ is an isomorphism

from $\{ u \in H^1(I);\ (1-x^2)u \in H^2(I) \}$ onto $L^2(I)$, (**)

from $C^\infty(\bar{I})$ onto $C^\infty(\bar{I})$,

from $\mathcal{A}(\bar{I})$ onto $\mathcal{A}(\bar{I})$,

from $G_s(\bar{I})$ " $G_s(\bar{I})$ for $s \geq 1$.

(*) The author's contribution to the following results was done in collaboration with M.S.Baouendi.

(**) For every $m \in \mathbb{N}$, we note $H^m(\Omega) = \{ u \in L^2(\Omega);\ D^\alpha u \in L^2(\Omega) \text{ for } |\alpha| \leq m \}$.

Proposition 2: For $u \in C^{\infty}(\bar{I})$ the following properties are equivalent:

$$\begin{cases} \text{i) } u \in \mathcal{A}(\bar{I}) \\ \text{ii) } \exists L>0;\ \forall k \in \mathbb{N};\ \|\mathcal{L}^k u\|_{L^2(I)} \leq L^{k+1}\, 2k! \end{cases}$$

It follows easily that the Fourier expansion $\mathcal{F}$ on the basis of Legendre's polynomials (normalized in $L^2(I)$) is an isomorphism:

$$\begin{cases} \text{from } C^{\infty}(\bar{I}) \text{ onto } \mathcal{S} \\ \text{from } \mathcal{A}(\bar{I}) \text{ onto } \varinjlim_{a>0} \ell^2_{a^{j^{1/2}}}\ . \end{cases} \quad (*)$$

These results are not very difficult, but the ideas of proof may be useful for the generalizations ; the first one is to consider the Laplacian Δ on the unit sphere S of $\mathbb{R}^3$ restricted to functions invariant by rotation with respect to a diameter; the projection on this diameter gives the operator $\mathcal{L}$. This idea reduces the problem to a similar one for an elliptic operator on the sphere S ; for example, proposition 2 appears to be a corollary of a theorem of [1] [8] about the iterates of an elliptic operator in an open set of $\mathbb{R}^n$.

Another method is to consider the operator $D_y y D_y$ on $]0,\infty[$, and to use some inequalities of Hardy's type ; for example :

for $u \in \mathcal{D}([0,\infty[)$ and $k \in \mathbb{N}$, we have:

$$(2) \begin{cases} \|D_y^k u\|_{L^2(0,\infty)} \leq \dfrac{2}{2k+1} \|D_y^{k+1} y u\|_{L^2(0,\infty)} \\ \|D_y^k u\|_{L^2(0,\infty)} \leq \dfrac{2^k}{k!} \|(D_y y D_y)^k u\|_{L^2(0,\infty)} . \end{cases}$$

(*) We note $\mathcal{S} = \{(f_j) \in \mathbb{C}^{\mathbb{N}};\ \forall \alpha \in \mathbb{N},\ (j^{\alpha} f_j)$ is a bounded sequence$\}$.

If $P(j)$ is a sequence of non-negative real numbers, we note

$$\ell^2_{P(j)} = \{(f_j) \in \mathbb{C}^{\mathbb{N}};\ \sum_{j=0}^{\infty} |f_j|^2 P(j) < \infty\}.$$

II. Generalizations for several variables

1°) Let Ω be a bounded open subset of $\mathbb{R}^n$ and φ a real analytic function defined in a neighborhood of $\overline{\Omega}$ such that

$$(3)\quad \begin{cases} \Omega = \{x \in \mathbb{R}^n;\ \varphi(x) > 0\} \\ \Gamma = \partial\Omega = \{x \in \mathbb{R}^n,\ \varphi(x) = 0\} \\ d\varphi(x) \neq 0 \ \text{for}\ x \in \Gamma. \end{cases}$$

Let us consider the following sesquilinear form, defined for u , v in $C_0^\infty(\Omega)$:

$$(4)\quad a(u,v) = \sum_{\substack{|\alpha| \leq 1 \\ |\beta| \leq 1}} \int_\Omega a_{\alpha\beta}(x)\varphi(x) D^\alpha u(x) \overline{D^\beta v(x)}\, dx \qquad (*)$$

We assume that the coefficients $a_{\alpha\beta}$ are analytic up to the boundary and there exists a constant $\lambda > 0$ such that for every $u \in C_0^\infty(\Omega)$ we have

$$(5)\quad \operatorname{Re} a(u,u) \geq \lambda \sum_{|\alpha| \leq 1} \|\sqrt{\varphi}\, D^\alpha u\|_{L^2(\Omega)} .$$

For $1 \leq j \leq n$ and $1 \leq k \leq n$, let Λ_{kj} denote the following first order operator

$$(6)\quad \Lambda_{kj} = \frac{\partial\varphi}{\partial x_k} D_j - \frac{\partial\varphi}{\partial x_j} D_k .$$

Finally, we consider the following second order degenerate elliptic operators:

$$(7)\quad \begin{cases} \mathcal{A}_1 = \mathcal{A}_1(x,D) = \sum_{\substack{|\alpha| \leq 1 \\ |\beta| \leq 1}} D^\beta a_{\alpha\beta}\, \varphi\, D^\alpha \\ \mathcal{A}_2 = \mathcal{A}_2(x,D) = \mathcal{A}_1 + \sum_{k,j} \Lambda_{kj}^* \Lambda_{kj} . \end{cases}$$

We have:

Theorem I. The operator $\mathcal{A}_i$ (i=1 or 2) realizes an isomorphism

from $C^\infty(\overline{\Omega})$ onto $C^\infty(\overline{\Omega})$

from $\mathcal{A}(\overline{\Omega})$ onto $\mathcal{A}(\overline{\Omega})$.

(*) We note $D_k = -i\frac{\partial}{\partial x_k}$ and $D^\alpha = D_1^{\alpha_1} \dots D_n^{\alpha_n}$ for $\alpha = (\alpha_1, \dots, \alpha_n) \in \mathbb{N}^n$.

For the proofs, we refer to [2][3]. The initial idea is the method of Morrey and Nirenberg [11] but with some modifications, for example an important use of the Hardy's inequalities (2).

Let us point out that the case $\mathcal{A}_2$ is simpler than the case $\mathcal{A}_1$ and can be reduced to an elliptic case by extending the number of variables.

<u>Remark I.</u> As a corollary of theorem I, we obtain some other analytic regularity for the operator $\mathcal{A}_1$; more preciely , let $\mathcal{O}$ be an open subset of $\mathbb{R}^n$ and assume that the coefficients of $\mathcal{A}_1$ are analytic in $\mathcal{O}$; then for $u \in C^\infty(\mathcal{O})$ and such that $\mathcal{A}_1 u \in a(\mathcal{O})$, we have $u \in a(\mathcal{O})$. But it is easy to see that $\mathcal{A}_1$ is not hypoelliptic in $\mathcal{O}$ if $\mathcal{O} \cap \Gamma \neq \emptyset$.

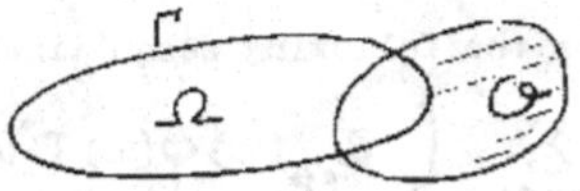

2)) Now, we want to generalize proposition 2, that is to characterize the analytic functions on a regular manifold with boundary, using powers of a differential operator: such a characterization in several variables needs a degenerate elliptic operator (we recall that, in the case of an analytic manifold without boundary, this sort of characterization was obtained in [8][1].

We consider first some subspaces of $C^\infty(\bar{\Omega})$ related to Gevrey's classes:

<u>Definition:</u> Let s be a real number ≥ 1; a function u is said to be in $a_s(\bar{\Omega})$ if and only if :

i) $u \in C^\infty(\bar{\Omega}) \cap G_s(\Omega)$,

ii) for every point in Γ , there exists a neighborhood V with local coordinates $x = (x_1, \dots, x_{n-1})$ (tangential) and y (normal), and a constant $M > 0$ such that for every $\alpha \in \mathbb{N}^{n-1}$ and $k \in \mathbb{N}$,

$$\| (D_y y D_y)^k D_x^\alpha u \|_{L^2(V)} \leq M^{|\alpha|+k+1} ((2k+|\alpha|)!)^s .$$

We can prove easily, using Hardy's inequalities (2):

$$G_s(\bar{\Omega}) \subsetneq a_s(\bar{\Omega}) \subsetneq G_{2s-1}(\bar{\Omega})$$

Therefore, for $s = 1$, $a_1(\bar{\Omega}) = a(\bar{\Omega})$.

(for $s > 1$, $G_s(\bar{\Omega}) \neq a_s(\bar{\Omega})$).

Theorem 2: Let s be a real number ≥ 1 and $u \in C^{\infty}(\bar{\Omega})$.

The following conditions are equivalent :

i) $u \in \mathcal{A}_s(\bar{\Omega})$

ii) There exists $L > 0$ such that, for every integer $k \geq 0$, $\|\mathcal{A}_2^k u\|_{L^2(\Omega)} \leq L^{k+1} (2k!)^s$.

For the proof, which is too long to be given here, we refer to [3].

3°) We can give some applications of theorem 2, using an operator $\mathcal{A}_2$ which we assume to be self adjoint.

We can describe the asymptotic behaviour of the eigenvalues (λ_j) of $\mathcal{A}_2$; more precisely, there exists a constant $C > 0$ such that

(9) $\lambda_j \sim C\, j^{2/n}$ when $j \to +\infty$.

Therefore, the Fourier expansion on an orthonormal basis of eigenfunctions is an isomorphism

$$\begin{cases} \text{from } C^{\infty}(\bar{\Omega}) \text{ onto } s \\ \text{from } \mathcal{A}_s(\bar{\Omega}) \text{ onto } \varinjlim_{a>0} \ell^2_{a^{j^{1/sn}}} \end{cases}.$$

It follows that the spaces $\mathcal{A}_s(\bar{\Omega})$ are interpolation spaces between $C^{\infty}(\bar{\Omega})$ and $\mathcal{A}(\bar{\Omega})$; they are characterized by their polynomial approximation; we can get by interpolation the results of regularity for elliptic boundary value problems with data in such spaces...etc.

Let us notice that the Gevrey spaces $G_s(\bar{\Omega})$ for $s > 1$ have not so good properties !

4°) It is not possible to characterize the analytic functions by the powers of an operator $\mathcal{A}_1$ (for $n \geq 2$). That can be proved by showing that the asymptotic behaviour of the eigenvalues of $\mathcal{A}_1$ (in the self-adjoint case) is not given by (9) (for $n \geq 3$, $\lambda_j(\mathcal{A}_1) \sim C_1\, j^{1/n-1}$); it is also possible to find a function

$$(10) \quad \begin{cases} u \in G_2(\bar{\Omega}) \text{ but } u \notin G_s(\bar{\Omega}) \text{ for } s < 2 \\ \text{such that } \|\mathcal{A}_1^k u\|_{L^2(\Omega)} \leq 2k! \text{ for } k \in \mathbb{N}. \end{cases}$$

5°) The characterization of analytic functions by the powers of a differential operator is done also for some irregular subsets of $\mathbb{R}^n$ (see [5]), but the problem is still open for some other reasonnable subsets, for example conic ones.

III. Remarks about analytic regularity

For an operator $\mathcal{A}$, the analytic regularity and the property about the iterates (theorem 2) are connected.

For example, we can prove that ii) implies i) in theorem 2 for $s=1$, using an idea of [9]:

We consider the series

$$v(t,x) = \sum_{k=0}^{\infty} t^k \frac{\mathcal{A}_2^k u(x)}{2k!} \quad ;$$

by assumption, this series defines a C^∞ function on $\Omega \times]-\varepsilon,\varepsilon[$ for some $\varepsilon > 0$; furthermore, we have

$$(D_t^2 + \mathcal{A}_2)\, v(t,x) = 0 \quad \text{in } \Omega \times]-\varepsilon,\varepsilon[\quad ,$$

but the operator $D_t^2 + \mathcal{A}_2$ is of the same type as $\mathcal{A}_2$ and by theorem 1 we know that v is analytic on $\overline{\Omega} \times]-\varepsilon,\varepsilon[$; then $u = v(\cdot,0)$ is analytic on $\overline{\Omega}$.

The same proof and (10) show that the operator $D_t^2 + \mathcal{A}_1$ has not the analytic regularity on $\overline{\Omega} \times \mathbb{R}$.

By the same way, we can find, for example, a function $v(t,x,y)$ which is in the Gevrey class G_2 near the origin in $\mathbb{R}^2 \times [0,\infty[$, and not in any class G_s near the origin for $s < 2$, and such that :

$$\text{(11)} \qquad D_t^2 + 4D_y\, y\, D_y + y\, D_x^2 + 2\, i\, D_y\,)\, v = 0 \quad .$$

Now, we define a function w near the origin in $\mathbb{R}^3$ by :

$$\text{(12)} \qquad w(t,x,z) = u(t,x,z^2)$$

and we see easily that w does not belong to any Gevrey class G_s for $s<2$ near the origin, and however

$$\text{(13)} \qquad \mathcal{A}_3 w = 0 \qquad \text{with } \mathcal{A}_3 = D_t^2 + D_z^2 + z^2 D_x^2 \quad .$$

In [6] it is proved for operators like $\mathcal{A}_3$ the following result : let $\mathcal{O}$ be an open subset of $\mathbb{R}^3$, s a real number ≥ 2, $u \in \mathcal{D}'(\mathcal{O})$ such that $\mathcal{A}_3 u \in G_s(\mathcal{O})$; then $u \in G_s(\mathcal{O})$.

This result and (12) are therefore complementary.

On the other hand, in [10] it is shown that the operator $\mathcal{A}_4 = D_z^2 + z^2 D_x^2$ is analytic hypoelliptic in $\mathbb{R}^2$; an open problem is then to characterize the degenerate elliptic operators with analytic coefficients which are analytic hypoelliptic, even when considering only operators of the form $\sum_{i=1}^{r} X_i^2$ where $(X_i)_{i=1,\dots,r}$ is a family of analytic vector fields (cf. [7]).

Bibliography

[1] N.Aronszajn ; Sur un theorème de la théorie des fonctions analytiques de plusieurs variables complexes; C.R. Acad. Sc. Paris,t.205(1937),16-18.

[2] M.S.Baouendi et C.Goulaouic;Etude de l'analyticité et de la régularité Gevrey pour une classe d'opérateurs elliptiques dégénérés ;Ann.SCIent.Ec.Norm.Sup. 4° série ,t.4 (1971),31-46.

[3] M.S.Baouendi et C. Goulaouic ; Régularité et itérés d'opérateurs elliptiques dégénérés ,applications;Journal of functional Analysis 9,(1972),208-248.

[4] M.S.Baouendi et C.Goulaouic ; Nonanalytic - hypoellipticity for some degenerate elliptic operators ; Bull.A.M.S. vol 78,number 3 (1972).

[5] M.S.Baouendi ,C.Goulaouic et B. Hanouzet; Caracterisation de classes de fonctions C^{∞} et analytiques sur une variété irrégulière à l'aide d'un opérateur différentiel;J.Math.pures et appl. (à paraitre).

[6] M.Derridj et C. Zuily; Régularité Gevrey d'opérateurs elliptiques dégénérés; C.R. Acad.Sc.Paris t. 273 (1971),720-723 . (et article à paraître).

[7] L.Hörmander; Hypoelliptic second order differential equations;Acta Math.(1968), 147-171.

[8] T. Kotake et N.S. Narasimhan; Fractional power of a linear elliptic operator; Bull.Soc.Math.France,90(1962),449-447 .

[9] J.L.Lions et E.Magenes ;Problèmes aux limites non homogènes t.3,Dunod (Paris) 1970.

[10] T.Matsuzawa; Sur les équations $u_{tt} + t^{\alpha} u_{xx} = f$ ($\alpha \geq 0$);Nagoya Math.J. vol 42 (1971),43-55.

[11] C.B. Morrey et L.Nirenberg; On the analyticity of the solutions of linear elliptic systems of partial differential equations;C.P.A.M.vol X (1957), 271-290.

Singular Elliptic Perturbations of Vanishing First-Order Differential Operators

E.M. de Jager

1. Introduction

In this paper we consider boundary value problems of the following type

$$L_\varepsilon[u_\varepsilon(x,y)] = \varepsilon L_2 u_\varepsilon(x,y) + L_1 u_\varepsilon(x,y) = 0 \text{ in a bounded domain } G \quad (1.1)$$

with

$$u_\varepsilon(x,y) = \phi_{+1}(x) \text{ along the upper boundary } \partial G_{+1}: y = \gamma_{+1}(x),$$
$$u_\varepsilon(x,y) = \phi_{-1}(x) \text{ along the lower boundary } \partial G_{-1}: y = \gamma_{-1}(x). \quad (1.2)$$

L_2 is a uniform elliptic differential operator of the second order

$$L_2 = a(x,y)\frac{\partial^2}{\partial x^2} + 2b(x,y)\frac{\partial^2}{\partial x\partial y} + c(x,y)\frac{\partial^2}{\partial y^2} + d(x,y)\frac{\partial}{\partial x} + e(x,y)\frac{\partial}{\partial y} + f(x,y) \quad (1.3)$$

and L_1 is a differential operator of the first order

$$L_1 = g(x,y)\frac{\partial}{\partial y} \quad (1.4)$$

ε is a small positive parameter.

Assuming $a,\dots,g$, ϕ_{-1} and ϕ_{+1} Hölder continuous, $a > 0$ and $f \leq 0$ in G, γ_- and $\gamma_+ \in C_{2+\alpha}$, the unique solvability of the b.v.p. is secured.

The coefficients $a,\dots,g$, the boundary values $\phi_{\pm 1}$ and the parametric representation of the boundary $y = \gamma_{\pm 1}(x)$ are called the parameters of the b.v.p.

Theorem I If the parameters of the b.v.p. belong to C_∞, $g(x,y) \neq 0$ in G and if G has the property that every line x = constant, intersecting the boundary, intersects this boundary exactly twice, it is possible to give an explicit asymptotic expansion of $u_\varepsilon(x,y)$ into powers of ε for ε sufficiently small.

This expansion is valid uniformly in all of G, with the exception of arbitrarely small fixed neighbourhoods $V(A)$ and $V(B)$ of the endpoints A and B. (see figure 1).

This theorem is proved in lit. ([1],[2]), where the differential equation (1.1) has even the more general form

$$(\varepsilon L_2+\bar{L}_1)[u_\varepsilon] = \varepsilon L_2[u_\varepsilon] + \left(\frac{\partial u_\varepsilon}{\partial y} + h(x,y)u_\varepsilon\right) = k(x,y) \quad (1.5)$$

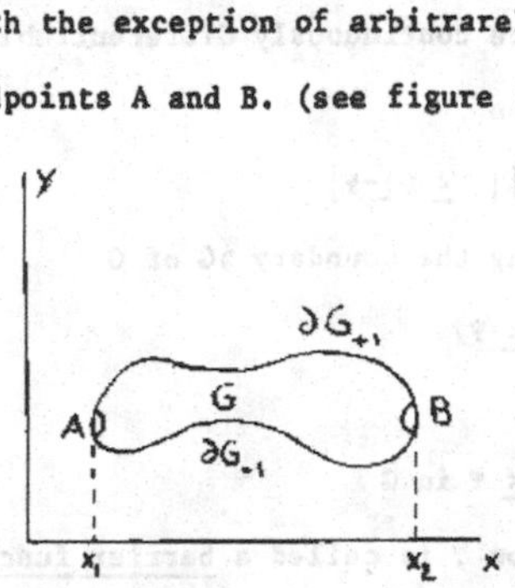

figure 1

In case the parameters are not infinitely differentiable an approximation of $u_\varepsilon(x,y)$ can be made up to order $O(\varepsilon^m)$, where m depends on the degree of differentiability of these parameters. For the precise conditions, see lit. [2], Theorem VI and VII. It is also possible to give a uniform approximation of $u_\varepsilon(x,y)$ in all of G, including V(A) and V(B); for this we refer the reader to lit. [3], [4].

In a first approximation $u_\varepsilon(x,y)$ is given by

$$u_\varepsilon(x,y) = w(x,y) + v(x,y;\varepsilon) + z(x,y;\varepsilon) \tag{1.6}$$

with $z(x,y;\varepsilon) = O(\varepsilon)$ in $G - V(A) - V(B)$ (1.7)

$w(x,y)$ is a solution of the socalled reduced equation

$$\bar{L}_1[w] = k(x,y) \tag{1.8}$$

satisfying only the boundary condition along ∂G_{-1} or ∂G_{+1}, say ∂G_i. The correction term $v(x,y;\varepsilon)$ has its support in a small neighbourhood along the boundary ∂G_{-i} of width $O(\varepsilon)$; further it satisfies along ∂G_{-i} the boundary condition

$$v(x,\gamma_{-i}(x);\varepsilon) = \phi_{-i}(x) - w(x,\gamma_{-i}(x)) \tag{1.9}$$

The function $v(x,y;\varepsilon)$ is a socalled <u>boundary layer</u> term.

The estimate $z(x,y;\varepsilon) = O(\varepsilon)$ is obtained by applying the maximum principle of E.Hopf. From this principle the following theorem can be derived (see lit. [2], lemma 1 and lit. [5], pp. 61-80).

<u>Theorem II</u> Let L be a second order linear differential operator

$$L = a(x,y)\frac{\partial^2}{\partial x^2} + 2b(x,y)\frac{\partial^2}{\partial x\partial y} + c(x,y)\frac{\partial^2}{\partial y^2} + d(x,y)\frac{\partial}{\partial x} + e(x,y)\frac{\partial}{\partial y} + f(x,y),$$

elliptic in a bounded domain G, with coefficients a, b etc. continuous in G; $a > 0$ and $f \leq 0$ in G.

If the twice continuously differentiable functions $\Phi(x,y)$ and $\Psi(x,y)$ satisfy in G the relation

$$|L[\Phi]| \leq L[-\Psi] \tag{1.10}$$

and if along the boundary ∂G of G

$$|\Phi| \leq \Psi, \tag{1.11}$$

then also

$$|\Phi| \leq \Psi \text{ in } \bar{G} \tag{1.12}$$

The function Ψ is called a <u>barrier function</u> for the function Φ.

By way of introduction we give in section 2 an asymptotic approximation of the solution of a b.v.p. of the type (1.1) - (1.2) with $g(x,y) \neq 0$ in G. In order not to grapple with many tedious calculations we give a representative example from which the method of lit. [1], [2] becomes fully clear.

The main purpose of this paper is to treat the case with $g(x,y) = 0$ along an arc ℓ in G. If $\frac{\partial g}{\partial y} > 0$ along ℓ there appears a socalled free boundary layer along ℓ; section 3 is devoted to the approximation of $u_\varepsilon(x,y)$ with $g(x,y) = 0$ along ℓ and $\frac{\partial g}{\partial y} > 0$ along ℓ.

In section 4 we discuss briefly the case, where $\frac{\partial g}{\partial y}$ is smaller than zero along ℓ.

2. The case $g(x,y) \neq 0$ in G

2.1 The boundary value problem and the location of the boundary layer.

We consider the boundary value problem

$$L_\varepsilon[u_\varepsilon] = \varepsilon\Delta u_\varepsilon + \frac{\partial u_\varepsilon}{\partial y} = 0, \text{ valid for } x^2+y^2 = r^2 < 1 \tag{2.1}$$

$$u_\varepsilon = \phi(\theta) \text{ for } r = 1;\ 0 \leq \theta \leq 2\pi.$$

Δ is the Laplace-operator, $\varepsilon > 0$ with $\varepsilon << 1$ and $\phi(\theta) \in C_3$.

The boundary for $y < 0$ is denoted by $y = \gamma_-(x)$ and for $y > 0$ by $y = \gamma_+(x)$; the pertaining boundary values are $u_{\varepsilon -} = \phi_-(\theta) = \tilde{\phi}_-(x)$ respectively $u_{\varepsilon +} = \phi_+(\theta) = \tilde{\phi}_+(x)$.

Assuming $|\phi(\theta)| < m$ (constant) one sees immediately that the function

$$w(x) = m + (1-x^2) \tag{2.2}$$

is a barrier function for $u_\varepsilon(x,y)$; application of theorem II gives that $|u_\varepsilon(x,y)|$ is uniformly bounded by the constant $M = 1+m$.

We introduce now the function

$$u_\varepsilon^*(x,y) = u_\varepsilon(x,y) - \tilde{\phi}_+(x)$$

u_ε^* satisfies the b.v.p.

$$\varepsilon\Delta u_\varepsilon^* + \frac{\partial u_\varepsilon^*}{\partial y} = -\varepsilon\tilde{\phi}_+''(x) \text{ for } x^2+y^2 < 1 \tag{2.3}$$

$$\begin{aligned} u_\varepsilon^* &= 0 \text{ along } y = \gamma_+(x) \\ u_\varepsilon^* &= \tilde{\phi}_-(x) - \tilde{\phi}_+(x) \text{ along } y = \gamma_-(x) \end{aligned} \tag{2.4}$$

Further we consider the function

$$\Psi(x,y) = \{\gamma_+(x)-y\}\chi(x) \qquad (2.5)$$

with $\chi(x) \in c_2$ in $[-1,+1]$ and

$$\min \chi(x) = \sup_{-1\leq x\leq +1} \frac{|\tilde{\phi}_+(x) - \tilde{\phi}_-(x)|}{\gamma_+(x) - \gamma_-(x)} + \sup_{-1\leq x\leq +1} \left|\frac{\tilde{\phi}''_+(x)}{\gamma''_+(x)}\right| +1$$

First of all we note that $\Psi(x,\gamma_-(x)) \geq |\tilde{\phi}_+(x) - \tilde{\phi}_-(x)|$ and

hence $\Psi(x,y) \geq |u^*_\varepsilon(x,y)|$ for $x^2+y^2 = 1$ (2.6)

We choose now $\chi(x)$ in such a way that the function $\Psi(x,y)$ becomes a barrier function for $u^*_\varepsilon(x,y)$.

Applying the differential operator to $\Psi(x,y)$ we get

$$\begin{aligned}\varepsilon\Delta\Psi + \frac{\partial\Psi}{\partial y} &= \varepsilon\left[(\gamma_+(x)-y)\chi''(x) + 2\gamma'_+(x)\chi'(x) + \gamma''_+(x)\chi(x)\right] - \chi(x)\\ &\leq -\tfrac{1}{2}N + 2\varepsilon\gamma'_+(x)\chi'(x) + \varepsilon\gamma''_+(x)\chi(x)\\ &\leq -\tfrac{1}{2}N - \varepsilon|\tilde{\phi}''_+(x)| + 2\varepsilon\gamma'_+(x)\chi'(x)\end{aligned}$$

for $x^2+y^2 < 1$ and ε sufficiently small, say $\varepsilon < \varepsilon_0$.

Taking finally $\chi'(x) \leq 0$ for $-1 \leq x \leq 0$ and $\chi'(x) \geq 0$ for $0 \leq x \leq 1$ we obtain

$$\varepsilon\Delta\Psi + \frac{\partial\Psi}{\partial y} \leq -\tfrac{1}{2}N - \varepsilon|\tilde{\phi}''_+(x)| < -\varepsilon|\tilde{\phi}''_+(x)| = -\left|\varepsilon\Delta u^*_\varepsilon + \frac{\partial u^*_\varepsilon}{\partial y}\right|$$

for $x^2+y^2 < 1$ and $\varepsilon < \varepsilon_0$ (2.7)

From (2.6) and (2.7) it follows that $\Psi(x,y)$ is a barrier function for $u^*(x,y)$, if $\varepsilon < \varepsilon_0$, and hence

$$|u_\varepsilon(x,y) - \tilde{\phi}_+(x)| \leq \{\gamma_+(x) - y\}\chi(x) \text{ for } r \leq 1 \text{ and } \varepsilon < \varepsilon_0 \qquad (2.8)$$

An important consequence of (2.8) is that $\frac{\partial u_\varepsilon}{\partial x}$ and $\frac{\partial u_\varepsilon}{\partial y}$ are uniformly bounded along the upper boundary $y = \gamma_+(x)$. If, instead of $+\frac{\partial u_\varepsilon}{\partial y}$, the term $-\frac{\partial u_\varepsilon}{\partial y}$ had appeared in the differential equation, then $\frac{\partial u_\varepsilon}{\partial x}$ and $\frac{\partial u_\varepsilon}{\partial y}$ were uniformly bounded along the lower boundary $y = \gamma_-(x)$.

It is to be expected that an approximation of $u_\varepsilon(x,y)$ can be obtained by a solution of the socalled <u>reduced</u> equation

$$\frac{\partial w}{\partial y} = 0$$

satisfying the boundary values along the upper boundary $y = \gamma_+(x)$.

The solution of this problem is simply

$$w = \tilde{\phi}_+(x) \tag{2.9}$$

However, w does not satisfy the boundary values along the lower boundary $y = \gamma_-(x)$; in order to make a uniformly good approximation of $u_\varepsilon(x,y)$ it is necessary to construct a correction term which adjusts the boundary values along the lower boundary.

2.2 The construction of the approximation.

Putting

$$u_\varepsilon(x,y) = \tilde{\phi}_+(x) + v(x,y;\varepsilon) \tag{2.10}$$

we obtain for $v(x,y;\varepsilon)$ the following boundary value problem

$$\varepsilon\Delta v + \frac{\partial v}{\partial y} = -\varepsilon\tilde{\phi}_+'' \text{ for } r < 1 \tag{2.11}$$

$$v = \tilde{\phi}_-(x) - \tilde{\phi}_+(x) \text{ along } y = \gamma_-(x) \tag{2.12}$$

$$v = 0 \qquad \text{along } y = \gamma_+(x)$$

In order to construct $v(x,y;\varepsilon)$ we develop (2.11) in local coordinates. Equation (2.11) reads in polar coordinates

$$L_\varepsilon[v] = \varepsilon\left(\frac{\partial^2 v}{\partial r^2} + \frac{1}{r}\frac{\partial v}{\partial r} + \frac{1}{r^2}\frac{\partial^2 v}{\partial \theta^2}\right) + \frac{\partial v}{\partial r}\sin\theta + \frac{1}{r}\frac{\partial v}{\partial \theta}\cos\theta = -\varepsilon\tilde{\phi}_+''$$

Substituting $r = 1 - \rho = 1 - \varepsilon t$ we obtain after a simple calculation the following expansion of the operator L_ε:

$$L_\varepsilon = \varepsilon^{-1}M_{-1} + M_0 + \varepsilon M_1 \tag{2.13}$$

with

$$M_{-1} = \frac{\partial^2}{\partial t^2} - \sin\theta\,\frac{\partial}{\partial t},$$

$$M_0 = -\frac{\partial}{\partial t} + \cos\theta\,\frac{\partial}{\partial \theta},$$

$$M_1 = \frac{1}{(1-\varepsilon t)^2}\frac{\partial^2}{\partial \theta^2} - \frac{t}{1-\varepsilon t}\left(\frac{\partial}{\partial t} - \cos\theta\,\frac{\partial}{\partial \theta}\right).$$

We consider now the functions v_0 and v_1 which satisfy the b.v.p.'s:

$$M_{-1}[v_0] = 0 \text{ for } 0 < t < \infty,\ \pi+\delta < \theta < 2\pi-\delta, \tag{2.14}$$

with $v_0 = \phi_-(\theta) - \phi_+(\theta)$ for $t = 0$,

$v_0 \to 0$ for $t \to +\infty$;

δ is an arbitrarely small, but fixed number (independent of ε).

$$M_{-1}[v_1] = -M_0[v_0] \text{ for } 0 < t < \infty,\ \pi+\delta < \theta < 2\pi-\delta \tag{2.15}$$

$v_1 = 0$ for $t = 0$

$v_1 \to 0$ for $t \to +\infty$.

From (2.14) it follows that

$$v_0 = \{\phi_-(\theta) - \phi_+(\theta)\}e^{t\sin\theta} = \{\phi_-(\theta) - \phi_+(\theta)\}e^{\frac{1-r}{\varepsilon}\sin\theta} \quad \pi+\delta<\theta<2\pi-\delta \tag{2.16}$$

and $v_1 = \{a(\theta)t^2 + b(\theta)t\}e^{t\sin\theta} =$

$$\{a(\theta)\ (\tfrac{1-r}{\varepsilon})^2 + b(\theta)\ (\tfrac{1-r}{\varepsilon})\}e^{\frac{1-r}{\varepsilon}\sin\theta}, \quad \pi+\delta < \theta < 2\pi-\delta \tag{2.17}$$

The functions $a(\theta)$ and $b(\theta)$, originating from the right hand side of (2.15), are twice continuously differentiable. ($\phi(\theta) \in C_3$!)

The functions v_0 and v_1 are typical boundary layer functions; they are asymptotically zero for $r \leq 1-\varepsilon^\mu$, $0 < \mu < 1$.

In order to define v_0 and v_1 uniquely and differentiable for $r = 0$ we introduce the cut-off factor $\chi(\rho)$, with $\chi(\rho) \in C_\infty$, $\chi(\rho) \equiv 1$ for $\rho < \rho_0$ and $\chi(\rho) \equiv 0$ for $\rho > 2\rho_0$; ρ_0 is chosen according to figure 2, where Δ is again a fixed arbitrarely small number, independent of ε.

Finally we set

$$u_\varepsilon(x,y) = \tilde{\phi}_+(x)+v(x,y;\varepsilon) =$$
$$= \tilde{\phi}_+(x)+\tilde{v}_0(x,y;\varepsilon)+\varepsilon\tilde{v}_1(x,y;\varepsilon)+$$
$$+ z(x,y;\varepsilon) \tag{2.18}$$

with $\tilde{v}_0 = \chi(\rho)v_0$ and $\tilde{v}_1 = \chi(\rho)v_1$.

Substituting (2.18) into (2.1) and using (2.13), (2.14) and (2.15) we get for the remainder term $z(x,y;\varepsilon)$ the b.v.p.

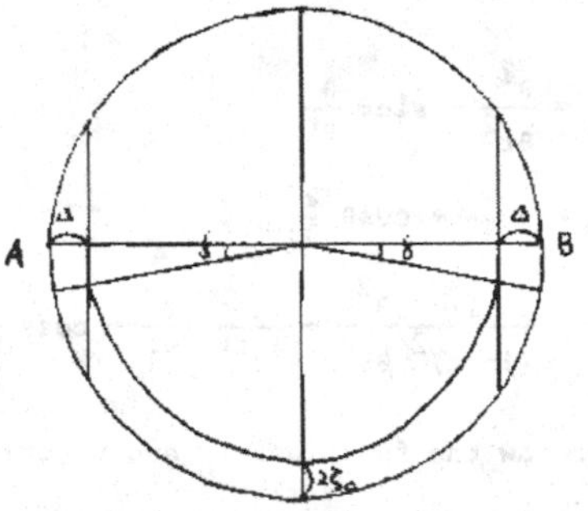

figure 2

$$L_\varepsilon[z] = L_\varepsilon[u_\varepsilon - \tilde{\phi}_+ - \tilde{v}_0 - \varepsilon\tilde{v}_1] =$$

$$-\varepsilon\tilde{\phi}''_+ - (\varepsilon^{-1}M_{-1}+M_0+\varepsilon M_1)[\tilde{v}_0+\varepsilon\tilde{v}_1] =$$

$$-\varepsilon\tilde{\phi}''_+ - \chi(\rho)(\varepsilon^{-1}M_{-1}+M_0+\varepsilon M_1)[v_0+\varepsilon v_1] + O(\varepsilon^N) =$$

$$-\varepsilon\tilde{\phi}''_+ - \varepsilon\chi(\rho)\{M_0[v_1] + M_1[v_0] + \varepsilon M_1[v_1]\} + O(\varepsilon^N) =$$

$$O(\varepsilon).$$

(N is an arbitrary positive number; $O(\varepsilon^N)$ denotes asymptotically zero) or $|L_\varepsilon[z]| < A\varepsilon$ for $x^2+y^2 < 1$, $-1+\Delta < x < 1-\Delta$ and for ε sufficiently small, say $\varepsilon < \varepsilon_0$; A is some fixed positive number, independent of ε.

$z = 0$ **for** $r = 1$, $-1+\Delta \le x \le 1-\Delta$;

$|z| \le B$ **for** $x = \mp 1 \pm \Delta$, **B some fixed positive number, independent of** ε. **(2.10)**

A barrier function for z in $\{x,y;\ x^2+y^2 \le 1,\ -1+\Delta \le x \le 1-\Delta\}$ **is given by**

$$\omega(x,y) = \varepsilon C(1-y) + B\{\chi_1(x) + \chi_2(x)\} \tag{2.11}$$

with $\chi_1(x) \equiv 1$ **for** $x \le -1+\Delta$, $\chi_1(x) \equiv 0$ **for** $x \ge -1+2\Delta$,

$\chi_2(x) \equiv 1$ **for** $x \ge +1-\Delta$, $\chi_2(x) \equiv 0$ **for** $x \le +1-2\Delta$,

$\chi_1(x)$ **and** $\chi_2(x) \in C_\infty$

and $C > \max\limits_{-1\le x\le +1} B\{\chi''_1(x) + \chi''_2(x)\} + A$.

Applying theorem II of the introduction yields

$|z| \le \omega$ **for** $x^2+y^2 \le 1$, $-1+\Delta \le x \le 1-\Delta$, $\varepsilon < \varepsilon_0$

or $|z| \le \varepsilon C(1-y) \le 2\varepsilon C$ **for** $x^2+y^2 \le 1$, $-1+2\Delta \le x \le 1-2\Delta$, $\varepsilon < \varepsilon_0$ **(2.12)**

Hence $|u-\tilde{\phi}_+-\tilde{v}_0-\varepsilon\tilde{v}_1| = O(\varepsilon)$ **for** $x^2+y^2 \le 1$, $-1+2\Delta \le x \le 1-2\Delta$

or

$$u = \tilde{\phi}_+(x) + \tilde{v}_0(x,y;\varepsilon) + O(\varepsilon) \tag{2.13}$$

uniformly in $x^2+y^2 \le 1$ with the exception of arbitrarely small fixed neighbourhoods of the endpoint (-1,0) and +1,0). (The number Δ is arbitrarely small, but fixed and independent of ε).

3. The case $g(x,y) = 0$ along an arc ℓ in G, $\frac{\partial g}{\partial y} > 0$ along ℓ

3.1 Construction of the approximation.

We consider in this section the b.v.p.

$$\varepsilon\Delta u_\varepsilon + g(x,y)\frac{\partial u_\varepsilon}{\partial y} = 0, \quad (x,y) \in G \tag{3.1}$$

G as stated in Theorem I of the introduction.

$$\begin{aligned} u_\varepsilon &= \phi_+(x) \text{ on the upper boundary } \partial G_+ : y = \gamma_+(x) \\ u_\varepsilon &= \phi_-(x) \text{ on the lower boundary } \partial G_- : y = \gamma_-(x) \end{aligned} \tag{3.2}$$

We assume $g(x,y) = 0$ along a simple arc ℓ in $\bar{G}$, $\frac{\partial g}{\partial y} > 0$ along ℓ (end points included) and $g(x,y) \neq 0$ elsewhere in $\bar{G}$; moreover $g(x,y)$ three times continuously differentiable in $\bar{G}$, ϕ_+ and ϕ_- twice continuously differentiable with respect to the arc length along ∂G_+ and ∂G_-.

In order not to struggle for the moment with tedious non-essential calculations we assume that the endpoints of ℓ are the endpoints A and B of the domain G and that the arc ℓ lies completely in G (see figure 3).

Similarly as has been shown in section 2 one can show that $u_\varepsilon(x,y)$ is uniformly bounded in G. $g(x,y)$ can be written as

$$g(x,y) = \{y-f(x)\}h(x,y). \tag{3.3}$$

with $h(x,y) > 0$ in $\bar{G}$.

A barrier function for $u_\varepsilon(x,y)$ is given by

$$\omega(x,y) = K - \{y-f(x)\}^2 - L\,x^2 \tag{3.4}$$

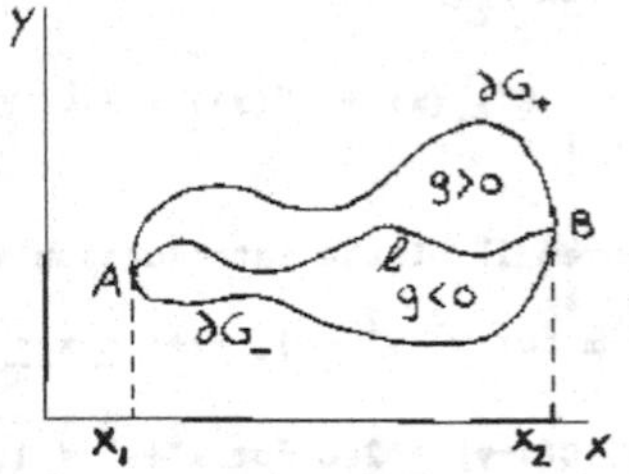

figure 3

with

$$L = \max_{\bar{G}} \left[\{ y-f(x)\}f''(x)\right] \text{ and } K > \max_{x_1\le x\le x_2}\{|\phi_+| + |\phi_-|\}$$

The coefficient of $\frac{\partial u_\varepsilon}{\partial y}$ in the differential equation is positive along the upper boundary and negative alon the lower boundary. In view of the considerations of section 2.1 we do not expect the appearance of boundary layers along the boundaries ∂G_- and ∂G_+.

Therefore it is a good guess to assume that a good approximation of $u_\varepsilon(x,y)$

obtained by putting

$$u_\varepsilon(x,y) \approx \phi_-(x) \text{ for } y < f(x)$$
$$u_\varepsilon(x,y) \approx \phi_+(x) \text{ for } y > f(x) \tag{3.5}$$

In order to obtain a smooth connection between these approximations at the arc ℓ we need a socalled free boundary layer along ℓ.

For constructing this boundary layer we introduce the transformation

$$\xi = x$$
$$\eta = y - f(x) \tag{3.6}$$

The differential operator becomes

$$\varepsilon\left[\frac{\partial^2}{\partial\xi^2} - 2f'(\xi)\frac{\partial^2}{\partial\xi\partial\eta} + \{1+(f'(\xi))^2\}\frac{\partial^2}{\partial\eta^2} - f''(\xi)\frac{\partial}{\partial\eta}\right] + \eta\tilde{h}(\xi,\eta)\frac{\partial}{\partial\eta}$$

with $\tilde{h}(\xi,\eta) = h(\xi,\eta + f(\xi)) > 0$.

Putting $\eta = t\sqrt{\varepsilon}$, we get

$$\{1+(f'(\xi))^2\}\frac{\partial^2}{\partial t^2} + t\tilde{h}(\xi,0)\frac{\partial}{\partial t} +$$

$$\sqrt{\varepsilon}\left[\left\{\frac{\tilde{h}(\xi,t\sqrt{\varepsilon})-\tilde{h}(\xi,0)}{t\sqrt{\varepsilon}}\,t^2 - f''(\xi)\right\}\frac{\partial}{\partial t} - 2f'(\xi)\frac{\partial^2}{\partial\xi\partial t}\right] + \varepsilon\frac{\partial^2}{\partial\xi^2} \tag{3.7}$$

Now we write:

$$u_\varepsilon(x,y) = \tfrac{1}{2}\{\phi_+(x) + \phi_-(x)\} + v(x,y;\varepsilon) + z(x,y;\varepsilon) \tag{3.8}$$

with

$$v(x,y;\varepsilon) = \tilde{v}(\xi,t)$$

and $\tilde{v}(\xi,t)$ satisfying the b.v.p.:

$$\{1 + (f'(\xi))^2\}\frac{\partial^2\tilde{v}}{\partial t^2} + t\tilde{h}(\xi,0)\frac{\partial\tilde{v}}{\partial t} = 0 \tag{3.9}$$

$$\lim_{t\to+\infty}\tilde{v}(\xi,t) = \tfrac{1}{2}\{\phi_+(\xi) - \phi_-(\xi)\}$$
$$\lim_{t\to-\infty}\tilde{v}(\xi,t) = -\tfrac{1}{2}\{\phi_+(\xi) - \phi_-(\xi)\} \tag{3.10}$$

The solution of this b.v.p. is

$$\tilde{v}(\xi,t) = \frac{\phi_+(\xi)-\phi_-(\xi)}{\sqrt{\pi}}\ \text{erf}\left[\left\{\frac{\frac{1}{2}\tilde{h}(\xi,0)}{1+(f'(\xi))^2}\right\}^{\frac{1}{2}} t\right]$$

or

$$v(x,y;\varepsilon) = \frac{\phi_+(x)-\phi_-(x)}{\sqrt{\pi}} \quad \text{erf}\left[\left\{\frac{\frac{1}{2}h(x,f(x))}{1+(f'(x))^2}\right\}^{\frac{1}{2}} \frac{y-f(x)}{\sqrt{\varepsilon}}\right] \tag{3.11}$$

The remainder term $z(x,y;\varepsilon)$ in (3.8) satisfies the differential equation

$$L_\varepsilon[z] = L_\varepsilon[u-\tfrac{1}{2}\{\phi_+(x)+\phi_-(x)\} - v(x,y;\varepsilon)]$$

$$= -\sqrt{\varepsilon}\left[\left\{\frac{\tilde{h}(\xi,t\sqrt{\varepsilon})-\tilde{h}(\xi,0)}{t\sqrt{\varepsilon}}\, t^2-f''(\xi)\right\} \frac{\partial\tilde{v}}{\partial t} - 2f'(\xi)\,\frac{\partial^2\tilde{v}}{\partial\xi\partial t}\right] - \varepsilon\,\frac{\partial^2\tilde{v}}{\partial\xi^2} - \frac{\varepsilon}{2}\{\varphi_+''+\varphi_-''\}$$

Remarking that $\frac{d}{dt}\,\text{erf}\; t = \frac{1}{2}\sqrt{\pi}\, e^{-t^2}$ we have

$$L_\varepsilon[z] = O(\sqrt{\varepsilon}) \text{ for } (x,y) \in G,\ x_1+\delta < x < x_2-\delta, \tag{3.12}$$

where δ is a fixed positive arbitrarely small number, independent of ε. (We exclude these small neighbourhoods of the endpoints A and B, because $\phi_+''(x)$ and $\phi_-''(x)$ may become infinite at A or B).

Along the boundaries we have

$$\begin{aligned} &z = O(\varepsilon^N)\ (z \lesssim 0) \text{ for } (x,y) \in \delta G,\ x_1+\delta \le x \le x_2-\delta, \\ &|z| < M \text{ for } x=x_1+\delta,\ x=x_2-\delta, \end{aligned} \tag{3.13}$$

with M some positive constant, independent of ε.

In the next section it will be shown that a barrier function can be constructed for z, leading to the result

$$|z| = O(\sqrt{\varepsilon}\log\varepsilon) \text{ for } (x,y) \in \bar{G}, x_1+2\delta \le x \le x_2-2\delta. \tag{3.14}$$

Hence we have

$$u_\varepsilon(x,y) = \tfrac{1}{2}\{\phi_+(x)+\phi_-(x)\}+v(x,y;\varepsilon)+O(\sqrt{\varepsilon}\log\varepsilon) \tag{3.15}$$

uniformly valid in $\bar{G}$, with the exception of arbitrarely small neighbourhoods of the endpoints A and B.

3.2 Construction of the barrier function.

Suppose $|L_\varepsilon[z]| \le N\sqrt{\varepsilon}$ for ε sufficiently small and $(x,y) \in G$, $x_1+\delta < x < x_2-\delta$.

We consider the particular solution

$$\omega(\eta) = -\frac{3N}{a\sqrt{\varepsilon}} \int_0^{\eta\sqrt{a}} \{\int_0^{\lambda} \exp\left[\frac{\mu^2-\lambda^2}{2\varepsilon}\right] d\mu\} d\lambda \qquad (3.16)$$

of the equation

$$\varepsilon \frac{d^2\omega}{\partial\eta^2} + \eta a \frac{d\omega}{\partial\eta} = -3N\sqrt{\varepsilon} \qquad (3.17)$$

with

$$a = \min_{(\xi,\eta)\in\bar{G}} \frac{\tilde{h}(\xi,\eta)}{1+\{f'(\xi)\}^2} > 0$$

Observing $\eta\frac{d\omega}{d\eta} \leq 0$ and $\frac{\partial\omega}{\partial\eta} = O(1)$ for all η we see that

$$L_\varepsilon[+\omega] = \varepsilon[\ \{1+(f'(\xi))^2\} \frac{\partial^2\omega}{\partial\eta^2} - f''(\xi)\ \frac{\partial\omega}{\partial\eta}] +$$

$$+ \eta\tilde{h}(\xi,\eta)\ \frac{\partial\omega}{\partial\eta} \leq -2N\sqrt{\varepsilon} \text{ for } \varepsilon \text{ sufficiently small and } (x,y) \in G. \qquad (3.18)$$

It is not difficult to make an estimate of the function $\omega(\eta)$ and a rather straigh tforward calculation yields

$$|\omega(\eta)| < -C\sqrt{\varepsilon}\ \log\ \varepsilon \qquad (3.19)$$

for ε sufficiently small, say $\varepsilon < \varepsilon_0$; C is a constant, independent of ε.

A barrier function for the remainder term $z(x,y;\varepsilon)$ is now readily obtained by taking

$$\tilde{\omega}(x,y) = M\{\chi_1(x)+\chi_2(x)\}-C\sqrt{\varepsilon}\ \log\ \varepsilon+\omega(y-f(x)) \qquad (3.20)$$

with $\chi_1(x) \equiv 1$ for $x \leq x_1+\delta$, $\chi_1(x) \equiv 0$ for $x \geq x_1+2\delta$

$\chi_2(x) \equiv 1$ for $x \geq x_2-\delta$, $\chi_2(x) \equiv 0$ for $x \leq x_2-2\delta$

Because the term $M\{\chi_1(x)+\chi_2(x)\}$ gives in the differential equation only a contribution of order $O(\varepsilon)$ we get from (3.18)

$$L_\varepsilon[\tilde{\omega}] \leq -N\sqrt{\varepsilon}$$

or

$L_\varepsilon[-\tilde{\omega}] \geq N\sqrt{\varepsilon} > |L_\varepsilon[z]|$ for $(x,y) \in G$, $x_1+\delta < x < x_2-\delta$, and ε sufficiently small, say $\varepsilon < \varepsilon_1$.

Moreover $\tilde{\omega} \geq |z|$ for $x = x_1+\delta, x = x_2-\delta$and for (x,y)on ∂G with $x_1+\delta \leq x \leq x_2-\delta$.

Application of theorem II of the introduction yields finally the result

$$|z| \le \tilde{\omega} = O(\sqrt{\varepsilon} \log \varepsilon) \text{ for all points in } \bar{G} \text{ with } x_1+2\delta \le x \le x_2 -2\delta.$$

3.3 General situation

We drop now the assumption that the arc $y = f(x)$ passes through the endpoints A and B and that the arc ℓ lies completely in G. The new situation is sketched in figure 4. The + sign indicates $g(x,y) > 0$, the − sign $g(x,y) < 0$. In the region ACC' with $u_\varepsilon(x,y)=\phi_+(x)$ along CC' we apply the theory of section 2; this gives rise to a boundary layer along AC due to the fact that $g(x,y) > 0$ in ACC'.

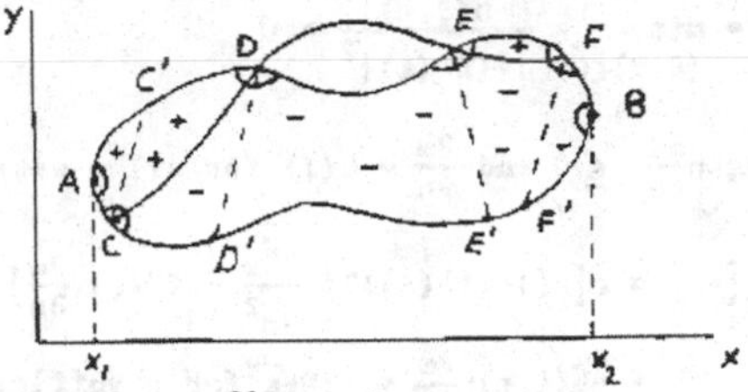

figure 4

Considering the region CC'DD' we take $u_\varepsilon = \phi_+(x)$ along CC' and $u_\varepsilon = \phi_-(x)$ along DD'; applying the theory of this section we get a boundary layer along CD. Considering the region D'DEE' we take $u_\varepsilon = \phi_-(x)$ along DD' and $u_\varepsilon = \phi_-(x)$ along EE'; applying again the theory of section 2 we get a boundary layer along DE, due to the fact that $g(x,y) < 0$ in D'DEE' etc.

Summing up we can give a uniform approximation of u_ε in all of G with the exception of arbitrarely small neighbourhoods of the points A,C,D,E,F and B. We obtain boundary layers along AC, CD, DE, EF and FB. Outside the boundary layers $u_\varepsilon(x,y)$ is uniformly approximated by $\phi_+(x)$ when $g(x,y) > 0$ and by $\phi_-(x)$ when $g(x,y) < 0$.

We have given here a method of approximation in case the perturbation operator is $\varepsilon\Delta$; it is expected that a generalization to εL_2, L_2 being an arbitrary linear uniform elliptic operator with sufficiently smooth coefficients, is possible without many difficulties.

4. The case $g(x, y) = 0$ along an arc ℓ in G, $\frac{\partial g}{\partial y} < 0$ along ℓ

This case cannot be treated by the method described in this paper.

Consider the boundary value problem:

$$\epsilon\{\Delta z + z\} - y\frac{\partial z}{\partial y} = \epsilon^{\alpha}\sin x$$

$$z(0, y) = z(\pi, y) = z(x, -1) = z(x, +1) = 0.$$

A calculation shows that $\lim_{\epsilon\to\infty} z(x, y) = \infty$ for all (x, y) within the boundary and for all finite values of α.

Hence it follows that a function $\tilde{u}_{\epsilon}$ satisfying the differential equation

$$\epsilon\{\Delta u_{\epsilon} + u_{\epsilon}\} - y\frac{\partial u_{\epsilon}}{\partial y} = 0$$

and the boundary conditions

$$u_{\epsilon}(0, y) = g_1(y),\ u_{\epsilon}(\pi, y) = g_2(y),\ u_{\epsilon}(x, -1) = f_1(x),\ u_{\epsilon}(x+1) = f_2(x)$$

up to the order ϵ^{α}, may have a remainder term z which is not of some order $O(\epsilon^{\beta})$; $u_{\epsilon} = \tilde{u}_{\epsilon} + z$. The application of the maximum principle to the estimation of z fails.

Acknowledgement The author thanks Mr P. P. N. de Groen for drawing his attention to the case $g(x, y) \equiv y$ and for working out this example.

Literature

1. Visik, M. I; Lyusternik, L. A. Regular Degeneration and Boundary Layers for Linear Differential equations with Small Parameter. Uspehi Mat. Nauk. 12 (1957) (A.M.S. Translation Series 2, 20 pp. 239-364 (1962)).

2. Eckhaus, W. de Jager, E.M. — Asymptotic Solutions of Singular Perturbation Problems for Linear Differential Equations of Elliptic Type. Arch.Rat.Mech. and An. 23,1. pp. 26-86 (1966).

3. Frankena, J.F. — A Uniform Asymptotic Expansion of the Solution of a Linear Elliptic Singular Perturbation Problem. Arch.Rat.Mech. and An. 31, no. 3. pp. 185-198 (1968).

4. Grasman, J. — On the Birth of Boundary Layers. Mathematical Centre Tract 36, 137 pp. Mathematisch Centrum, Amsterdam, 1971.

5. Protter, M.H. Weinberger, H.F. — Maximum Principles in Differential Equations, pp. 261. Prentice Hall, Partial Differential Equations Series; Englewood Cliffs, N.J., 1967.

Perturbations of the Dirac Operator

K. Jörgens

Let T_o be the Dirac operator for a free electron, and let V_o be a symmetric perturbation representing a field of force acting on the electron. It is of considerable importance for relativistic quantum mechanics to know conditions on V_o such that $S_o = T_o + V_o$ is essentially selfadjoint. Such conditions have been found by T. Kato [7], F. Rellich [10], J. Schillemeit [11], R.T. Prosser [8], W.D. Evans [1], P.A. Rejto [9], J. Weidmann [13], and W. Schmincke [12]. They are all concerned with operators V_o of potential type:

$$(V_o u)(x) = q(x)u(x)$$

where q is a function on R^3 with values in the 4by4 Hermitean matrices. The present paper treats general perturbations V_o and, at the same time, more general unperturbed operators T_o ; in particular none of the two operators is required to be "local" in the sense, that the support of $T_o u$ is always contained in the support of u . Most of the previous criteria are contained in our main result (theorem 5.6). In the final section 7 we give conditions for V_o to be T_o-compact and a stronger result on invariance of the essential spectrum adapted from [5] . The essential spectrum of perturbed Dirac operators has also been studied by W.D. Evans [2] who gave conditions for the essential spectrum to fill the entire real line.

1. NOTATION

For $c \in K$, the field of complex numbers, denote by c^* the complex conjugate number. Let K^p denote the space of column vectors u with p components $u_j \in K$, and let u^* be the row vector with components u_j^* , so that $u^* v$ is the inner product and $|u| = (u^* u)^{1/2}$ is the length of vectors in K^p. The Hilbert space $H = [L_2(R^m)]^p$ is the space of functions $u : R^m \longrightarrow K^p$ with components $u_j \in L_2(R^m)$; the inner product and the norm in H are defined by

$$\langle u|v\rangle = \int u(x)^* v(x)\,dx \quad \text{and} \quad \|u\| = \left(\int |u(x)|^2 dx\right)^{1/2}$$

respectively. For every non-void open set $\Omega \subset R^m$ we define the subspace $D_o(\Omega) = [C_o^\infty(\Omega)]^p$ of H ; in particular $D_o = D_o(R^m)$ is a dense subspace of H. Let $H_{loc} = [L_{2,loc}(R^m)]^p$ be the space of functions $u : R^m \longrightarrow K^p$ with locally square-integrable components. H_{loc} is a locally convex topological vector space with the topology generated by the family of seminorms $\|u\|_\Omega = (\int_\Omega |u(x)|^2 dx)^{1/2}$ for bounded open sets $\Omega \subset R^m$; obviously H_{loc} is metrizable and complete (a Fréchet space). For $u \in H_{loc}$ and $\varphi \in C_o^\infty(R^m)$ define $\varphi u : R^m \longrightarrow K^p$ as the function with components φu_j ; then φu has compact support and hence $\varphi u \in H$.

2. THE UNPERTURBED OPERATOR

Let T_o be a linear operator in H with domain $D(T_o) = D_o$ having the following properties:

(2.1) T_o is essentially selfadjoint.

(2.2) For every $\varphi \in C_o^\infty(R^m)$ the operator $A_o(\varphi)$ defined by $A_o(\varphi)u = T_o\varphi u - \varphi T_o u$ for $u \in D_o$ is bounded.

The best-known example is the Dirac operator: Here $m = 3$, $p = 4$ and

$$T_o u = i \sum_{k=1}^{3} \alpha_k \partial_k u - \mu \alpha_4 u \quad \text{for } u \in D_o , \tag{2.3}$$

where $\partial_k u$ is the function with components $\partial_k u_j$, μ a real number, and where α_k are 4by4 Hermitean matrices satisfying $\alpha_j\alpha_k + \alpha_k\alpha_j = 2\,\delta_{jk}\,\alpha_o$ for $j, k = 1,\dots,4$, α_o the unit matrix. Property (2.1) is well known for the Dirac operator (cf. T. Kato [6] V.5.4); (2.2) follows from $A_o(\varphi)u = i(\sum_{k=1}^{3} \alpha_k \partial_k \varphi)u$ for $u \in D_o$. The Dirac operator is a special case of the following type of operators: Let $K^{p,p}$ be the set of p-by-p complex matrices and, for $\alpha \in K^{p,p}$, let $|\alpha|$ denote the norm of α as an operator in K^p. Let $\hat{u}$ be the Fourier transform of $u \in H$ and define the unitary operator U in H by $Uu = \hat{u}$. T_o is given by

(2.4) $$T_o u = U^{-1} k\, U u \quad \text{for } u \in D_o ,$$

where $k : R^m \longrightarrow K^{p,p}$ is such that $k(y)$ is Hermitean for all y and

(2.5) $$|k(y) - k(z)| \leq c|y - z| \quad \text{for all } y, z \in R^m$$

with $c > 0$ and independent of y and z. Since this inequality implies $|k(y)| \leq c_1 + c_2|y|$, the operator T_o is well defined and symmetric. Its closure T is given by $Tu = U^{-1} k\, U u$ for all $u \in D(T) = \{u \mid u \in H,\ k\, U u \in H\}$; hence T is selfadjoint, and T_o is essentially selfadjoint. A computation shows that $A_o(\varphi)u = U^{-1}GUu$ for $u \in D_o$ where G is the integral operator with kernel $G(y,z) = \hat{\varphi}(y - z)[k(y) - k(z)]$. By (2.5) we have $|G(y,z)| \leq c|y-z|\,|\hat{\varphi}(y-z)|$ and hence (cf. [6] III.2.1)

(2.6) $$\|A_o(\varphi)\| = \|G\| \leq c \int |z|\,|\hat{\varphi}(z)|\,dz < \infty ,$$

which proves (2.2). The Dirac operator is obtained by putting $k(y) = -\sum_{j=1}^{3} y_j \alpha_j - \mu\alpha_4$; here $|k(y)|^2 = \mu^2 + |y|^2$ and hence $D(T) = [H^1(R^3)]^4$, where $H^1(R^3)$ is the Sobolev space of order 1.

3. EXTENSIONS OF THE UNPERTURBED OPERATOR

Let T denote the unique selfadjoint extension of T_o; as is well known, T is equal to the closure $\overline{T}_o$ and to the adjoint T_o^* of T_o. Let $D(T)$ be the domain of T, and for every $\varphi \in C_o^\infty(R^m)$ let $A(\varphi)$ be the unique bounded extension of $A_o(\varphi)$ to all of H.

(3.1) Lemma. For every $u \in D(T)$ and for every $\varphi \in C_o^\infty(R^m)$ we have $\varphi u \in D(T)$ and $T\varphi u = \varphi Tu + A(\varphi)u$.

This follows at once from the definition of $A(\varphi)$ and of T.

(3.2) Definition. Let $F(T)$ be the vector space of all $u \in H$ such that $\varphi u \in D(T)$ for all $\varphi \in C_o^\infty(R^m)$; $F(T)$ is given the locally convex topology generated by the family of norms p_φ, $\varphi \in C_o^\infty(R^m)$, defined by $p_\varphi(u) = \|u\| + \|T\varphi u\|$ for $u \in F(T)$.

(3.3) Lemma. $F(T)$ is metrizable and complete (hence $F(T)$ is a Fréchet space).

Proof. Let $\varphi_n \in C_o^\infty(R^m)$ satisfy $0 \leq \varphi_n \leq 1$, $\varphi_n(x) = 1$ for $|x| \leq n$ and $\varphi_n(x) = 0$ for $|x| \geq n+1$. The norms $p_n = p_{\varphi_n}$ generate the topology of $F(T)$, i.e. for every finite subset M of $C_o^\infty(R^m)$ there exists a positive integer n and a number $c > 0$ such that $p_\varphi(u) \leq c\, p_n(u)$ for all $\varphi \in M$ and for all $u \in F(T)$. To see this, choose n such that the support of φ is contained in the closed ball of radius n for all $\varphi \in M$. Then $\varphi\varphi_n = \varphi$ for all $\varphi \in M$ and by (3.1), for all $u \in F(T)$, we have $p_\varphi(u) = \|u\| + \|T\varphi\varphi_n u\| \leq \|u\| + \|\varphi T\varphi_n u\| + \|A(\varphi)\varphi_n u\| \leq \|u\| + |\varphi|_\infty \|T\varphi_n u\| + \|A(\varphi)\| \|\varphi_n u\|$ where $|\ |_\infty$ is the maximum-norm. Since $|\varphi_n| \leq 1$ the estimate follows with $c = \max\{1 + |\varphi|_\infty + \|A(\varphi)\| \mid \varphi \in M\}$. This proves that $F(T)$ is metrizable. The completeness is an easy consequence of the fact that T is closed.

(3.4) Corollary. There are positive numbers c_n such that the norms $q_n = c_n p_n$ form an increasing sequence and generate the topology of $F(T)$.

(3.5) Lemma. D_o is dense in $F(T)$.

Proof (due to J.Voigt). Let $u \in F(T)$; by (3.4) it suffices to show that for every $\varphi \in C_o^\infty(R^m)$ there is a sequence (u_j) in D_o such that $p_\varphi(u_j-u) \longrightarrow 0$. Choose $\psi \in C_o^\infty(R^m)$ such that $\varphi\psi = \varphi$; then $\psi u \in D(T)$ and hence there is a sequence (v_j) in D_o such that $v_j \longrightarrow \psi u$ and $Tv_j \longrightarrow T\psi u$ in H. By (3.1) we have $T\varphi v_j \longrightarrow T\varphi\psi u = T\varphi u$. Now choose a sequence (w_j) in D_o such that $w_j \longrightarrow u$ and put $u_j = v_j + (1-\psi)w_j$. It follows $u_j \longrightarrow u$ and $T\varphi u_j = T\varphi v_j \longrightarrow T\varphi u$, hence $p_\varphi(u_j-u) \longrightarrow 0$, q.e.d.

(3.6) Lemma. The operator T has a unique extension $T_1 : F(T) \longrightarrow H_{loc}$ such that $\varphi T_1 u = T\varphi u - A(\varphi)u$ for all $\varphi \in C_o^\infty(R^m)$ and for all $u \in F(T)$. T_1 is linear and continuous.

Proof. Let $u \in F(T)$; by (3.5) there is a sequence (u_j) in D_o such that $p_\varphi(u_j-u) \longrightarrow 0$ for every $\varphi \in C_o^\infty(R^m)$. Then $\varphi Tu_j = T\varphi u_j - A(\varphi)u_j \longrightarrow T\varphi u - A(\varphi)u$ in H for every φ and hence (Tu_j) converges in H_{loc} to some element $T_1 u$, say, and we have $\varphi T_1 u = T\varphi u - A(\varphi)u$ for every φ . Hence T_1 is well defined and linear. Continuity follows from the estimate $\|T_1 u\|_{\Omega_n} \le \|\varphi_n T_1 u\| \le (1+\|A(\varphi_n)\|)\, p_n(u)$ where φ_n and p_n have been defined in the proof of (3.3) and where Ω_n is the open ball of radius n and center 0 .

4. THE PERTURBATION

Let T_o satisfy (2.1) and (2.2) and let V_o be a perturbation satisfying the conditions:

(4.1) V_o has domain D_o and is symmetric.

(4.2) For every $\varphi \in C_o^\infty(R^m)$ the operator $B_o(\varphi)$ defined by $B_o(\varphi)u = V_o\varphi u - \varphi V_o u$ for $u \in D_o$ is bounded.

(4.3) For every non-void open bounded set $\Omega \subset R^m$ there are constants $a > 0$ and $b \in]0,1[$ such that $\|V_o u\|_\Omega \le a\|u\| + b\,\|T_o u\|$ for all $u \in D_o(\Omega)$.

We postpone the discussion of the assumptions on V_o to section 6 where examples will be given. Let $V = \overline{V}_o$ denote the closure of V_o , and for every $\varphi \in C_o^\infty(R^m)$ let $B(\varphi)$ be the unique bounded extension of $B_o(\varphi)$ to all of H.

(4.4) <u>Lemma</u>. For every $u \in D(V)$ and for every $\varphi \in C_o^\infty(R^m)$ we have $\varphi u \in D(V)$ and $V\varphi u = \varphi Vu + B(\varphi)u$. If $u \in D(T)$ has compact support, then $u \in D(V)$ and $\|Vu\|_\Omega \leq a\|u\| + b\|Tu\|$ for every bounded open set Ω containing the support of u with the same constants a and b as in (4.3).

<u>Proof</u>. The first assertion follows from (4.2) and from the definition of V and $B(\varphi)$. Let $u \in D(T)$ have compact support; choose a bounded open set Ω containing the support of u and choose $\varphi \in C_o(\Omega)$ such that $\varphi u = u$. Then there is a sequence (u_j) in D_o such that $u_j \longrightarrow u$ and $T_o u_j \longrightarrow Tu$ in H. It follows that $\varphi u_j \longrightarrow \varphi u = u$ and $T_o \varphi u_j \longrightarrow Tu$ by (3.1). Since $\varphi u_j \in D_o(\Omega)$ we get from (4.3)

$$\|V_o\varphi u_j - V_o\varphi u_k\|_\Omega \leq a\|\varphi u_j - \varphi u_k\| + b\|T_o\varphi u_j - T_o\varphi u_k\|$$

and this tends to zero for $j, k \longrightarrow \infty$, i.e. the sequence $(V_o\varphi u_j)$ converges in $[L_2(\Omega)]^p$. Also by (4.2) we have $(V_o\varphi u_j)(x) = (B_o(\varphi)u_j)(x)$ for all $x \in \Omega' = R^m \setminus \Omega$; hence $(V_o\varphi u_j)$ converges in $[L_2(\Omega')]^p$ and therefore it converges in H. This proves $u = \lim \varphi u_j \in D(V)$. The inequality follows from (4.3) by inserting φu_j and passing to the limit $j \longrightarrow \infty$.

(4.5) <u>Corollary</u>. Let $u \in D(T)$ have compact support. Then there is a sequence (v_j) in D_o such that $v_j \longrightarrow u$, $T_o v_j \longrightarrow Tu$ and $V_o v_j \longrightarrow Vu$.

(4.6) <u>Lemma</u>. The operator V_o has a unique extension $V_1 : F(T) \longrightarrow H_{loc}$ such that $\varphi V_1 u = V\varphi u - B(\varphi)u$ for all $u \in F(T)$ and for all $\varphi \in C_o^\infty(R^m)$. V_1 is linear and continuous. For $u \in F(T) \cap D(V)$ we have $V_1 u = Vu$.

<u>Proof</u>. Using (3.5) the proof is almost the same as for lemma (3.6); the inequality of (4.4) is used, however, in proving the continuity

of V_1. The last statement also follows from (4.4).

So far the fact that $b < 1$ in (4.3) has not been used; the next lemma, however, requires the full strength of assumption (4.3).

(4.7) Lemma. For every $\varphi \in C_o^\infty(R^m)$ such that $0 \leq \varphi \leq 1$, the operator $V_o(\varphi)$ defined by $V_o(\varphi)u = \varphi V_o \varphi u$ for $u \in D_o$ is symmetric and T_o-bounded with T_o-bound smaller than 1.

Proof. Obviously $V_o(\varphi)$ is symmetric. Choose a bounded open set Ω containing the support of φ. Then for every $u \in D_o$ we have $\varphi u \in D_o(\Omega)$, and by (4.3) we get $\|V_o(\varphi)u\| \leq \|V_o \varphi u\|_\Omega \leq a\|\varphi u\| + b\|T_o \varphi u\| \leq a\|u\| + b\|\varphi T_o u + A_o(\varphi)u\| \leq (a + b\|A_o(\varphi)\|)\|u\| + b\|T_o u\|$. Hence $V_o(\varphi)$ is T_o-bounded with T_o-bound < 1.

(4.8) Lemma. For every $\varphi \in C_o^\infty(R^m)$ such that $0 \leq \varphi \leq 1$, the operator $S_o(\varphi) = T_o + V_o(\varphi)$ is essentially selfadjoint; its unique selfadjoint extension $S(\varphi) = \overline{S_o(\varphi)}$ has domain $D(T)$, and $S(\varphi)u = Tu + \varphi V \varphi u$ holds for every $u \in D(T)$.

Proof. By (4.7) and a theorem of Rellich (cf. [6] V, th.4.4) $S_o(\varphi)$ is essentially selfadjoint and $S(\varphi) = \overline{S_o(\varphi)}$ has domain $D(T)$ and is equal to $T + \overline{V_o(\varphi)}$. Now (4.4) implies that, for every $u \in D(T)$ and for every (u_j) in D_o such that $u_j \longrightarrow u$ and $T_o u_j \longrightarrow Tu$, we have $V_o(\varphi)u_j \longrightarrow \varphi V \varphi u$; hence $\overline{V_o(\varphi)}u = \varphi V \varphi u$ for every $u \in D(T)$.

5. THE PERTURBED OPERATOR

Let T_o and V_o satisfy the assumptions of sections 2 and 4 respectively. Now we want to investigate the perturbed operator $S_o = T_o + V_o$ defined on D_o ; evidently S_o is symmetric. Let S_o^* denote its adjoint operator.

(5.1) Lemma. For every $u \in D(S_o^*)$ and for every $\varphi \in C_o^\infty(R^m)$ we have $\varphi u \in D(S_o^*)$ and $S_o^* \varphi u = \varphi S_o^* u + A(\varphi)u + B(\varphi)u$.

Proof. Let $u \in D(S_o^*)$; by definition of the adjoint operator we have $\langle S_o^* u | v\rangle = \langle u | S_o v\rangle$ for all $v \in D_o$. Hence for every $\varphi \in C_o^\infty(R^m)$ we get $\langle \varphi S_o^* u | v\rangle = \langle S_o^* u | \varphi^* v\rangle = \langle u | S_o \varphi^* v\rangle = \langle u | \varphi^* S_o v + A_o(\varphi^*)v + B_o(\varphi^*)v\rangle = \langle \varphi u | S_o v\rangle + \langle A_o(\varphi^*)^* u + B_o(\varphi^*)^* u | v\rangle$ where we have used (2.2) and (4.2). It is easy to see that $A_o(\varphi^*)^* = - A(\varphi)$ and similarly $B_o(\varphi^*)^* = - B(\varphi)$. Hence $\langle \varphi u | S_o v\rangle = \langle \varphi S_o^* u + A(\varphi)u + B(\varphi)u | v\rangle$ for all $v \in D_o$, i.e. $\varphi u \in D(S_o^*)$ and $S_o^* \varphi u = \varphi S_o^* u + A(\varphi)u + B(\varphi)u$, q.e.d.

(5.2) Lemma. For every $u \in D(S_o^*)$ and for every $\varphi \in C_o^\infty(R^m)$ we have $\varphi u \in D(T)$ and $S_o^* \varphi u = (T + V)\varphi u$.

Proof. Let $u \in D(S_o^*)$ and $\varphi \in C_o^\infty(R^m)$; then $\varphi u \in D(S_o^*)$ by lemma (5.1) and hence $\langle S_o^* \varphi u | v\rangle = \langle \varphi u | S_o v\rangle$ for all $v \in D_o$. Choose $\psi \in C_o^\infty(R^m)$ such that $0 \leq \psi \leq 1$ and $\psi = 1$ on the support of φ . Replacing v by ψv and using $\varphi^* \psi = \varphi^*$ we get $\langle \psi S_o^* \varphi u | v\rangle = \langle u | \varphi^* S_o \psi v\rangle = \langle u | \varphi^* \psi T_o v + \varphi^* A_o(\psi)v + \varphi^* \psi V_o \psi v\rangle = \langle u | \varphi^* S_o(\psi)v + \varphi^* A_o(\psi)v\rangle = \langle \varphi u | S_o(\psi)v\rangle - \langle A(\psi)\varphi u | v\rangle$ for all $v \in D_o$ (recall the definition of $S_o(\psi)$ in lemma (4.7)). Hence $\varphi u \in D(S_o(\psi)^*)$ and $S_o(\psi)^* \varphi u = \psi S_o^* \varphi u + A(\psi)\varphi u$. By lemma (4.7) we have $D(S_o(\psi)^*) = D(T)$ and $S_o(\psi)^* w = Tw + \psi V \psi w$ for all $w \in D(T)$. It follows that $\varphi u \in D(T)$ and $\psi S_o^* \varphi u = T\varphi u + \psi V \psi \varphi u - A(\psi)\varphi u = T\psi\varphi u - A(\psi)\varphi u + \psi V \varphi u = \psi (T + V)\varphi u$. This holds for every $\psi \in C_o^\infty(R^m)$ such that $0 \leq \psi \leq 1$ and $\psi = 1$ on the support of φ ; hence $S_o^* \varphi u = (T + V)\varphi u$, q.e.d.

(5.3) Theorem. $D(S_o^*) = \{u \mid u \in F(T), (T_1 + V_1)u \in H\}$ and $S_o^* u = (T_1 + V_1)u$ for $u \in D(S_o^*)$.

Proof. a) For $u \in D(S_o^*)$ we have $u \in H$ and $\varphi u \in D(T)$ for every $\varphi \in C_o^\infty(R^m)$ by lemma (5.2); hence $u \in F(T)$. Lemmas (5.1), (5.2), (3.6) and (4.6) give $\varphi S_o^* u = (T + V)\varphi u - A(\varphi)u - B(\varphi)u = \varphi (T_1 + V_1)u$ for every $\varphi \in C_o^\infty(R^m)$, hence $S_o^* u = (T_1 + V_1)u \in H$.

b) Let $u \in F(T)$ satisfy $(T_1 + V_1)u \in H$. By lemma (3.5) there is a sequence (u_j) in D_o which converges to u in $F(T)$. The mapping $T_1 + V_1 : F(T) \longrightarrow H_{loc}$ is continuous according to (3.6) and (4.6). Hence $S_o u_j = (T_1 + V_1)u_j \longrightarrow (T_1 + V_1)u$ in H_{loc} and consequently for every $v \in D_o$ we get $\langle (T_1 + V_1)u | v \rangle = \lim \langle S_o u_j | v \rangle = \langle u | S_o v \rangle$. This proves $u \in D(S_o^*)$ and $(T_1 + V_1)u = S_o^* u$, q.e.d.

Our main problem is essential selfadjointness of S_o; apparently this can not be derived from the previous assumptions on T_o and V_o. The following additional assumption is sufficient:

(5.4) <u>Condition</u>. There exists a sequence (φ_n) in $C_o^\infty(R^m)$ such that

(i) $0 \leq \varphi_n \leq 1$;

(ii) $\varphi_n(x) \longrightarrow 1$ for $n \longrightarrow \infty$ and for every $x \in R^m$;

(iii) for every $u \in D(S_o^*)$ the sequence $(\|A(\varphi_n)u + B(\varphi_n)u\|)$ is bounded.

(5.5) <u>Remark</u>. Let $\varphi \in C_o^\infty(R^m)$ satisfy $0 \leq \varphi \leq 1$, $\varphi(x) = 1$ for $|x| \leq 1$ and $\varphi(x) = 0$ for $|x| \geq 2$; put $\varphi_n(x) = \varphi(x/n)$ for $x \in R^m$ and $n = 1, 2, \ldots$ Then conditions (i), (ii) of (5.4) are satisfied. The Fourier transform $\hat{\varphi}_n$ of φ_n is given by $\hat{\varphi}_n(y) = n^m \hat{\varphi}(ny)$; hence for T_o defined by (2.4), (2.5) we have by (2.6) $\|A(\varphi_n)\| \leq c\, n^m \int |z|\, |\hat{\varphi}(nz)|\, dz = c/n \int |y|\, |\hat{\varphi}(y)|\, dy \longrightarrow 0$. Therefore, in order to fulfill (5.4) in this case, it suffices to require that the sequence $(\|B(\varphi_n)\|)$ is bounded.

(5.6) <u>Theorem</u>. If (5.4) holds (in addition to (2.1), (2.2), (4.1), (4.2) and (4.3)), then S_o is essentially selfadjoint.

<u>Proof</u>. Let $u \in D(S_o^*)$ and let (φ_n) be a sequence satisfying (5.4); put $u_n = \varphi_n u$. Then $u_n \in D(T)$ by lemma (5.2) and u_n has compact support, hence $u_n \in D(\overline{S}_o)$ by corollary (4.5) and $\overline{S}_o u_n = (T + V)u_n = \varphi_n S_o^* u + A(\varphi_n)u + B(\varphi_n)u$ by (5.1) and (5.2). It follows from (5.4) that the sequence $(\|\overline{S}_o u_n\|)$ is bounded. Furthermore $u_n \longrightarrow u$ and

hence $\langle S_o^* u \mid u_n \rangle \longrightarrow \langle S_o^* u \mid u \rangle$. Now $\overline{S}_o$ is symmetric and $\overline{S}_o^* = S_o^*$; therefore $\langle u_n \mid \overline{S}_o u_n \rangle - \langle S_o^* u \mid u_n \rangle = \langle u_n - u \mid \overline{S}_o u_n \rangle \longrightarrow 0$ and hence $\langle S_o^* u \mid u \rangle = \lim \langle u_n \mid \overline{S}_o u_n \rangle$ is a real number. Since this holds for all $u \in D(S_o^*)$, S_o^* is symmetric, and consequently S_o is essentially selfadjoint, q.e.d.

6. EXAMPLES

In this section we take $m = 3$, $p = 4$ and we consider the Dirac operator T_o defined by (2.3). Let the perturbation V_o be of the form $(V_o u)(x) = q(x)u(x)$, where $q : R^3 \longrightarrow K^{4,4}$ is a function on R^3 with values in the 4-by-4 Hermitean matrices with components in $L_{2,loc}(R^3)$. Then (4.1) is satisfied and (4.2) holds with $B_o(\varphi) = 0$ for all $\varphi \in C_o^\infty(R^3)$. Several sufficient conditions for (4.3) are known:

(6.1) Coulomb potentials: $q(x) = \sum_{n=1}^{N} \gamma_n |x - x_n|^{-1}$ with n different points $x_n \in R^3$ and with Hermitean matrices $\gamma_n \in K^{4,4}$ satisfying $|\gamma_n| < 1/2$. Here (4.3) holds even for $\Omega = R^3$ and with $b = \max\{2|\gamma_n| \mid n = 1,\dots,N\}$. The proof is essentially contained in [6] V.5.4 ; for $N > 1$ one uses a C^∞-partition of unity.

(6.2) $L_{3,loc}$-potentials: $|q| \in L_{3,loc}(R^3)$; this condition is due to L.Gross [3]. Using a decomposition $q = q_1 + q_2$ such that q_1 is bounded on Ω and $\int_\Omega |q_2(x)|^3 dx$ is small, and the Sobolev inequality $\int |u(x)|^6 dx \leq C \|T_o u\|^6$ for $u \in D_o$, one proves (4.3) for arbitrary $b \in]0,1[$ with suitable $a > 0$. If $|q| \in L_3(R^3)$ then (4.3) holds for $\Omega = R^3$ and arbitrary $b \in]0,1[$.

(6.3) Stummel-potentials: For some $\delta > 0$ the function $x \longmapsto M(x) = \int_{|x-y| \leq 1} |q(y)|^2 |x - y|^{-1-\delta} dy$ is locally bounded. This condition has been used by W.D. Evans [1] . Here (4.3) holds for arbitrary $b \in]0,1[$ with suitable $a > 0$. If the function M is bounded, then (4.3) holds for $\Omega = R^3$.

Evidently a sum $q = q_1 + q_2 + q_3$, where q_j is of type (6.j), also satisfies (4.3). Finally (5.4) holds according to remark (5.5) because $B(\varphi) = 0$ for all $\varphi \in C_o^\infty(R^3)$. Hence $S_o = T_o + V_o$ is essentially selfadjoint by theorem (5.6); if $|q_2| \in L_3(R^3)$ and if M in (6.3) is bounded, then V_o is T_o-bounded with T_o-bound smaller than 1 , and the result follows from Rellich's theorem with the additional information that $D(\overline{S}_o) = D(T)$. Stronger results are known for the special class of potentials $q(x) = r(x)\alpha_o$ where α_o is the unit matrix and r is real-valued (cf.[9],[10],[11],[12] and [13]).

(6.4) Non-local perturbations. Let ρ be positive and $\Sigma = \{(x,y) | x, y \in R^3, |x-y| \leq \rho\}$. We consider a function $q : \Sigma \longrightarrow K^{4,4}$ with the properties:

(i) $q(x,y) = q(y,x)^*$, $(x,y) \in \Sigma$;

(ii) q is continuously differentiable ;

(iii) $|q(x,y)| \leq c(1 + |x| + |y|)$, $(x,y) \in \Sigma$ with c independent of x and y . Let $\delta \in]0,1[$; define

$$(V_\varepsilon u)(x) = \int_{\varepsilon \leq |x-y| \leq \rho} |x-y|^{\delta-4} q(x,y)u(y)dy - \frac{4\pi\varepsilon^{\delta-1}}{3-\delta} q(x,x)u(x)$$

for $x \in R^3$, $u \in D_o$ and $\varepsilon \in]0,\rho[$; we want to define $(V_o u)(x) = \lim (V_\varepsilon u)(x)$ for $\varepsilon \longrightarrow 0$. This limit exists (it is the "finite part" in the sense of Hadamard of the integral over $|x-y| \leq \rho$): Let $h: [0,\rho] \longrightarrow R$ be continuously differentiable, $0 \leq h \leq 1$, $h(t) = 1$ for $t \in [0,\eta]$ and $h(t) = 0$ for $t \in [2\eta,\rho]$ where $\eta \in]0,\rho/2[$. In the integral we insert $1 = h(|x-y|) + [1 - h(|x-y|)]$ and we write the integral as a sum of two integrals accordingly; then we use $|x-y|^{\delta-4} = (3-\delta)^{-1} \operatorname{grad}_y (x-y)|x-y|^{\delta-4}$ in the first integral and perform partial integration. Passing to the limit $\varepsilon \longrightarrow 0$ we get

$$(V_o u)(x) = \frac{-1}{3-\delta} \int_{|x-y| \le 2\eta} |x-y|^{\delta-4}(x-y)\,\mathrm{grad}_y h(|x-y|)\, q(x,y)u(y)\,dy + \int_{\eta \le |x-y| \le \rho} |x-y|^{\delta-4}[1-h(|x-y|)]\, q(x,y)u(y)\,dy$$

and it is easy to see that the limit is uniform with respect to x ; hence $V_o u$ is a continuous function. Obviously $V_o u$ has compact support; consequently V_o is defined on D_o. V_ε is symmetric for every $\varepsilon \in]0,\rho[$ because of (i) ; hence V_o is symmetric. Let Ω be a bounded open set and let $u \in D_o(\Omega)$. By (ii) we have $c_1 = \max \{|q(x,y)| \mid (x,y) \in \Sigma \cap \Omega^2\} < \infty$, $c_2 = \max\{|\mathrm{grad}_y q(x,y)| \mid (x,y) \in \Sigma \cap \Omega^2\} < \infty$ and we define $c_o(\eta) = \max\{|h'(t)| \mid t \in [0,\rho]\}$. For $x \in \Omega$ we get

$$|(V_o u)(x)| \le \frac{c_1}{3-\delta} \int_{|x-y| \le 2\eta} |x-y|^{\delta-3} |\mathrm{grad}\, u(y)|\,dy + \frac{c_o(\eta)c_1+c_2}{3-\delta} \int_{|x-y| \le 2\eta} |x-y|^{\delta-3} |u(y)|\,dy + c_1 \eta^{-1} \int_{|x-y| \le \rho} |x-y|^{\delta-3} |u(y)|\,dy$$

$$\le \frac{c_1}{3-\delta} \int_{|x-y| \le 2\eta} |x-y|^{\delta-3} |T_o u(y)|\,dy + c_3(\eta) \int_{|x-y| \le \rho} |x-y|^{\delta-3} |u(y)|\,dy$$

A well-known estimate for the norm of an integral operator (cf. [6] III.2.1) now gives

$$\|V_o u\|_\Omega \le \frac{c_1}{3-\delta} \int_{|z| \le 2\eta} |z|^{\delta-3} dz \, \|T_o u\| + c_3(\eta) \int_{|z| \le \rho} |z|^{\delta-3} dz \, \|u\|$$

$= c_4 \eta^\delta \|T_o u\| + c_5(\eta)\|u\|$. Hence V_o satisfies (4.3) for arbitrary $b \in]0,1[$ with suitable $a > 0$. Furthermore we have for $\varphi \in C_o^\infty(R^3)$

$$(B_o(\varphi)u)(x) = \int_{|x-y| \le \rho} |x-y|^{\delta-4} [\varphi(y) - \varphi(x)]\, q(x,y)u(y)\,dy \; , \text{ hence}$$

$$\|B_o(\varphi)\| \le \sup\left\{ \int_{|x-y| \le \rho} |x-y|^{\delta-4} |\varphi(y) - \varphi(x)|\, |q(x,y)|\,dy \;\middle|\; x \in R^3 \right\}$$

and this is finite for every $\varphi \in C_o^\infty(R^3)$, i.e. (4.2) holds. Finally, according to (5.5), the condition (5.4) is satisfied if the sequence $(\|B(\varphi_n)\|)$ is bounded for the functions φ_n defined in (5.5).

Using the estimate $|\varphi_n(y) - \varphi_n(x)| \leq \frac{1}{n}|y-x| \max\{|\text{grad}\,\varphi(z)| \; z \in R^3\} = \frac{1}{n} c_6 |x-y|$ and property (iii) of q we get

$$\|B(\varphi_n)\| \leq \frac{1}{n} c_7 \sup\left\{ \int_{|z| \leq \rho} |z|^{\delta-3}(1+|x|+|x+z|)\,dz \;\Big|\; |x| \leq 2n+\rho \right\} \leq c_8 .$$

All assumptions of theorem (5.6) are fulfilled, hence S_o is essentially selfadjoint.

7. THE ESSENTIAL SPECTRUM OF S

Let $S_o = T_o + V_o$ be essentially selfadjoint and let S be the unique selfadjoint extension of S_o. The essential spectrum $\sigma_e(S)$ of S is the set of all points in the spectrum $\sigma(S)$ which are not isolated eigenvalues of finite multiplicity. The essential spectrum has remarkable stability properties; the best-known result is the theorem of H.Weyl which says that $\sigma_e(S) = \sigma_e(T)$ if V_o is T_o-compact (cf. [6] IV.1.3 for the definition and IV.5.6 for the theorem). For certain operators in $L_2(R^m)$ there is a necessary and sufficient condition for T_o-compactness due to J.Weidmann and the author ([5] theorem 3.2) which evidently also holds for operators in H :

(7.1) Theorem. Let T_o and V_o be closable operators in H with domain D_o and let the operator $u \longmapsto \varphi u$ be T_o-compact for every $\varphi \in C_o^\infty(R^m)$. Then V_o is T_o-compact if and only if for every $\varepsilon > 0$ there is a function $\varphi_\varepsilon \in C_o^\infty(R^m)$ such that $\|V_o u\| \leq \varepsilon(\|u\| + \|T_o u\|) + \|\varphi_\varepsilon u\|$ for all $u \in D_o$.

(7.2) Lemma. Let T_o be defined by (2.4) and (2.5); assume that $1 + |k(y)|^2 \geq c_o(1+|y|^2)^r$ for all $y \in R^m$ with positive constants c_o and r . Then the operator $u \longmapsto \varphi u$ is T_o-compact for every $\varphi \in C_o^\infty(R^m)$.

Proof. Denote by $\|\;\|_r$ the norm of the Sobolev space $[H^r(R^m)]^p$; then for $\varphi \in C_o^\infty(R^m)$ and $u \in D_o$ we have

$\|\varphi u\|_r^2 = \int |\widehat{\varphi u}(y)|^2(1+|y|^2)^r dy \leq c_o^{-1} \int |\widehat{\varphi u}(y)|^2(1+|k(y)|^2)dy =$ $c_o^{-1}(\|\varphi u\|^2 + \|T_o\varphi u\|^2) \leq c_1(\|u\|^2 + \|T_o u\|^2)$ by (2.2), where c_1 is independent of u. Let (u_n) be a sequence in D_o such that $(\|u_n\| + \|Tu_n\|)$ is bounded; it follows that $(\|\varphi u_n\|_r)$ is bounded and hence by the Rellich-Gårding theorem (cf. [14] X.3) there is a subsequence (u_{n_j}) such that (φu_{n_j}) converges in H, q.e.d.

(7.3) Theorem. Let T_o satisfy (2.1), (2.2); assume that the sequence $(\|A_o(\varphi_n)\|)$ is bounded for the functions φ_n defined in (5.5) and let the operator $u \longmapsto \varphi u$ be T_o-compact for every $\varphi \in C_o^\infty(R^m)$. Let V_o satisfy (4.1) and (4.2). Then V_o is T_o-compact if and only if the following two conditions hold:

(i) (4.3) holds for arbitrary $b \in]0,1[$ with suitable $a > 0$.

(ii) For every $\varepsilon > 0$ there is an open neighbourhood of ∞, say Ω_ε, such that $\|V_o u\| \leq \varepsilon(\|u\| + \|T_o u\|)$ for all $u \in D_o(\Omega_\varepsilon)$.

Proof. If V_o is T_o-compact then (i) and (ii) follow directly from theorem (7.1). Assume that (i) and (ii) hold. By assumption there is a constant $c > 0$ such that $\|A_o(\varphi_n)\| \leq c$ for all n and for the functions φ_n defined in (5.5). Let $\varepsilon > 0$ be given and let $\eta = (2+2c)^{-1}\varepsilon$; choose Ω_η according to (ii), then choose n such that all x with $|x| \geq n$ are contained in Ω_η. Then $1-\varphi_n$ has support in Ω_η and $(1-\varphi_n)u \in D_o(\Omega_\eta)$ for all $u \in D_o$. From (ii) we get $\|V_o(1-\varphi_n)u\| \leq \eta(\|(1-\varphi_n)u\| + \|T_o(1-\varphi_n)u\|) \leq \eta(\|u\| + \|A_o(\varphi_n)u\| + \|T_o u\|)$ for all $u \in D_o$. Let Ω be the open ball of radius $2n+1$ and center 0; then $\varphi_n \in C_o^\infty(\Omega)$ and hence $\varphi_n u \in D_o(\Omega)$ for all $u \in D_o$. Apply (4.3) for this Ω and choose $b \leq \eta$ according to (i); then $\|V_o\varphi_n u\| \leq a\|\varphi_n u\| + \eta\|T_o\varphi_n u\| \leq a\|\varphi_n u\| + \eta\|A_o(\varphi_n)u\| + \eta\|T_o u\|$ and hence $\|V_o u\| \leq \eta(1+2\|A_o(\varphi_n)\|)\|u\| + 2\eta\|T_o u\| + a\|\varphi_n u\|$ $\leq \varepsilon(\|u\| + \|T_o u\|) + \|a\varphi_n u\|$ for all $u \in D_o$; it follows from theorem (7.1) that V_o is T_o-compact, q.e.d.

If T_o is of the form (2.4) and satisfies the assumption of lemma (7.2) then, by this lemma and by (5.5), T_o satisfies the assumptions of theorem (7.3). In particular this is true for the Dirac operator, because $|k(y)|^2 = \mu^2 + |y|^2$ in this case. Let us check the examples of perturbations of the Dirac operator given in section 6: The Coulomb potentials (6.1) do not satisfy (i) of theorem (7.3); the other perturbations (6.2), (6.3) and (6.4) satisfy this condition. The perturbation (6.2) satisfies (ii) if $|q| \in L_3(R^3)$; the proof makes use of the Sobolev inequality mentioned in (6.2). The Stummel potentials (6.3) satisfy (ii), if $\int_{|x-y| \leq 1} |q(y)|^2 dy$ goes to zero for $|x| \to \infty$; this is proved with the methods of [4] §4 and §5. Finally the integral operator (6.4) satisfies (ii) if $q(x,y)$ goes to zero for $|x|+|y| \to \infty$ and if $\mathrm{grad}_y q(x,y)$ is uniformly bounded; this follows from the estimates in (6.4). By theorem (7.3) and by Weyl's theorem we get $\sigma_e(S) = \sigma_e(T) = \{\lambda \mid \lambda \in R, |\lambda| \geq |\mu|\}$ if V_o is a sum of three operators satisfying the conditions given above.

Actually T_o-compactness of V_o is not necessary for T and S to have the same essential spectrum. The following theorem is adapted from [5] (theorem 3.15, corollary 3.19, remark 3.20) and therefore is given without proof:

(7.4) <u>Theorem</u>. Let T_o satisfy (2.1), (2.2) and let $u \mapsto \varphi u$ be T_o-compact for every $\varphi \in C_o^\infty(R^m)$. Let V_o satisfy (4.1), (4.2) and let V_o be T_o-bounded with T_o-bound smaller than 1 (i.e. (4.3) holds for $\Omega = R^m$) and satisfy (ii) of theorem (7.3). Then $S = \overline{S}_o$ (which is selfadjoint by Rellich's theorem) and $T = \overline{T}_o$ have the same essential spectrum.

This theorem applies in particular to the Dirac operator T_o and to the Coulomb perturbation (6.1). Note that the assumptions of theorem

(7.3) (even with the condition on $A_o(\varphi_n)$ omitted) imply those of theorem (7.4). It follows that a perturbation $V_o = \sum_{j=1}^{4} V_j$ does not change the essential spectrum, if V_1 is a Coulomb perturbation, V_2 is given by a L_3-potential, V_3 is given by a Stummel potential q_3 such that $\int_{|x-y| \leq 1} |q_3(y)|^2 dy \longrightarrow 0$ for $|x| \longrightarrow \infty$, and V_4 is the operator (6.4) with $q(x,y) \longrightarrow 0$ for $|x|+|y| \longrightarrow \infty$ and $\operatorname{grad}_y q(x,y)$ uniformly bounded.

REFERENCES

[1] Evans, W.D. Proc. London Math. Soc.(3)20,537-557(1970).

[2] " " Math.Zeitschr.121,1-23(1971).

[3] Gross, L. Comm.pure appl.Math. 19,1-15(1966).

[4] Jörgens, K. Math.Zeitschr. 96,355-372(1967).

[5] " " and J.Weidmann. Spectral theory of Hamiltonian operators (to appear).

[6] Kato, T. Perturbation theory for linear operators.Springer 1966.

[7] Kato, T. Trans.Amer.Math.Soc. 70,195-211(1951).

[8] Prosser, R.T. J.Math.Phys. 4,1048-1054(1963).

[9] Rejto, P.A. Israel J.Math. 9,144-171(1971).

[10] Rellich, F. Eigenwerttheorie partieller Differentialgleichungen II. Lecture Notes, Göttingen 1953.

[11] Schillemeit, J. Thesis, Univ. Göttingen 1954.

[12] Schmincke, U.-W. Essential selfadjointness of Dirac operators with a strongly singular potential (to appear).

[13] Weidmann, J. Math.Zeitschr. 119,349-373(1971).

[14] Yosida, K. Functional analysis. Springer 1968.

On Optimal Control Problems with Bounded State Variables

H.W. Knobloch

1. Introduction

We consider control problems which are described by the following data: (i)a differential equation

$\dot{x} = dx/dt = f(x,u), \; f = (f^1,\ldots,f^n),$

where $x = (x^1,\ldots,x^n)$ are the state variables and $u = (u^1,\ldots,u^m)$ the control variables, (ii) an arbitrary non-empty set U in u-space (control region), (iii) a scalar function g(x). By a control function we mean a piecewise continuous function u(t) which is such that $u(t \pm 0) \in U$ for every t. A solution of the control problem on some interval $[0,\tilde{t}]$ is a pair (x(t), u(t)), where u(t) is a control function and x(t) solution of $\dot{x} = f(x, u(t))$ and where x(t) is required to satisfy the inequality restraint $g(x(t)) \leq 0$ and some kind of boundary condition. An optimal solution is one which minimizes a given integral functional (which always can assumed to be a component of the state vector x(t)).

The main result of the paper is theorem 4.1 which contains four first-order necessary conditions for optimal solutions. The first two conditions constitute the maximum principle for the problem in question and are essentially known. The remaining conditions however seem to be new and of some significance for applications, since they allow certain a-priori-statements about the Lagrange multipliers. The knowledge of these relations may reduce the number of cases to be discussed in a practical situation, as will be illustrated in section 5 by the example of time-optimal control of an ore unloader.

At first glance our approach is similar to the ones used by Hestenes and Pontryagin (see e.g. [1], [2], [5]). However there is a considerable difference concerning the notion of the "cone of attainability". This cone will be defined in section 3 by means of generating elements a n d relations and thereby becomes larger than the usual one.

A major portion of the material presented in section 2 can be found in a slightly different form in the unpublished thesis of G.Plate [4]. All definitions and theorems used from [4] are repeated here in full but proofs are merely sketched or omitted.

Matrix notation is used exclusively, in particular matrices of one row will be called vectors and denoted by x,u,w etc. with $|x|$ signifying the maximum norm. The transpose of a matrix A is denoted by A^T. For convenience we will assume that the value of a function of t always equals the left-hand limit. Several statements involve a positive ϵ and are valid if ϵ is sufficiently small, a premise which will not always be mentioned explicitely. Finally we introduce the notion of "property L". If $\{M_r\}$, $0 \le r \le r_o$, is a family of subsets of some Euclidean space and φ a function defined on $\cup M_r$, then we say that φ has property L on $\{M_r\}$ if φ satisfies a Lipschitz-condition on M_r with a Lipschitz-constant L_r and if $\lim_{r\to o} L_r = 0$. We will assume (which is more than actually required) that f(x,u) and g(x) are defined and continuously differentiable everywhere. Furthermore - and this is essentially needed - we consider only those x(t),u(t) which satisfy the following condition with respect to the function $G(x,u) = f(x,u)\cdot\mathrm{grad}(g(x))^T$

$$(\partial G/\partial u^1,\ldots,\partial G/\partial u^m) \neq (0,\ldots,0) \text{ for } (x,u) = (x(t_1),u(t_1\pm 0))$$

whenever t_1 is a limit point of the zeros of g(x(t)). (1.1)

2. Variation of Controls and Trajectories

In this section the variational technique will be developed. Throughout the following considerations we will assume that an interval $[0,\tilde{t}]$ and a certain solution $(u(t),x(t))$ of the control problem has been fixed once and for all. Hence $u(t)$ is a control function on $[0,\tilde{t}]$ and $x(t)$ solution of the corresponding equation $\dot{x} = f(x,u(t))$. The condition $g(x(t)) \leq 0$ is not required for the present. We put $x_o = x(0)$ and denote by D the (finite) set of all $t \in [0,\tilde{t}]$ for which u becomes discontinuous. Furthermore let

$$\dot{y} = - yF(t)^T \tag{2.1}$$

be the adjoint variational equation. $F(t)$ is thus the matrix having $\partial f/\partial x^i(x(t),u(t))$ as row vectors.

Let there be given finitely many t_i, $0 < t_1 \leq t_2 \leq .... \leq t_N < \tilde{t}$, which are such that

$$t_i \notin D \text{ if } t_i = t_{i+1}. \tag{2.2}$$

Also let there be given the same number of dements $v_i \in U$. We put

$$\begin{aligned} k_i &= f(x(t_i),v_i) - f(x(t_i),\ u(t_i+0)),\\ \tilde{k}_i &= f(x(t_i),u(t_i)) - f(x(t_i),u(t_i+0)). \end{aligned} \tag{2.3}$$

The t_i and v_i will also be kept fixed throughout this section. We next choose control functions $u_i(t)$, $i = 0,\dots,N$, $v_i(t)$, $i = 1,\dots,N$ which are defined on $[0,\tilde{t}]$ and satisfy the following conditions

$u_i(t) = u(t)$ for $t \in (t_i,t_{i+1})$, u_i continuous for $t \in (t_i,t_{i+1})$,
$u_i(t_{i+1}) = u(t_{i+1})$, $i = 1,\dots,N-1$,

$u_o(t) = u(t)$ for $t < t_1$, u_o continuous for $t \geq t_1$,

$u_N(t) = u(t)$ for $t > t_N$, u_N continuous for $t \leq t_N$,

v_i continuous for all t, $v_i(t_i) = v_i$.

Such functions exist, take e.g. $v_i(t) = \text{const} = v_i$ for all t and $u_i(t)$ = const outside (t_i,t_{i+1}).

We are now going to explain what one may call "variation of the control function". To this purpose let $W \subset \mathbb{R}^{2N}$ be the set of all $w = (w^1,\dots,w^{2N})$ satisfying $w^j \le w^{j+1}$, $j = 1,\dots,N-1$ and let $\tilde{w}$ be the element $(\tilde{w}^1,\dots,\tilde{w}^{2N})$ with $\tilde{w}^{2i-1} = \tilde{w}^{2i} = t_i$.

If $w \in W$ does not differ too much from $\tilde{w}$ then the following definition makes sense and yields a controlfunction $u_w(t)$ on $[0,\tilde{t}]$.

$$u_w(t) = \begin{cases} u_o(t) \text{ if } t \le w^1, \\ v_i(t) \text{ if } w^{2i-1} < t \le w^{2i},\ i = 1,\dots,N, \\ u_i(t) \text{ if } w^{2i} < t \le w^{2i+1},\ i = 1,\dots,N-1, \\ u_N(t) \text{ if } t > w^{2N}. \end{cases}$$

Let $x(t,w)$ then be the solution of the initial-value problem

$$\dot{x} = f(x,u_w(t)),\ x(0) = x_o. \tag{2.4}$$

Since $u_{\tilde{w}}(t) = u(t)$ we have $x(t,\tilde{w}) = x(t)$. The family $x(t,w)$ may be regarded as the corresponding "variation of the trajectory".

<u>Theorem 2.1.</u> If $|w-\tilde{w}|$ is sufficiently small, then $x(t,w)$ exists on $[0,\tilde{t}]$. Furthermore, if $X(t)^T$ is the matrix solution of (2.1) satisfying $X(\tilde{t}) = I$ (= identity matrix), then the function

$$\Phi(t,w) = x(t,w) - x(\tilde{t}) - f(x(\tilde{t}),u(\tilde{t}))(t-\tilde{t}) - \sum_{i=1}^{N}(w^{2i}-w^{2i-1})k_i X(t_i) - \sum_{i=1}^{N}(w^{2i-1}-t_i)\tilde{k}_i X(t_i)$$

has property L on the family $M_r = \{(t,w):\ |t-\tilde{t}| \le r,\ |w-\tilde{w}| \le r\}$, r sufficiently small (for the definition of $k_i,\tilde{k}_i$ see (2.3)).

The proof is given in [4] (Satz (2.1)). We now wish to establish the analogous result for the control problem stated in the introduction, that is the control problem which includes the condition $g(x) \le 0$ for the trajectory. So we assume that the given $x(t)$ satisfies the condition $\gamma(t) = g(x(t)) \leq 0$ for all t and we ask how the parameter w has to be chosen in order that $g(x(t,w)) \leq 0$ for all $t \in [0,\tilde{t}]$. The answer is given first in two special cases, namely when the set of the zeros of $\gamma(t)$ reduces to a point or to an interval.

<u>Lemma 2.1.</u> Let t_i be one of the numbers fixed in the beginning and let t', t'' be numbers such that $0 \leq t' < t_i < t'' \leq \tilde{t}$, $\gamma(t) < 0$ if $t \neq t_i$ and $t \in [t', t'']$, $\gamma(t_i) = 0$, $\dot{\gamma}(t_i) > 0$, $\dot{\gamma}(t_i+0) < 0$. Let $v_i = u(t_i+0)$.

Then, for $|w-\tilde{w}|$ sufficiently small, the inequality $g(x(t,w)) \leq 0$ on $[t', t'']$ follows from

$$w^{2i-1} \leq \Delta_i(w'), \tag{2.5}$$

where $w' = (w^1, \ldots, w^{2i-2})$ and Δ_i is a certain function defined on the set

$$W'_\epsilon = \left\{ w' = (w^1, \ldots, w^{2i-2}) \; : \; w^{\nu+1} \geq w^\nu, \; |w^\nu - \tilde{w}^\nu| \leq \epsilon \right\}.$$

Furthermore $\Delta_i(\tilde{w}') = t_i$ and the function

$$\Delta_i(w') - t_i + (\dot{\gamma}(t_i))^{-1} \left\{ \sum_{\nu=1}^{i-1} (w^{2\nu} - w^{2\nu-1}) k_\nu X_i(t_\nu) + \right.$$

$$\left. + \sum_{\nu=1}^{i-1} (w^{2\nu-1} - t_\nu) \tilde{k}_\nu X_i(t_\nu) \right\} \cdot \text{grad}\,(g(x(t_i))^T$$

has property L on the family $\{W'_r\}$, $r \leq \epsilon$. Here $X_i^T(t)$ denotes the solution of (2.1) satisfying $X_i(t_i) = I$.

The <u>proof</u> of this lemma can be found in [4], section 5. $\Delta(w')$ represents, roughly speaking, the moment when the trajectory $x(t,w)$ would reach the hypersurface $g(x) = 0$, if the corresponding control function would not switch from $u_{2i-1}(t)$ to $v_{2i-1}(t)$ at $t = w^{2i-1}$.

For the remaining part of the paper we will assume that $G = f(x,u) \cdot \text{grad}(g(x))^T$ is a function of u only, which will be denoted by G(u) henceforth. As will be pointed out in section 6 this restriction actually can be removed and is introduced merely to avoid lengthy explanations which may obscure the basic line of our approach.

<u>Lemma 2.2.</u> Let there be given numbers t', t'', t_i, t_j, where t_i, t_j belong to the set introduced in the beginning. Assume that the following conditions are satisfied ($\overset{\circ}{U}$ is the interior of U)

(i) $0 \le t' < t_i < t_j < t'' \le \tilde{t}$,

(ii) $\gamma(t) < 0$ for $t \notin [t_i, t_j]$, $\gamma(t) = 0$ for $t \in [t_i, t_j]$,

(iii) $\dot{\gamma}(t_i - 0) = G(u(t_i)) > 0$, $\dot{\gamma}(t_j + 0) = G(u(t_j + 0)) < 0$,

(iv) $v_i = u(t_i + 0)$,

(v) $u(t \pm 0) \in \overset{\circ}{U}$ if $t \in [t_i, t_j]$.

Then for $|w - \tilde{w}|$ sufficiently small the inequality $g(x(t,w)) \le 0$ will hold on $[t', t'']$ provided

$w^{2i-1} \le \Delta_i(w')$ and

$$\sum_{\nu=i+1}^{m} (w^{2\nu} - w^{2\nu-1}) G(v_\nu) \le 0, \quad m = i+1, \ldots, j. \tag{2.6}$$

Δ_i is the function introduced in lemma 2.1.

<u>Proof.</u> The proof of the lemma requires an appropriate choice of the functions $u_i(t)$ and $v_i(t)$ which are used to define the variation $u_w(t)$ of the control. It follows from (1.1) and from the hypotheses (iii) - (v) that one can choose $u_\nu(t)$, $v_\mu(t)$ in accordance with previous requirements such that we have for some $\epsilon_1 > 0$

$$\begin{aligned} &G(u_\nu(t)) = 0 \text{ on } [t_\nu - \epsilon_1, t_\nu + \epsilon_1], \ \nu = i, \ldots, j-1, \\ &G(u_j(t)) < 0 \text{ on } [t_j - \epsilon_1, t_j + \epsilon_1], \\ &G(v_i(t)) = 0 \text{ on } [t_i - \epsilon_1, t_i + \epsilon_1] \end{aligned} \tag{2.7}$$

and $v_\mu(t)$ = const for $\mu \neq i$. As a consequence of this choice the function $G(u_w(t))$ is piecewise constant on $[w^{2i-1},w^{2j}]$ and takes the values $G(v_i(t)) = 0,0,\ G(v_{i+1}),\ 0,\dots,0,\ G(v_j)$ on the intervals $[w^{2i-1},w^{2i}]$, $[w^{2i},w^{2i+1}],\dots,$ $[w^{2j-1},w^{2j}]$ respectively if $|w-\tilde{w}| \leq \epsilon$. It follows then from (2.6), (2.7) that

$$g(x(t,w)) = g(x(w^{2i-1},w)) + \int_{w^{2i-1}}^{t} G(u_w(\tau))d\tau$$

$$\leq g(x(w^{2i-1},w)) \text{ if } w^{2i-1} \leq t \leq t_j + \epsilon_1.$$

On the other hand $g(x(t,w)) \leq 0$ for $t' \leq t \leq w^{2i-1}$ if $w^{2i-1} \leq \Delta_i(w')$ and $|w-\tilde{w}| \leq \epsilon$, as we have shown in lemma 2.1. Finally, for reasons of continuity, $g(x(t,w)) < 0$ for $t_j+\epsilon_1 \leq t \leq t''$ and $|w-\tilde{w}|$ sufficiently small, since the inequality holds for $w = \tilde{w}$. Thereby the lemma is proved.

We now enter into the discussion of the general case. Let us assume that the set of zeros of the function $\gamma(t)$ consists of finitely many intervals (or isolated points) $[t_\mu^*,t_\mu^{**}]$ such that $0 < t_\mu^* \leq t_\mu < \tilde{t}$, $\mu = 1,\dots,q$. Furthermore we assume, that all t_μ^*, t_μ^{**} appear among the t_i. That means that there exist numbers $i(\mu)$, $j(\mu)$ such that $t_\mu^* = t_{i(\mu)}$, $t_\mu^{**} = t_{j(\mu)}$. Finally we assume that the following conditions are satisfied for $\mu = 1,\dots,q$

(i) $\dot{\gamma}(t_\mu^*) = G(u(t_\mu^*)) > 0$, $\dot{\gamma}(t_\mu^{**}+0) = G(u(t_\mu^{**}+0)) < 0$,

(ii) $v_{i(\mu)} = u(t_\mu^*+0)$,

(iii) $u(t\pm 0) \in \overset{\circ}{U}$ if $t \in [t_\mu^*,t_\mu^{**}]$ and $t_\mu^* < t_\mu^{**}$. (2.8)

In view of the two lemmas we can then state the following preliminary result. If the $u_\nu(t)$, $v_\mu(t)$ have been chosen suitably and if $|w-\tilde{w}|$ is sufficiently small the inequality $g(x(t,w)) \leq 0$ is true for all $t \in [0,\tilde{t}]$ provided (2.5) and (2.6) hold for $i = i(\mu)$, $j = j(\mu)$ and $\mu = 1,\dots,q$.

We next wish to put condition (2.5) in a more convenient form. To this purpose we introduce a vector-valued function $z = z(w) = (z^1(w),\dots,z^{2N}(w))$ by this definition of $z^\nu = z^\nu(w)$

$$z^\nu = w^\nu \text{ if } \nu \neq 2i(\mu)-1,\ \mu = 1,\dots,q,$$

$$z^{2i(\mu)-1} = w^{2i(\mu)-1} - \Delta_{i(\mu)}(w^1,\dots,w^{2i(\mu)-2}) + t^*_\mu,\ \mu = 1,\dots,q \quad (2.9)$$

$w \to z(w)$ is a one-to-one mapping of $W_\epsilon = \{w \in W : |w-\tilde{w}| \le \epsilon\}$ onto some subset Z_ϵ of the $\mathbb{R}^{2N}$. $z(w)$ satisfies a Lipschitz-condition with respect to w on W_ϵ and $z(\tilde{w}) = \tilde{w}$. The conditions which we imposed an w to enforce $g(x(t,w) \le 0$ can then be expressed in terms of $z = z(w)$ in this way

$$z^{2i(\mu)-1} \le t^*_\mu$$

$$\sum_{\nu=i(\mu)+1}^{m} (z^{2\nu} - z^{2\nu-1})G(v_\nu) \le 0,\ m = i(\mu)+1,\dots,j(\mu) \quad (2.10)$$

for $\mu = 1,\dots,q$.

For notational convenience we introduce the following abbreviations

$$g_\mu = \operatorname{grad} f(x(t^*_\mu)),\ \gamma_\mu = \dot{\gamma}(t^*_\mu).$$

For the same reason the notation $\Psi_1 \sim \Psi_2$ is used to indicate that the functions Ψ_1, Ψ_2 are defined on W_ϵ and that $\Psi_1 - \Psi_2$ has property L on the family $\{W_r\}$, $0 < r \le \epsilon$. The second statement of lemma 2.1 can then be given this form

$$w^{2i(\mu)-1} \sim z(w)^{2i(\mu)-1} - \gamma_\mu^{-1} \Big\{ \sum_{\nu=1}^{i(\mu)-1} (w^{2\nu} - w^{2\nu-1}) k_\nu X_{i(\mu)}(t_\nu)$$

$$+ \sum_{\nu=1}^{i(\mu)-1} (w^{2\nu} - t_\nu)\tilde{k}_\nu X_{i(\mu)}(t_\nu)\Big\} \cdot g^T_\mu . \quad (2.11)$$

This follows immediately from the definition of $z(w)$. We are now going to express $\Phi(t,w)$ (cf. Theorem 1) in terms of the components

$(z^1,\dots,z^{2N})$ of $z(w)$. Thereby a new kind of solutions $X(t)$ of (2.1) will appear, namely solutions which are continuous except at $t = t_\mu$ and satisfy the "jump-condition"

$$X(t_\mu^*) = (I - \gamma_\mu^{-1} g_\mu^T \cdot \tilde{k}_\mu^*) X(t_\mu^* + 0), \text{ where } \tilde{k}_\mu^* = \tilde{k}_{i(\mu)} \qquad (2.12)$$

<u>Lemma 2.3.</u> Given a matrix solution $X(t)$ of (2.1) which is continuous on $[0,\tilde{t}]$, and an integer m such that $0 \le m \le N$. Let $r = r(m)$ be the number of t_μ^* which belong to the set $\{t_1,\dots,t_m\}$ and let $v_{i(\mu)} = u(t_\mu^* + 0)$ for $\mu = 1,\dots,r$. Then

$$\sum_{\nu=1}^{m} (w^{2\nu} - w^{2\nu-1}) k_\nu X(t_\nu) + \sum_{\nu=1}^{m} (w^{2\nu-1} - t_\nu)\tilde{k}_\nu X(t_\nu) \sim$$

$$\sim \sum_{\nu=1}^{m} (z^{2\nu} - z^{2\nu-1}) k_\nu X^*(t_\nu + 0) + \sum_{\nu=1}^{m} (z^{2\nu-1} - t_\nu)\tilde{k}_\nu X^*(t_\nu + 0), \qquad (2.13)$$

where $X^*(t)$ is the solution of (2.1) satisfying the jump-conditions (2.12) for $\mu = 1,\dots,r$ and the condition $X^*(t) = X(t)$ for $t > t_r^*$. The z^j appearing on the right-hand of (2.13) are the components of $z(w)$.

<u>Proof.</u> Let us denote for the moment the expressions which occur on both sides in the relation (2.13) by $S_w(m,X(t))$ and $S_z(m,X^*(t))$ respectively. If $m = 0$ nothing has to be proved. So we proceed by induction and assume that the assertion is correct for sums $S_w(m',X(t)),S_z(m',X^*(t))$ provided $m' < m$. If $m > i(r)$ then $w^j = z^j$ for $j = 2m$ and $j = 2m-1$, $X(t_m) = X^*(t_m) = X^*(t_m + 0)$ and it is clear that (2.13) follows from $S_w(m-1,X(t)) \sim S_z(m-1,X^*(t))$. Let us now consider the case $m = i(r)$. Since $v_{i(r)} = u(t_r^* + 0)$ we have $k_m = 0$ and therefore

$$S_w(m,X(t)) = S_w(m-1,X(t)) + (w^{2m-1} - t_m)\tilde{k}_m X(t_m). \qquad (2.14)$$

By assumption of induction

$$S_w(m-1,X(t)) \sim S_z(m-1,X'(t)). \tag{2.15}$$

From (2.11) ($\mu = r$) we obtain, also by assumption of induction,

$$w^{2m-1} - t_m \sim z^{2m-1} - t_m - \gamma_r^{-1} S_z(m-1,X^*_{i(r)}(t))\cdot g_r^T. \tag{2.16}$$

Here $X'(t), X^*_{i(r)}(t)$ satisfy the jump-condition (2.12) for $\mu = 1,\dots,r-1$ and the condition

$$X'(t) = X(t),\ X^*_{i(r)}(t) = X_{i(r)}(t) \text{ for } t > t^*_{r-1}. \tag{2.17}$$

From (2.14) - (2.16)

$$\begin{aligned} S_w(m,X(t)) &\sim S_z(m-1,X'(t)) + (z^{2m-1}-t_m)\tilde{k}_m X(t_m) \\ &\quad -\gamma_r^{-1} S_z(m-1,X^*_{i(r)}(t)) g_r^T \tilde{k}_m X(t_m) \\ &= S_z(m,X^*(t)), \end{aligned}$$

where $X^*(t) = X'(t)-\gamma_r^{-1}X^*_{i(r)}(t)(g_r^T\cdot\tilde{k}_m X(t_m))$ for $t < t_m$ and $X^*(t) = X'(t)$ for $t > t_m$. $X^*(t)$ satisfies the jump-condition for $t = t^*_\mu$, $\mu = 1,\dots,r-1$, since this is true for $X'(t)$ and for $X^*_{i(r)}(t)$. Furthermore $X'(t_m) = X(t_m)$, $X^*_{i(r)}(t_m) = X_{i(r)}(t_m) = I$ because of (2.17) and $t_m = t_{i(r)}$. So $X^*(t_m) = (I-\gamma_r^{-1}g_r^T\cdot\tilde{k}_m)X'(t_m)$ and this is the jump-condition for $\mu = r$, since $X'(t_m) = X^*(t_m+0)$.

We summarize the results we have obtained so far.

<u>Theorem 2.2.</u> Let $(x(t),u(t))$ be a solution of the control problem and let (2.8) be satisfied for $\mu = 1,\dots,q$. If $|w-\tilde{w}|$ is small enough, then the following conditions, which are expressed in terms of $z = z(w)$, are sufficient in order that $g(x(t,w)) \le 0$.

(i) $z^{2i(\mu)-1} \le t^*_\mu$

(ii) $\sum\limits_{\nu=i(\mu)+1}^{m} (z^{2\nu} - z^{2\nu-1})G(v_\nu) \le 0,\ m = i(\mu)+1,\dots,j(\mu)$

and $\mu = 1,\dots,q$.

Furthermore, let $X^*(t)^T$ be the matrix solution of (2.1), which is continuous if $t \neq t_\mu^*$ and satisfies the jump-conditions (2.12) for $\mu = 1,\dots,q$ and the initial condition $X^*(\tilde{t}) = I$. Then the function

$$\Phi^*(t,w) = x(t,w) - x(\tilde{t}) - f(x(\tilde{t}),u(\tilde{t}))(t-\tilde{t}) - \sum_{i=1}^{N}(z^{2i}-z^{2i-1})k_i X^*(t_i+0) - \sum_{i=1}^{N}(z^{2i-1}-t_i)\tilde{k}_i X^*(t_i+0)$$

has property L on the family $\{M_r\}$ (cf. Theorem 1).

3. The Cone of Attainability

We now associate with $(x(t),u(t))$ a convex cone $K \subseteq \mathbb{R}^n$ with vertex in 0*). K will be the sum of cones $K_0, K_\mu, \mu = 1,\dots,q$, K^* which we have to define first.

K_0 is the cone generated by the vectors of the form

$$(f(x(t),v) - f(x(t),u(t)))X^*(t) \tag{3.1}$$

where $v \in U$, $t \in (0,\tilde{t})$, $t \notin D$, $t \notin (t_\mu^*, t_\mu^{**})$ for $\mu = 1,\dots,q$.

K^* is the cone generated by the vectors

(i) $\pm f(x(\tilde{t}),u(\tilde{t}))$,

(ii) $\pm(f(x(t),u(t))-f(x(t),u(t+0)))X^*(t)$, $t \neq t_\mu^*$, $\mu = 1,\dots,q$,

(iii) $-(f(x(t_\mu^*),u(t_\mu^*))-f(x(t_\mu^*),u(t_\mu^*+0)))X^*(t_\mu^*+0)$, $\mu = 1,\dots,q$. (3.2)

The cone K_μ is defined in a more complicated way by specifying its elements as follows

$a \in K_\mu$ if there exist finitely many $t_i \in (t_\mu^*, t_\mu^{**})$, $v_i \in U$, $\sigma_i \geq 0$, $i = 1,\dots,s$, such that $t_1 \leq t_2 \leq \dots \leq t_s$,

$$a = \sum_{\nu=1}^{s} \sigma_\nu (f(x(t_\nu),v_\nu)-f(x(t_\nu),u(t_\nu)))X^*(t_\nu) \text{ and}$$

$$\sum_{\nu=1}^{m} \sigma_\nu G(v_\nu) \leq 0,\ m = 1,\dots,s. \tag{3.3}$$

*) that is a set of vectors such that $\alpha a+\beta b \in K$ whenever $a \in K$, $b \in K$ and $\alpha \geq 0$, $\beta \geq 0$.

We then put $K = K_0+K^*+K_1+\ldots+K_q$. $x(\tilde{t})+K$ is then the "cone of attainability". As it will become clear from the next theorem $x(\tilde{t})+K$ represents so to speak a first approximation to the set of reachable points in the neighbourhood of $x(\tilde{t})$.

<u>Theorem 3.1.</u> Let a_o be an interior point of K (regarded as subset of $\mathbb{R}^n$). Then there exists for every sufficiently small $\epsilon > 0$ a number t_ϵ and a control function $u_\epsilon(t)$ such that the solution $x_\epsilon(t)$ of the initial value problem

$\dot{x} = f(x,u_\epsilon(t)),\ x_\epsilon(0) = x_o\ (= x(0))$

exists on $[0,t_\epsilon]$ and satisfies these conditions

(i) $g(x_\epsilon(t)) \leq 0$ for all $t \in [0,t_\epsilon]$,

(ii) $x_\epsilon(t_\epsilon) = x(\tilde{t})+\epsilon a_o$.

The proof of this theorem, which is the basic tool in dealing with the control problems in question, is given in [4] (Satz 3.1).

4. The Multiplier Rule

The statement of theorem 3.1 can be given this form: If a_o is interior point of K, then, for sufficiently small ϵ, $x(\tilde{t})+\epsilon a_o$ belongs to the reachable set (that is the set of endpoints of those trajectories which start at x_o and which arise from solutions of the control problem). Conversely, if $x(\tilde{t})+\epsilon a_o$ is not reachable, then a_o cannot be interior point of K. Therefore, by the separation properties for convex cones, one can find a non-zero vector b (the multiplier) such that

$$a\cdot b^T \leq 0 \text{ for } a \in K,\ a_o\cdot b^T \geq 0. \tag{4.1}$$

One can use this "multiplier rule" to derive necessary conditions for optimal solutions. This procedure is well known and need not repeated here. Instead we wish to discuss the consequences of

(4.1) by substituting for a the vectors (3.1) - (3.3) in turn. First we introduce the following notations

$$\psi(t) = bX^*(t)^T,\ H(t,v) = f(x(t),v)\cdot\psi(t)^T,\ h(t) = H(t,u(t)). \qquad (4.2)$$

We then obtain these inequalities

$$h(t) \geq H(t,v) \text{ if } v \in U,\ t \notin D,\ g(x(t)) < 0 \quad \text{(from (3.1))} \qquad (4.3)$$

$$h(\tilde{t}) = 0 \quad \text{(from (3.2),(i))} \qquad (4.4)$$

$$h(t+0) = h(t) \text{ if } t \neq t^*_\mu \text{ (from 3.2),(ii))} \qquad (4.5)$$

$$h(t^*_\mu+0) \leq H(t^*_\mu+0,u(t^*_\mu)),\ \mu = 1,\dots,q, \text{ from (3.2),(iii))} \qquad (4.6)$$

$$\sum_{\nu=1}^{s} \sigma_\nu h(t_\nu) \geq \sum_{\nu=1}^{s} \sigma_\nu H(t_\nu,v_\nu)$$

if $t^*_\mu < t_1 \leq t_2 \leq \dots \leq t_s < t^{**}_\mu$, $\sigma_\nu \geq 0$ and

$$\sum_{\nu=1}^{m} \sigma_\nu G(v_\nu) \leq 0,\ m = 1,\dots,s \quad \text{(from 3.3))} \qquad (4.7)$$

We obtain an additional relation by multiplying the jump-condition (2.12) from the left with $f(x(t^*_\mu),u(t^*_\mu))$ and from the right with b^T. This gives (note that $\gamma_\mu = \dot{\gamma}(t^*_\mu) = f(x(t^*_\mu),u(t^*_\mu))\cdot g^T_\mu$)

$$h(t^*_\mu) = (f(x(t^*_\mu),u(t^*_\mu)) - \tilde{k}^*_\mu)\psi(t^*_\mu+0) = h(t^*_\mu+0) \text{ (cf.(2.3)).}$$

Hence, in view of (4.5), h(t) is continuous everywhere.

We are now going to showthat dh/dt exists except at points of D and is equal to zero. This will be done as usually in three steps, which we merely sketch. First one proves the inequalities

$$H(t,u(t_o))-H(t_o,u(t_o)) \leq h(t)-h(t_o) \leq H(t,u(t))-H(t_o,u(t)) \qquad (4.8)$$

if $t_o \notin D$, $|t-t_o| \leq \epsilon$.

Next one concludes from (4.8) by elementary arguments that dh/dt exits at $t = t_o$ and is equal to $\partial/\partial t H(t_o, u(t_o))$. That the partial derivative $\partial/\partial t H$ vanishes for $u = u(t)$ then turns out to be a formal identity.

In our case (4.8) follows immediately from (4.3) if $g(x(t_o)) < 0$, and it follows from (4.7) if $t_o \in (t^*_\mu, t^{**}_\mu)$. Note that $G(u(t)) = 0$ for all $t \in (t^*_\mu, t^{**}_\mu)$ and that for this reason $h(t_1) \geq H(t_1, u(t))$ if $t, t_1 \in (t^*_\mu, t^{**}_\mu)$(take $s = 1$, $v_1 = u(t)$). So we know that $h(t) = 0$ for all $t \in [0, \tilde{t}]$ and are thus arrived at our final result.

<u>Theorem 4.1.</u> Let b satisfy (4.1) and let $H(t,v)$ be defined as in (4.2). Then

(i) $H(t,u(t)) = 0$ for all $t \in [0,\tilde{t}]$,

(ii) $H(t,v) \leq 0$ if $v \in U$ and $t \notin (t^*_\mu, t^{**}_\mu]$, $\mu = 1,\dots,q$,

(iii) $H(t^*_\mu + 0, u(t^*_\mu)) \geq 0$, $\mu = 1,\dots,q$, equality holds if $t^*_\mu = t^{**}_\mu$.

(iv) $\sum_{\nu=1}^{s} \sigma_\nu H(t_\nu, v_\nu) \leq 0$ if $t^*_\mu < t_1 \leq t_2 \leq \dots \leq t_s < t^{**}_\mu$, $\sigma_\nu \geq 0$

and $\sum_{\nu=1}^{m} \sigma_\nu G(v_\nu) \leq 0$, for $m = 1,\dots,s$.

<u>Proof.</u> In view of what has been said before, only (ii) (iii) need an explanation. If $t_o \in D$, $t_o \notin (t^*_\mu, t^{**}_\mu]$ then (4.3) is valid in a certain left neighbourhood of t_o and hence also for $t = t_o$, for reasons of continuity. We have for the same reason $0 = h(t^*_\mu) \geq H(t^*_\mu + 0, v)$ in case $t^{**}_\mu = t^*_\mu$. Hence (iii) follows from (4.6).

<u>5. Example</u>

We consider the differential equation $\dot{x}^1 = x^2$, $\dot{x}^2 = -x^1 + u$, $\dot{x}^3 = x^4$, $\dot{x}^4 = u$, where u is a scalar control variable. If one takes the control region U to be the interval $[-1,1]$ and imposes on the state variable the constraint $g(x) = x^4 - \alpha \leq 0$, one has obtained a control problem which may serve as a model for the operation of an ore unloader (see [3],[4]). We will briefly describe the kind of information which can be gained from theorem 4.1 for the corresponding time-optimal control problem with the H-function $H(t,u) = \Psi^0(t)+x^2(t)\Psi^1(t)-x^1\Psi^2(t)+x^4(t)\Psi^3(t)+u[\Psi^2(t)+\Psi^4(t)]$. Since $h(t) = H(t,u(t)) = 0$ the function H can also be written in this form

$$H(t,u) = (u-u(t))\omega(t) \text{ where } \omega = \Psi^2+\Psi^4. \tag{5.1}$$

If $g\,(x(t)) < 0$ and $\omega(t) \neq 0$, then $u(t) = \text{sign}\,\omega(t)$, because of the maximum principle (ii). It follows from the jump-conditions (2.12), that $\Psi^i(t)$ is continuous everywhere, for $i = 0,\dots,3$, whereas ω satisfies the conditions $\omega(t_\mu^*) = -\omega(t_\mu^*+0)$ if $t_\mu^{**} = t_\mu^*$, $\omega(t_\mu^*) = 0$ if $t_\mu^{**} > t_\mu^*$ On the other hand $\omega(t_\mu^*+0) = 0$ if $t_\mu^{**} = t_\mu^*$, because of (5.1) and statement (iii) of theorem 4.1. So ω is continuous and vanishes at every isolated zero of $g(x(t))$. If $t_\mu^* < t_\mu^{**}$, then $\sum_{\nu=1}^{s}\sigma_\nu v_\nu \omega(t_\nu) \leq 0$ whenever $\sum_{\nu=1}^{m}\sigma_\nu v_\nu \leq 0$ for $m = 1,\dots,s$, and $t_\mu^* < t_1 \leq \dots \leq t_s < t_\mu^{**}$ according to (iv). As a consequence of this property ω must satisty $\dot\omega(t_\mu^*) \leqq 0$, $\omega\,(t_\mu^{**}) = 0$, $\dot\omega < 0$ and $\omega > 0$ on (t_μ^*,t_μ^{**}). From these results one can obtain an a-priori estimate (in terms of $\tilde t$)of the number of switching points of the control function.

6.Remark

If G depends upon x one has to be more careful in defining the adjoint variational equation (2.1). Theorem 4.1 remains true if $G(v_\nu)$ is replaced by $G(x(t_\nu),v_\nu)$ and if v_ν satisfies $v_\nu \in \overset{o}{U}$ and $(\partial G/\partial u^1,\dots,\partial G/\partial u^m) \neq (0,\dots,0)$ for $(x,u) = (x(t),v_\nu)$.

References

[1] ANOROV, V. P.: Maximum principle for processes with constraints of general form I,II. Automat. Remote Control 1967, no.3, 357-367, 533,543

[2] HESTENES, M.R.: Calculus of variations and optimal control theory. John Wiley and Sons, Inc., New York - London - Sydney 1966

[3] HIPPE, P.: Zeitoptimale Steuerung eines Erzentladers. Reglungstechnik und Prozeß- Datenverarbeitung, Heft 8, (18. Jahrgang 1970) 346-350

[4] PLATE, G.: Über Optimierungsprobleme mit eingeschränkten Phasenveränderlichen. Dissertation Technische Universität Berlin 1971

[5] PONTRJAGIN, L. S.; BOLTYANSKII, V. G.; GAMKRELIDZE, R. V.; MISHCHENKO, E.F.: The mathematical theory of optimal processes. John Wiley and Sons, 1962

Addendum

Condition (v) in Lemma 2.2 is actually not required and can be cancelled, also (iii) in (2.8). In the general case (when G depends upon x) the condition $u(t\pm 0) \in \overset{\circ}{U}$, $v_\nu \in \overset{\circ}{U}$ can be replaced by the regularity conditions (2) and (3) as introduced in [5], p. 266. Note that we do not need the regularity condition (1), which would take the form $G(x(t_\nu),v_\nu) = 0$ in our terminology. Instead we have condition (4.7).

Quelques Propriétés 'Génériques' des Systèmes à Commande

C. Lobry

Le but de ce papier est de décrire un certain nombre de propriétés vérifiées "génériquement" par les systèmes à commande (ou systèmes guidables). Pour la terminologie classique concernant les systèmes à commande on pourra consulter [1] ou encore [2] en langue française. On dit qu'une propriété P sur un espace topologique est "générique" lorsqu'elle est vraie sur un ensemble contenant une intersection dénombrable d'ouverts denses ; cette notion est intéressante dans le cas des espaces qui possèdent la propriété de Baire (i.e. une intersection dénombrable d'ouverts denses est dense). En fait, les propriétés qui seront décrites ici sont vraies sur un ouvert dense.

L'idée de s'intéresser aux propriétés "généralement vérifiées" par les systèmes à commande n'est pas neuve. Kalman [3] , Boltyanskii [4] remarquent déjà que la propriété essentielle des systèmes linéaires autonomes:

$$\frac{dx}{dt} = Ax + Bu \quad ; \quad x \in \mathbb{R}^n \qquad u \in \mathbb{R}^p ,$$

à savoir que le rang de la matrice :

$$(B \, , \, AB \, , \, A^2B \, , \, \dots \, , \, A^{n-1}B)$$

soit égal à n, est vérifiée par "presque tous les systèmes" (c'est une condition "ouverte" sur les coefficients des matrices A et B). Dans [1] Lee et Markus précisent ce qu'il faut entendre par "presque tous les systèmes". Enfin Dauer [5] démontre précisément que la propriété de "controlabilité" est générique.

Nous allons voir ici que les propriétés essentielles des systèmes linéaires non autonomes sont génériquement vérifiées ; les résultats proposés ici ne généralisent pas ceux de [3], [4] et [5] ; ils sont disjoints.

Quelques résultats concernant les propriétés génériques des systèmes non linéaires sont publiés dans [6], [7] , [8].

I - Introduction

Considérons le système linéaire :

$$\frac{dx}{dt} = A(t)\,x + u\,V(t) \quad ; \quad x \in \mathbb{R}^n \quad ; \quad u \in [-1\,;+1]$$

On suppose les application $t \mapsto A(t) \in \mathcal{L}(R^n, R^n)$; $t \mapsto V(t) \in \mathbb{R}^n$ **de classe** C^∞. **Etant donné l'application continue par morceaux** $t \mapsto \mathcal{U}(t)$ $(t_o \leqslant t \leqslant t_1)$ **(appelée commande) on appelle réponse d'origine** x_o **l'unique solution**

$$t \mapsto x(t, x_o - t_o, \mathcal{U}) \quad ; \quad t_o \leqslant t \leqslant t_1$$

du problème de Cauchy :

$$\frac{dx}{dt}(t) = A(t)\,x(t) + \mathcal{U}(t)\;V(t)$$

$$x(t_o) = x_o$$

Posons-nous le problème du retour à l'origine en temps minimum ; il s'agit de déterminer, pour un x_o **donné, une commande dont la réponse atteigne le plus vite possible l'origine. C'est un problème classique de régulation automatique. Les questions naturelles sont les suivantes :**

1) Soit $\mathcal{R}(t_o)$ **l'ensemble des points x de** $\mathbb{R}^n$ **tels qu'il existe une commande** $t \mapsto \mathcal{U}_x(t)$ $(t_o \leqslant t \leqslant t_1)$ **telle que :**

$$0 = x(t_1\,,\, x - t_o\,,\, \mathcal{U}_x)$$

L'ensemble $\mathcal{R}(t_o)$ **est-il un voisinage de 0 ?**

2) Soit x_o **un élément de** $\mathcal{R}(t_o)$ **; ce point peut par définition être "recallé" à l'origine en un certain temps ; posons :**

$$t^*_{x_o} = \mathrm{Inf}\ \{\, t_1 \geqslant t_o \ :\ x_o \in \mathcal{R}(t_o\,,\, t_1)$$

où $\mathcal{R}(t_o\,,\, t_1)$ **désigne l'ensemble des x de** $\mathbb{R}^n$ **tels qu'il existe une commande** $t \mapsto \mathcal{U}_x$ $(t_o \leqslant t \leqslant t_1)$ **telle que :**

$$0 = x(t_1\,,\, x - t_o\,,\, \mathcal{U}_x)\ .$$

L'ensemble $\mathcal{R}(t_o\,,\, t_1)$ **désigne donc l'ensemble des points qui peuvent être recallés à l'origine en un temps** t_1.

Le point x_o **appartient-t-il à** $\mathcal{R}(t_o\,,\, t^*_{x_o})$ **? En d'autres termes, existe-t-il une commande en temps minimum chaque fois qu'il existe une commande assurant le retour à l'origine ?**

3) Considérons la fonction de Hamilton :

$$(x, \psi, t, u) \mapsto H(x, \psi, t, u) = \langle \psi, A(t)x + u\,V(t) \rangle$$

Posons :

$$(1) \qquad \mathcal{U}(\psi, t) = \begin{cases} +1 & \text{si } \langle \psi, V(t) \rangle \geqslant 0 \\ -1 & \text{si } \langle \psi, V(t) \rangle < 0 \end{cases}$$

on a évidemment la relation :

$$H(x, \psi, t, \mathcal{U}(\psi, t)) = \underset{u \in [-1; +1]}{\operatorname{Max}} H(x, \psi, t, u)$$

Le principe du maximum de Pontriaguine exprime que si $t \mapsto \mathcal{U}_{x_o}(t)$ ($t_o \leqslant t \leqslant t_1$) est une commande qui assure le retour à l'origine en temps minimum alors il existe une application $t \mapsto \psi(t)$ (non identiquement nulle) absolument continue, telle que le couple :

$$t \mapsto x(t, x_o - t_o, \mathcal{U}_{x_o})$$

$$t \mapsto \psi(t)$$

soit solution du système hamiltonien

$$\frac{dx}{dt}(t) = \frac{\partial H}{\partial \psi}(x(t), \psi(t), t, \mathcal{U}(\psi(t), t)) \qquad \text{p.p.}$$

$$\frac{d\psi}{dt}(t) = -\frac{\partial H}{\partial x}(x(t), \psi(t), t, \mathcal{U}(\psi(t), t)) \qquad \text{p.p.}$$

Réciproquement, si $t \mapsto (x(t), \psi(t))$ ($t_o \leqslant t \leqslant t_1$) est une solution des équations de Hamilton telle que $x(t_o) = x_o$ et $x(t_1) = o$, la commande $t \mapsto \mathcal{U}(\psi(t), t)$ définie par (1) est-elle continue par morceaux ? Si oui, elle admet la réponse :

$$x(t, x_o - t_o, \mathcal{U}(\psi(t), t)) = x(t) \;;$$

cette commande est-elle optimale ? En d'autres termes, les conditions nécessaires de Pontriaguine sont-elles suffisantes ? (Il n'est pas absurde de se poser une telle question dans le cas d'un problème linéaire).

2 - La condition H. W

A l'application $t \mapsto A(t)$ on associe l'opérateur D_A sur $C^\infty(\mathbb{R}, \mathbb{R}^n)$ qui à l'application φ associe $D_A\varphi$ définie par :

$$t \mapsto D_A\varphi(t) = -A(t)\varphi(t) + \frac{d}{dt}\varphi(t).$$

2 - 1 Définition : On dit que le système :

$$\frac{dx}{dt} = A(t)x + u\,V(t) \quad ; \quad x \in \mathbb{R}^n \quad ; \quad u \in [-1\,;\,+1]$$

satisfait la condition H. W si quel que soit t dans $\mathbb{R}$ le rang du système des n + 1 vecteurs :

$$V(t)\,,\ D_A\,V(t)\,,\ D_A^2\,V(t)\,,\ \ldots\ldots,\ D_A^n\,V(t)$$

est égal à n.

Cette condition apparaît dans Hermann [9] ; elle est exploitée plus systématiquement par Weiss dans [10].

2-2 Théorème : Si le système

$$\frac{dx}{dt} = A(t)x + u\,V(t) \quad ; \quad x \in \mathbb{R}^n \quad ; \quad u \in [-1\,;\,+1]$$

satisfait la condition H. W. la réponse aux questions 1), 2) et 3) de l'introduction est positive :

Ce théorème n'est pas original encore qu'il n'ait jamais, à ma connaissance, été exprimé sous cette forme ; pour le montrer, on procède ainsi :

Dans [10] Weiss montre que si le système satisfait H. W., il est alors "différentiablement contrôlable" ou en "expansion", ce qui veut dire que si t_1 est strictement plus petit que t_2 alors $\mathcal{R}(t_o\,,\,t_1)$ est inclus dans l'intérieur de $\mathcal{R}(t_o\,,\,t_2)$. Ceci prouve que $\mathcal{R}(t_o)$ est un ouvert contenant l'origine et répond donc à la question 1. Un calcul très facile montre que si la condition H. W. est satisfaite, si $t \mapsto \psi(t)$ est solution de :

$$\frac{d\psi}{dt}(t) = -\frac{\partial H}{\partial x}\left(x(t)\,,\,\psi(t)\,,\,\mathcal{U}(\psi(t)\,,\,t)\right) = -{}^tA(t)\,\psi(t)$$

alors la fonction :

$$t \mapsto \langle \psi(t)\,,\,V(t) \rangle = h(t)$$

n'a que des zéros isolés ; (ceci provient de ce que grâce à la condition H. W. un au moins des nombres $h(t)\,,\ \frac{dh}{dt}(t)\,,\ \frac{d^2h}{dt^2}(t)\,,\ \ldots,\ \frac{d^n h(t)}{dt^n}$ est toujours différent de 0). La fonction $t \mapsto \mathcal{U}(\psi(t)\,,\,t)$ est donc continue par morceaux (en fait constante par morceaux) et par suite il est possible de parler de sa réponse. A partir de là, la théorie classique telle qu'on peut la trouver dans [1] ou encore dans [11] permet de répondre aux questions 2 et 3.

2 - 3 Remarques : 1) On sait que dans le cas linéaire la réponse à la question 2) est toujours oui si on accepte de considérer des commandes mesurables et non plus constantes par morceaux. La condition H. W. est donc une condition de régularité qui est à rapprocher de la condition de [12] : $A(t)$ et $V(t)$ analytiques par morceaux ; sous cette condition, on montre également que les commandes optimales sont constantes par morceaux.

2) On peut généraliser le théorème 2-2 aux systèmes :

$$\frac{dx}{dt} = A(t)x + B(t)u \quad ; \quad x \in \mathbb{R}^n \quad ; \quad u \in U \subset \mathbb{R}^p$$

$$t \mapsto A(t) \in \mathcal{L}(\mathbb{R}^n, \mathbb{R}^n)$$

$$t \mapsto B(t) \in \mathcal{L}(\mathbb{R}^p, \mathbb{R}^n)$$

où U est un polyhèdre de $\mathbb{R}^p$ tel que l'origine soit un point intérieur d'un segment contenu dans U. La condition H. W. est alors plus compliquée à exprimer mais les phénomènes sont essentiellement les mêmes.

3 - La condition H. W. est générique

Le but de ce numéro est de montrer le

3 - 1 Théorème : L'ensemble $\mathcal{R}(A)$ des applications $t \mapsto V(t)$ de $C^\infty([t_o, t_1] \to \mathbb{R}^n)$ telles que :

$$\text{rang}\ (V(t), D_A V(t), D_A^2 V(t), \dots, D_A^n V(t)) = n \ ; \ t \in [t_o, t_1]$$

est un ouvert dense pour la topologie de la convergence uniforme à l'ordre n :

La démonstration de ce théorème se fait très simplement à l'aide de la technique standard de la théorie de la transversalite (cf. [13] par exemple) ; elle n'utilise d'ailleurs qu'un théorème de Sard "trivial" à savoir :

"Si φ est une application continuement différentiable de $\mathbb{R}^p$ dans $\mathbb{R}^q$, si p est strictement plus petit que q, alors l'image de φ est de mesure nulle, donc de complémentaire dense".

Montrons pour commencer le lemme :

3 - 2 Lemme : Sur $C^{n+1}([t_o ; t_1], \mathbb{R}^n)$ muni de la topologie de la convergence des fonctions et de leurs dérivées à l'ordre n, l'ensemble $\mathcal{R}$ des fonctions φ telles que le rang du système :

$$\varphi(t), \frac{d\varphi}{dt}(t), \frac{d^2\varphi}{dt^2}(t), \dots, \frac{d^n\varphi}{dt^n}(t)$$

soit égal à n pour tout t est dense

Démonstration : Soit φ un élément de $C^{n+1}([t_o, t_1] \to \mathbb{R}^n)$

Considérons l'espace vectoriel P :

$$P = E_o \times E_1 \times \ldots \times E_n$$

avec $E_o = E_1 = \ldots = E_n = \mathbb{R}^n$. L'espace vectoriel P est un espace de perturbations, on note $u = (u_o, u_1, \ldots, u_i, \ldots, u_n)$ ($u_i \in \mathbb{R}^n$) le point "courant" de P. On définit maintenant la famille à n (n + 1) paramètres d'applications de $[t_o ; t_1]$ dans $\mathbb{R}^n$.

$$t \mapsto \varphi_u(t) = \varphi(t) + u_o + tu_1 + \frac{t^2}{2!} u_2 + \ldots + \frac{t^n}{n!} u_n \quad .$$

Soit f l'application de $[t_o ; t_1] \times P$ dans P définie par :

$$f(t, u_o, u_1, \ldots, u_n) = \begin{cases} \varphi(t) + u_o + tu_1 + \frac{t^2}{2!} u_2 + \ldots + \frac{t^n}{n!} u_n \\ \frac{d\varphi}{dt}(t) + u_1 + tu_2 + \ldots + \frac{t^{n-1}}{(n-1)!} u_n \\ \text{-----------------------} \\ \frac{d^{n-1}\varphi}{dt^{n-1}}(t) + u_{n-1} + tu_n \\ \frac{d^n}{dt^n}(t) + u_n \end{cases}$$

Un calcul immédiat prouve que l'application f est de rang maximum n (n + 1) en tout point. Soit S l'ensemble des $u = (u_o, u_1, \ldots, u_n)$ de P tels que le rang des n + 1 vecteurs de $\mathbb{R}^n$, $u_o, u_1, \ldots, u_n$ ne soit pas égal à n. L'ensemble S est une variété algébrique réelle ; donc d'après un théorème de Whitney [14] on a :

$$S = \bigcup_{i=1}^{r} S_i$$

où chaque S_i est une variété connexe régulièrement plongée de P. Il est facile de voir que chaque S_i est de codimension supérieure à 2 (puisque en particulier sur chaque S_i les deux polynômes det $(u_o, u_1, \ldots, u_{n-1})$ et det $(u_1, u_2, \ldots, u_n)$ doivent être nuls. Puisque f est de rang maximum n (n + 1) l'ensemble $f^{-1}(S) = \bigcup f^{-1}(S_i)$ est une réunion dénombrable de sous variétés connexes régulièrement plongées $(\Gamma_j)_{j \in N}$ de dimensions inférieures à n (n + 1) - 1.

Dire qu'en tout point de $[t_o ; t_1]$ le rang du système :

$$\varphi_u(t), \frac{d\varphi_u}{dt}(t), \ldots, \frac{d^n\varphi_u}{dt^n}(t)$$

est égal à n , équivaut à dire, par définition de f, que l'image de $[t_o ; t_1]$ par:

$$t \mapsto f(t, u)$$

ne rencontre pas S, soit encore que le segment

$$[t_o ; t_1] \times \{u\}$$

ne rencontre pas $f^{-1}(S)$. Soit π_j l'application de Γ_j dans P qui à un point de Γ_j associe sa projection ; c'est une application différentiable de Γ_j dans $R^{n(n+1)}$; Γ_j étant de dimension strictement plus petite que n (n + 1) son image, $\pi_j(\Gamma_1)$ est de mesure nulle et par suite l'union

$$\bigcup_{j \in N} \pi_j(\Gamma_j)$$

est de mesure nulle. L'ensemble des u de P n'appartenant pas à cette union, donc tels que $[t_o ; t_1] \times \{u\}$ ne rencontre pas $f^{-1}(S)$, est partout dense dans P. Si on choisit U suffisamment voisin de O l'application φ_u est voisine de φ au sens de la convergence à l'ordre n sur $[t_o ; t_1]$. Ceci démontre donc la densité et achève la démonstration du lemme.

<u>Démonstration du théorème 1 - 3 - 1</u>: L'ensemble $\mathfrak{R}(A)$ est évidemment ouvert ; montrons sa densité. Soit T_A l'application de $C^{n+1}([t_o ; t_1], R^n)$ dans lui-même qui à la fonction φ associe $T_A\varphi$ définie par :

$$t \mapsto T_A\varphi(t) = \emptyset(t, t_o)\varphi(t) \qquad t_o \leq t \leq t_1$$

où $\Phi(t, t_o)$ est l'unique application $t \mapsto \Phi(t, t_o) \in \mathcal{L}(\mathbb{R}^n, \mathbb{R}^n)$ solution du système :

$$\frac{d\Phi}{dt}(t) = A(t)\,\Phi(t)$$

$$\Phi(t_o) = \text{identité}$$

On sait que pour tout t la matrice $\Phi(t, t_o)$ est inversible ; l'opérateur T_A est donc inversible et on vérifie aisément que c'est un homéomorphisme pour la convergence uniforme à l'ordre n. Soit $t \mapsto V(t)$ un élément de $C^{n+1}([t_o ; t_1], \mathbb{R}^n)$ tel que le rang de

$$V(t), D_A V(t), \ldots, D_A^n V(t)$$

soit égal à n ; si on dérive successivement $t \mapsto T_A^{-1} V(t)$ soit :

$$\Phi^{-1}(t, t_o)\, V(t)$$

compte tenu de ce que :

$$\frac{d\,\Phi^{-1}}{d\,t}(t, t_o) = -\,\Phi^{-1}(t, t_o)\, A(t)$$

on obtient :

$$T_A^{-1}\, V(t) = \Phi^{-1}(t, t_o)\, V(t)$$

$$\frac{d\,T_A^{-1}}{d\,t}\, V(t) = \Phi^{-1}(t, t_o)\,(-A(t)\, V(t) + V'(t)) = \Phi^{-1}(t, t_o)\, D_A V(t)$$

$$\frac{d^n T_A^{-1} V(t)}{d\,t} = \Phi^{-1}(t, t_o)\, D_A^n\, V(t)$$

Ceci prouve l'égalité :

$$C^{n+1}([t_o\, ;\, t_1], \mathbb{R}^n) \cap \mathcal{R}(A) = T_A(\mathcal{R})$$

et achève la démonstration du théorème 3 - 1 car $C^{n+1}([t_o\, ;\, t_1], \mathbb{R}^n)$ est dense dans $C^n([t_o\, ;\, t_1], \mathbb{R}^n)$.

Sans plus de difficultés, on peut obtenir le théorème suivant valable sur $\mathbb{R}$ tout entier :

3 - 3 <u>Théorème</u> : <u>L'ensemble $\mathcal{R}(A)$ des applications $t \mapsto V(t)$ de $C^n(\mathbb{R}, \mathbb{R}^n)$ telles que :</u>

$$\text{rang}\ (V(t)\, ,\, D_A V(t)\, ,\, D_A^2 V(t)\, ,\ldots,\, D_A^n V(t)) = n$$

<u>est un ouvert dense de $C^n(\mathbb{R}, \mathbb{R}^n)$ pour la C^n - Topologie de Whitney</u>.

3 - 4 <u>Remarques</u> :

On peut envisager également des théorèmes pour lesquels les "perturbations" se feront par rapport à $t \mapsto A(t)$ ou encore à certains coefficients de $A(t)$. C'est en fonction de chaque situation pratique que l'on décidera du meilleur théorème à démontrer ; par exemple si on considère dans $\mathbb{R}$ le système :

$$\frac{d^2 x}{d\,t^2} = \alpha\,\frac{d\,x}{d\,t} + \beta\,x + u\,V(t)$$

ce qui donne, si on pose $y = \dfrac{d\,x}{d\,t}$

$$\begin{pmatrix} \frac{dx}{dt} \\ \\ \frac{dy}{dt} \end{pmatrix} = \begin{pmatrix} o & 1 \\ \beta & \alpha \end{pmatrix} \begin{pmatrix} x \\ y \end{pmatrix} + u \begin{pmatrix} o \\ V(t) \end{pmatrix}$$

Les seules perturbations raisonnables doivent porter sur α, β et $V(t)$; dans ce cas particulier, qui n'est pas une conséquence du théorème 1 - 3 - 1 on peut facilement vérifier que la condition H. W. peut être réalisée en perturbant arbitrairement $t \mapsto V(t)$

4 - Systèmes non linéaires

On peut montrer cf. [8] dans le cas non linéaire la proposition suivante qui est à rapprocher d'un résultat de Markus [15]

Soit M une variété Riemannienne compacte, considérons les systèmes

$$\frac{dx}{dt} = f(x, u) \qquad x \in M \quad ; \quad u \in \{1 ; 2\}$$

où les champs de vecteurs $x \mapsto f(x, u)$ $(u = 1, 2)$ sont conservatifs.

4 - 1 Proposition : Génériquement, l'ensemble des états accessibles à partir du point x_o (i. e. l'ensemble des extrémités des réponses issues de x_o) est la variété M toute entière.

Ce résultat repose sur une extension d'un théorème de Chow [16] ; ce théorème a déjà été utilisé en théorie de la commande par Hermann [9] et Hermes [17].

Remerciements : Je désire remercier J. MARTINET pour l'intérêt constant qu'il a apporté à cette recherche. Je remercie également le Professeur L. W. MARKUS qui m'a suggéré de démontrer la proposition 4 - 1.

[1] LEE MARKUS : "Foundations of optimal Control Theory" John Witney & Sons 1967

[2] PALLU DE LA BARRIERE :"Cours d'automatique théorique" Dunod 1966

[3] R. E. KALMAN, Y. C. HO, K. S. NARENDRA : "Controlability of linear dynamical systems" ; Contribution to Differential Equations (1963) p. p. 189 - 213

[4] V. G. BOLTYANSKII : "Mathematical Methods of Optimal Control" Balskrishan - Neustadt Series - Holt. Rinehart Winston 1970

[5] J. P. DAUER : "Perturbations of linear control systems" S. I. A. M. J. on control Vol 9 n° 3 Aout 1971

[6] C. LOBRY : 'Controlabilité des systèmes non linéaires" S. I. A. M. J. on control Vol 8 n° 4 novembre 1970

[7] C. LOBRY : "Une propriété générique des couples de champs de vecteurs". Journal Mathématique Tchecoslovaquie Janvier 1972

[8] C. LOBRY : Thèse Sciences Mathématiques.

[9] R. HERMANN : "On the accessibility problem in control theory" Int - Symp. Non linear differential equations and non linear mechanics - Academic Dress N. Y. 1963 p. p. 325 - 332

[10] L. WEISS : "Controlability and Observability". Notes de lectures, école d'été du C. I. M. E. - Juin 1968 - Bologne

[11] HERMES-LA SALLE :"Functional Analysis and time optimal control" Academic press 1969

[12] H. HALKIN : "A generalisation of La Salle'S Bang Bang Principle" S. I. A. M. Journal on Control 1 - 1962 p. p. 76 - 84

[13] ABRAHAM-ROBBIN : Transversal mappings and Flows W. A. Benjamin . Inc. 1967

[14] WHITNEY : "Elementary Structure of Real Algebraic Varieties" Annals of Mathematics 66 - p. p. 545 - 556 (1967)

[15] L. W. MARKUS : Control Dynamical Systems Mathematical Systems Theory . Vol. 3 - n° 2

[16] W. I. CHOW : "Uber Systeme von linearen partiellen differential glechungen erster ordnung". Math. Ann. , 117 (1939) p.p. 98 - 105

[17] H. HERMES : "Controlability and the singular problem", S. I. A. M. Journal on Control, 2 - 1964 p.p. 241 - 260

Green's Functions for Pairs of Formally Selfadjoint Ordinary Differential Operators

Å. Pleijel

SUMMARY. A spectral theory is deduced for differential eigenvalue problems related to a formally selfadjoint differential equation $Su = \lambda Tu$, where u is complex-valued and λ is the eigenvalue parameter. The equation or rather the differential operators S and T are considered on an arbitrary open interval of the real axis. The lower order operator T is assumed to have a positive definite Dirichlet integral which serves as scalar product in spectral theorems determined by symmetric boundary conditions. The theory is given in terms of ordered pairs $u|\dot{u}$ of functions. Thus symmetric boundary conditions are certain subrelations of $\{u|\dot{u} : Su = T\dot{u}\}$. If for instance T is the identity operator the boundary conditions are equally well described as conditions on u only. As far as the spectral theorem is concerned the method of the paper is easily transferred to the case when S instead of T

has a positive definite Dirichlet integral. In the here considered case with T positive a kernel representation of the resolvent is deduced and used to prove the regularity of the elements of eigenspaces belonging to finite intervals of the spectral axis. The theory was worked out independently of the investigations by F.W. Schäfke, A. Schneider and H.-D. Niesser of systems of first order equations to which it seems related in different respects.

1. The differential operators

$$S = \sum_{k=0}^{m} (D^k a_k D^k + D^k b_k D^{k+1} + D^{k+1} b_k D^k) \,, \tag{1.1}$$

$$T = \sum_{k=0}^{n} D^k c_k D^k \,. \tag{1.2}$$

will be considered in which $D = id/dx$ and a_k, b_k, c_k denote functions of x with continuous derivatives in $I = \{x: a < x < b\}$ of the orders $k, k+1, k$ respectively. These functions shall be real-valued in which case S and T are formally selfadjoint. The operators shall have orders which are independent of x namely $M = 2m+1$ or $M = 2m$ (b_m identically 0) for S, and $N = 2n$ for T. It is supposed that $M > N$. With the further condition on T that not only $c_n(x) \neq 0$ but also

$$\text{all } c_k(x) \geq 0 \text{ on } I, \quad c_o \text{ not identically } 0, \tag{1.3}$$

a spectral theory will be deduced for the pair S, T under general symmetric boundary conditions ([3], [1])

2. For any integer $k \geq 0$ let $C^{(k)}(I)$ be the linear space of all complex-valued functions u for which $D^{k-1}u$ is absolutely continuous on every compact subinterval J of I and $D^k u \in L^2(J)$ for such intervals. For $k = 0$ only $u \in L^2(J)$ is required. If $u \in C^{(k)}(I)$ with $k = 0$ the relation $u = 0$ means $u = 0$ a.e., if $k > 0$ it means identically 0.

If $U = u|\dot{u}$ and $V = v|\dot{v}$ belong to the linear space of ordered pairs

$$E(I) = \{u|\dot{u} \in C^{(M)}(I) \times C^{(N)}(I) \colon Su = T\dot{u} \ \text{ a.e.}\}, \tag{2.1}$$

partial integrations of $Su\,\overline{v} - u\,\overline{Sv} = T\dot{u}\,\overline{v} - u\,\overline{T\dot{v}}$ over $J = \{x\colon \alpha \leq x \leq \beta\}$, $a < \alpha < \beta < b$, lead to a formula

$$q_\beta(U,V) - q_\alpha(U,V) = i^{-1}((\dot{u},v)_J - (u,\dot{v})_J). \tag{2.2}$$

Here

$$(u,v)_J = \int_J \sum_{k=0}^{n} c_k D^k u\,\overline{D^k v} \tag{2.3}$$

and

$$q(U,V) = \sum_{j=1}^{M-m} ((D^{-j}Su - D^{-j}T\dot{u})\,\overline{D^{j-1}v} + D^{j-1}u(\overline{D^{-j}Sv - D^{-j}T\dot{v}})), \tag{2.4}$$

where $D^{-j}S$, $D^{-j}T$ are obtained by formal multiplication of (1.1), (1.2) by D^{-j} and suppression of all terms with negative powers of D before $a_k D^k$, $b_k D^{k+1}$, $b_k D^k$ and $c_k D^k$. Observe that the left hand side of (2.2)

$$Q_J(U,V) = q_\beta(U,V) - q_\alpha(U,V) \tag{2.5}$$

is hermitean in U,V.

The Dirichlet integral $(u,v) = (u,v)_I$ defined by (2.3) but extended over I will serve as the scalar product of Hilbert spaces in which the spectral theory will take place. A positive definite character of this Dirichlet integral is secured by the assumptions (1.3). Indeed, because of (1.3), $(u,u)_J$ will be positive definite on $C^{(n)}(J)$ provided the compact subinterval J contains part of the support of c_0. Consequently (u,u) will be positive definite on the space $\overset{\circ}{C}{}^{(n)}(I)$ if we define

$$\overset{\circ}{C}{}^{(k)}(I) = \{u \in C^{(k)}(I): (u,u) < \infty\} \tag{2.6}$$

when $k \geq n$. Then $H = \overset{\circ}{C}{}^{(n)}(I)$ is a Hilbert space with scalar product (u,v). Any $\overset{\circ}{C}{}^{(k)}(I)$ with $k > n$ is dense in H.

The hermitean form

$$Q(U,V) = i^{-1}((\dot{u},v) - (u,\dot{v})) \tag{2.7}$$

exists when $U = u|\dot{u}$, $V = v|\dot{v}$ belong to $H \times H = H^2$ and in particular if they belong to the linear space

$$\overset{\circ}{E}(I) = \{u|\dot{u} \in E(I): (u,u) + (\dot{u},\dot{u}) < \infty\}. \tag{2.8}$$

Important finite-dimensional subspaces of $E(I)$, $\overset{\circ}{E}(I)$ are $E(\lambda,I) = \{u|\lambda u \in E(I)\}$ and $\overset{\circ}{E}(\lambda,I) = \{u|\lambda u \in \overset{\circ}{E}(I)\}$, where λ is a number which in general will be non-real. The elements of these subspaces correspond to solutions of $Su = \lambda Tu$. If $u|\lambda u \in \overset{\circ}{E}(\lambda,I)$, the solution u has a finite Dirichlet integral $(u,u) < \infty$. The dimension of $E(\lambda,I)$ is M and the dimension of $\overset{\circ}{E}(\lambda,I)$ is $\leq M$. According to the existence theorem for ordinary differential equations an element of $E(\lambda,I)$ is uniquely determined by its restriction to any subinterval of I.

3. By completion of squares in $q_\beta(U,U)$ and $q_\alpha(U,U)$ defined by (2.4) it follows that the Sylvester signature of (2.5) i.e. the pair of its positive and negative inertia indices satisfies the inequality

$$\operatorname{sign} \underset{J}{Q} \leq (M,M). \tag{3.1}$$

According to (2.2) the form (2.5) reduces to $\underset{J}{Q}(U,V) = c\,\underset{J}{(u,v)}$, $c = i^{-1}(\lambda - \overline{\lambda})$, on $E(\lambda,I)$ and to $-c\underset{J}{(u,v)}$ on $E(\overline{\lambda},I)$. Thus $c\underset{J}{Q}$ is positive and negative definite on these spaces provided $\operatorname{Im}\lambda \neq 0$ and J contains part of the support of c_o. Since $E(\lambda,I)$ is M-dimensional, $c\underset{J}{Q}$ cannot be positive definite on the linear hull of $E(\lambda,I)$ and any element outside this space. Because of this, $E(\lambda,I)$ is called maximal positive definite with respect to $c\underset{J}{Q}$ in $E(I)$. Similarly $E(\overline{\lambda},I)$ is maximal negative definite. The direct sum (λ non-real)

$$F(\lambda,I) = E(\lambda,I) \dot{+} E(\overline{\lambda},I) \tag{3.2}$$

is regular in the sense that $\underset{J}{Q}$ is not degenerate on $F(\lambda,I)$. An element of $F(\lambda,I)$ is uniquely determined by its restriction to J. The maximality of $E(\lambda,I)$, $E(\bar{\lambda},I)$ implies that $F(\lambda,I)$ is maximal as a regular subspace of $E(I)$ with respect to $\underset{J}{Q}$. A consequence is that

$$E(I) = E(I)^{perp} \dotplus F(\lambda,I) \,, \tag{3.3}$$

where $E(I)^{perp} = \{U \in E(I) : \underset{J}{Q}(U,E(I)) = 0\}$. We finally state that for all $U \in E(I)$

$$c \underset{J}{Q} (U - U(J),\, U - U(J)) \leq 0 \tag{3.4}$$

if $U(J)$ is the $\underset{J}{Q}$-projection of U on $E(\lambda,I)$ defined by

$$U(J) \in E(\lambda,I),\ \underset{J}{Q}(U - U(J),\, E(\lambda,I)) = 0 \,. \tag{3.5}$$

Such an inequality (3.4) is a general characteristicum for a positive definite finite dimensional subspace which is maximal with respect to a hermitean form, in our case $c \underset{J}{Q}$. The maximality of a negative definite finite dimensional subspace is of course similarly characterized by a reversed inequality.

4. If $v \in C^{(N)}(I)$ the equation $Su - \lambda Tu = Tv$ can be solved with $u \in C^{(M)}(I)$ i.e. there exist elements $U = u|(\lambda u + v)$ in $E(I)$. If $v \in \overset{\circ}{C}{}^{(N)}(I)$ and $U = u|(\lambda u + v) \in E(I)$ for a non-real λ, it can be shown by the help of (3.4) that when J tends to I the projections $U(J)$ tend to a limit $U_1 \in E(\lambda,I)$ such that $U - U_1 \in \overset{\circ}{E}(I)$. In the case when $U = u|(\lambda u + v)$ already belongs to $\overset{\circ}{E}(I)$ the limit U_1 coincides with the Q-projection $U(I)$ of U on $\overset{\circ}{E}(\lambda,I)$ which is defined by

$$U(I) \in \overset{\circ}{E}(\lambda,I),\ Q(U - U(I),\, \overset{\circ}{E}(\lambda,I)) = 0 \,. \tag{4.1}$$

In the last statement Q is defined by (2.7). It is readily seen that $c\,Q$ is positive and negative definite on $\overset{\circ}{E}(\lambda,I)$ and $\overset{\circ}{E}(\bar{\lambda},I)$. As it can be shown, the inequality (3.4) is carried over into

$$c\, Q(U - U_1,\, U - U_1) \leq 0 \tag{4.2}$$

by the transition to the limit $J \to I$. Since $U_1 = U(I)$ and (4.2) holds for any $U \in \mathring{E}(I)$ it follows that the subspace $\mathring{E}(\lambda,I)$ is maximal positive definite with respect to $c\,Q$ in $\mathring{E}(I)$. Similarly $\mathring{E}(\bar{\lambda},I)$ is maximal negative definite with respect to $c\,Q$. Hence, the direct sum

$$\mathring{F}(\lambda,I) = \mathring{E}(\lambda,I) \dotplus \mathring{E}(\bar{\lambda},I) \tag{4.3}$$

is maximal regular in $\mathring{E}(I)$ with respect to Q. In the same way as (3.3) it follows that

$$\mathring{E}(I) = \mathring{E}(I)^{perp} \dotplus \mathring{F}(\lambda,I) \tag{4.4}$$

with $\mathring{E}(I)^{perp} = \{U \in \mathring{E}(I) : Q(U, \mathring{E}(I)) = 0\}$.

5. The linear space $\mathring{E}(I)^{perp}$ is a Q-nullspace in the sense that $Q(U,V) = 0$ for all $U,V \in \mathring{E}(I)^{perp}$. So is

$$Z = \mathring{E}(I)^{perp} \dotplus Z' \tag{5.1}$$

provided Z' is a Q-nullspace in $\mathring{F}(\lambda,I)$. As a Q-nullspace (5.1) is maximal in $\mathring{E}(I)$, i.e. not extendable as such, if and only if Z' is a maximal Q-nullspace in $\mathring{F}(\lambda,I)$. A criterion for this is that

$$\dim Z' = \min(\dim \mathring{E}(\lambda,I), \dim \mathring{E}(\bar{\lambda},I)) . \tag{5.2}$$

A maximal Q-nullspace Z in $\mathring{E}(I)$ is a symmetric boundary condition. If u satisfies Z there is a $\dot{u}$ such that $u \mid \dot{u} \in Z$ so that u belongs to the linear space Z_1 of all first components of elements in Z .

In Section 4 it was stated that if $\operatorname{Im}\lambda \neq 0$, $U = u \mid (\lambda u + v) \in E(I)$ with v in $\mathring{C}^{(N)}(I)$, then U can be "compensated" by an element U_1 in $E(\lambda,I)$ so that $U - U_1 \in \mathring{E}(I)$. If $\dim \mathring{E}(\lambda,I) \geq \dim \mathring{E}(\bar{\lambda},I)$ it follows from (5.2) that $\mathring{F}(\lambda,I) = \mathring{E}(\lambda,I) \dotplus Z'$. A corresponding subdivision of the projection of $U - U_1$ on $\mathring{F}(\lambda,I)$ leads to

<u>THEOREM 5.1</u>. <u>If</u> Z <u>is a symmetric boundary condition</u>, $\operatorname{Im}\lambda \neq 0$, $\dim \mathring{E}(\lambda,I) \geq \dim \mathring{E}(\bar{\lambda},I)$ <u>and if</u> $U = u \mid (\lambda u + v) \in E(I)$ <u>with</u> $v \in \mathring{C}^{(N)}(I)$, <u>then there is a unique</u> $U_0 \in E(\lambda,I)$ <u>such that</u> $U - U_0 \in Z$.

6. A corollary of the last theorem is that if Z is a symmetric boundary condition, $\operatorname{Im}\lambda \neq 0$, $\dim \overset{\circ}{E}(\lambda,I) \geq \dim \overset{\circ}{E}(\bar{\lambda},I)$ and $v \in \overset{\circ}{C}^{(N)}(I)$, then there is a unique $u = R_\lambda v$ such that $u|(\lambda u+v) \in Z$. The mapping R_λ is linear and bounded in terms of the norm $|u| = (u,u)^{1/2}$ of H and maps $\overset{\circ}{C}^{(N)}(I)$ onto Z_1. Its closure $\bar{R}_\lambda$ in H is then also bounded and maps H onto $(\bar{Z})_1$, where $\bar{Z}$ is the closure of Z in H^2 and $(\bar{Z})_1$ is obtained from $\bar{Z}$ in the same way as Z_1 from Z. The range $(\bar{Z})_1$ of $\bar{R}_\lambda$ is a subset of

$$H_o = \bar{Z}_1 = \overline{(Z_1)} \tag{6.1}$$

and dense in this subhilbert of H. The statement $u = \bar{R}_\lambda v$ is equivalent to $u|(\lambda u+v) \in \bar{Z}$.

The linear set

$$H_\infty = \{v \in H\colon \bar{R}_\lambda v = 0\} \tag{6.2}$$

coincides with $H_\infty = \{v \in H\colon 0|v \in \bar{Z}\}$ and can be shown to be the orthogonal complement of H_o in H so that

$$H = H_o \oplus H_\infty . \tag{6.3}$$

The restriction A_λ of $\bar{R}_\lambda$ to H_o is a bounded and invertible linear operator with domain H_o and with range $(\bar{Z})_1 \subset H_o$.

A study of the question, when a linear subrelation of the relation defined by $u|\dot{u} \in \overset{\circ}{E}(I)$ is a symmetric hilbertspace operator, shows that the domain of this operator must be a subset of a set D related to a Z

$$D = \{u\colon u|\dot{u} \in Z \quad \text{for a} \quad \dot{u} \in H_o\}. \tag{6.4}$$

But this linear subset D of Z_1 actually determines a function K i.e. there is for every $u \in D$ only one $\dot{u} = Ku$ which satisfies the condition in (6.4). This K is symmetric on account of the definition (2.7) of Q so that $(Ku,v) = (u,Kv)$ for all $u,v \in D$. It maps D into H_o. Since $\dot{u} = Ku$ implies $Su = T\dot{u}$ and $u|\dot{u} \in Z$ it is reasonable to write $K = T^{-1}S$ determined by Z.

The closure $\overline{K}$ of K can be shown to coincide with $A_\lambda^{-1} + \lambda$. Thus $\overline{K} - \lambda$ maps $(\overline{Z})_1$ onto the entire space H_o and $\overline{K}$ is a maximal symmetric operator on H_o with domain $(\overline{Z})_1$. It also follows that the domain D of K is dense in H_o.

Recall that these results are obtained under the condition that $\dim \overset{o}{E}(\lambda,I) \geq \dim \overset{o}{E}(\overline{\lambda},I)$. If the reversed inequality holds the closure $\overline{K}$ is still maximal, this time since $\overline{K} - \overline{\lambda}$ has the range H_o. The condition

$$\dim \overset{o}{E}(\lambda,I) = \dim \overset{o}{E}(\overline{\lambda},I) \tag{6.5}$$

is necessary and sufficient in order that the operator K (for any Z) shall have a selfadjoint extension.

7. If (6.5) holds let $H_o(\Delta)$ be an eigenspace of $\overline{K}$ corresponding to a finite real interval Δ. It can be proved that all elements of $H_o(\Delta)$ are equivalent to functions in $\overset{o}{C}{}^M(I)$ i.e. to functions with finite Dirichlet integrals over I and with continuous derivatives of orders $\leq M$. In this way $Su = T\dot{u}$ (and $\dot{u} = Ku$) when $\dot{u} = \overline{K}u$ in $H_o(\Delta)$. The proof is accomplished according to the following observations. If λ is non-real $\overline{K} - \lambda$ is an invertible mapping of $H_o(\Delta)$ onto itself. By iterated application of $(\overline{K}-\lambda)^{-1}$ to an element v of $H_o(\Delta) \subset \overset{o}{C}{}^{(n)}(I)$, the regularity is successively increased so that $(\overline{K}-\lambda)^{-s}v$ equals a function in $\overset{o}{C}{}^M(I)$ if the integer s is sufficiently large. The last statement can for instance be deduced from an integral representation of $u = \overline{R}_\lambda v$ by the help of kernels related to the elements of the Green's space that will be introduced in Section 10.

8. By its nature a boundary condition should be related to values taken by functions close to a and b only. The consideration of discontinuities in the interior of I is needed to define Green's functions.

Because of this we introduce spaces $C^{(M)}(a\cdot\cdot b)$, $C^{(N)}(a\cdot\cdot b)$, $E(a\cdot\cdot b)$ etc. of functions and pairs defined in subintervals $a-$ and $-b$ of I

which have one endpoint at a or one at b respectively, but may be different for different elements. For simplicity a- and -b can be taken non-overlapping. If they have one point ξ in common we then distinguish between $\xi - 0 \in$ a- and $\xi + 0 \in$ -b.

For an element $U = u|\dot{u}$ of $E(a\cdot\cdot b)$ the conditions in the beginning of Section 2 concerning the derivatives of u, $\dot{u}$ shall be fulfilled for any compact subinterval J of the domain $(a-) \cup (-b)$. Also $Su = T\dot{u}$ a.e. in this domain.

The subspace $\overset{\circ}{E}(a\cdot\cdot b)$ of $E(a\cdot\cdot b)$ is determined by the condition that the Dirichlet integrals of u and $\dot{u}$ in $U = u|\dot{u} \in \overset{\circ}{E}(a\cdot\cdot b)$ shall be finite when extended over a set $(a-) \cup (-b)$. The subspaces $E(\lambda, a\cdot\cdot b)$ of $E(a\cdot\cdot b)$ and $\overset{\circ}{E}(\lambda, a\cdot\cdot b)$ of $\overset{\circ}{E}(a\cdot\cdot b)$ are defined by the condition that $\dot{u} = \lambda u$.

The new spaces are linear under otherwise evident definitions if $u|\dot{u} = 0$ means that $u = 0$, $\dot{u} = 0$ in a set $(a-) \cup (-b)$ and equality between elements is correspondingly defined. Elements of our previous spaces can be considered as elements of the new ones so that $E(I) \subset$ $\subset E(a\cdot\cdot b)$ etc.

According to the existence theorem of $Su = \lambda Tu$ a solution u in an interval is determined by the values of $D^j u$, $j = 0, 1, \ldots, M-1$, at one point of the interval. Thus in $E(\lambda, a\cdot\cdot b)$ and $\overset{\circ}{E}(\lambda, a\cdot\cdot b)$ it can be assumed that all elements are defined in arbitrarily chosen intervals a- and -b. The dimension of $E(\lambda, a\cdot\cdot b)$ is $2M$ and the dimension of $\overset{\circ}{E}(\lambda, a\cdot\cdot b)$ is $\leq 2M$.

If $U, V \in E(a\cdot\cdot b)$ let

$$\underset{J}{Q}(U,V) = q_\beta(U,V) - q_\alpha(U,V) \tag{8.1}$$

for any $J = \{x\colon \alpha \leq x \leq \beta\}$ which has its endpoints α, β in (not overlapping) intervals a-, -b in which U,V are defined. The inequality (3.1) subsists. If $J \supset L = \{x\colon A \leq x \leq B\}$ and already $A \in$ a-, $B \in$ -b the formula

$$Q_J(U,V) = Q_L(U,V) + i^{-1}((\dot{u},v)_{J-L} - (u,\dot{v})_{J-L}) \qquad (8.2)$$

is obtained if (2.2) is applied to the two intervals which constitute $J - L$. When $U,V \in \overset{o}{E}(a\cdot\cdot b)$ the transition to the limit $J \to I$ in (8.2) leads to the definition

$$Q(U,V) = Q_L(U,V) + i^{-1}((\dot{u},v)_{I-L} - (u,\dot{v})_{I-L}) \, . \qquad (8.3)$$

In (8.2), (8.3) the choice of L does not influence the values of Q_J and Q. Of course L together with the domain $(a-) \cup (-b)$ of U and V must cover I. If $U,V \in E(I)$ or $\overset{o}{E}(I)$ one can let A and B coincide ($A = \xi - 0$, $B = \xi + 0$, $q_x(U,V)$ continuous at $\xi \in I$) which reduces (8.2), (8.3) to the previous forms Q_J and Q.

Corresponding to the result in Section 3, $c\,Q_J$ cannot be positive definite for sufficiently large J on any linear hull $\{U, E(\lambda,I)\}$ in which U is in $E(a\cdot\cdot b)$ but is outside $E(\lambda,I)$. The condition on J is that it shall have its endpoints in $a-$ and $-b$, where U is defined and that it contains part of the support of c_o. In this sense $E(\lambda,I)$ is maximal positive definite in $E(a\cdot\cdot b)$ with respect to $c\,Q_J$. Similarly $E(\bar{\lambda},I)$ is negative definite. It follows that

$$c\,Q_J(U-U(J),\, U-U(J)) \leq 0 \qquad (8.4)$$

for every $U \in E(a\cdot\cdot b)$ and for all sufficiently large J when $U(J)$ is defined by

$$U(J) \in E(\lambda,I), \; Q_J(U-U(J),\, E(\lambda,I)) = 0 \, .$$

It is furthermore trivial that if $v \in C^{(N)}(a\cdot\cdot b)$ the equation $Su - \lambda\, Tu = Tv$ a.e. in a domain $(a-) \cup (-b)$ can be solved for u so that there exist elements $U = u|(\lambda u + v) \in E(a\cdot\cdot b)$.

Thus all prerequisites for a reasoning parallel to the one in Section 4 are available. The results are similar. In particular $\overset{o}{E}(\lambda,I)$, $\overset{o}{E}(\bar{\lambda},I)$ and $\overset{o}{F}(\lambda,I)$ are maximal in $\overset{o}{E}(a\cdot\cdot b)$ as subspaces on which the form cQ defined by (8.3) is positive and negative definite or Q is regular.

9. Corresponding to the results of Section 5 the general expression for a maximal Q-nullspace (Q defined in (8.3)) in $\overset{\circ}{E}(a\cdot\cdot b)$ is

$$Z(a\cdot\cdot b) = \overset{\circ}{E}(a\cdot\cdot b)^{perp} \dotplus Z' , \tag{9.1}$$

where Z' is a maximal Q-nullspace in $\overset{\circ}{F}(\lambda,I)$. Evidently (9.1) can be considered to be the same symmetric boundary condition as $Z = Z(I)$ in Section 5 if $Z(a\cdot\cdot b)$ and $Z(I)$ are determined by one and the same Z'. Actually

$$Z(I) = Z(a\cdot\cdot b) \cap \overset{\circ}{E}(I). \tag{9.2}$$

Finally, in the same way as Theorem 5.1, one deduces

<u>THEOREM 9.1</u>: <u>If</u> $Z(a\cdot\cdot b)$ <u>is a symmetric boundary condition in</u> $\overset{\circ}{E}(a\cdot\cdot b)$, <u>if</u> $\operatorname{Im}\lambda \neq 0$ <u>and</u> $\dim \overset{\circ}{E}(\lambda,I) \geq \dim \overset{\circ}{E}(\overline{\lambda},I)$ <u>and if</u> $U = u|(\lambda u + v) \in E(a\cdot\cdot b)$ <u>with</u> $v \in \overset{\circ}{C}^{(N)}(a\cdot\cdot b)$, <u>then there is a unique</u> $U_o \in E(\lambda,I)$ <u>such that</u> $U - U_o \in Z(a\cdot\cdot b)$.

It should be observed that the case $v = 0$ is of consequence. To any $U = u|\lambda u \in E(\lambda, a\cdot\cdot b)$ it gives a unique $U_o \in E(\lambda,I)$ such that $U - U_o$ satisfies Z. In Theorem 5.1 this case has no interest since there $U - U_o$ reduces to 0.

10. The mapping $\overline{R}_\lambda$ of Section 6 and consequently its restrictions $(\overline{K} - \lambda)^{-1}$ and $(K - \lambda)^{-1}$ have an integral representation with a certain kernel. In studying this question we shall restrict ourselves to the case when

$$\dim \overset{\circ}{E}(\lambda,I) = \dim \overset{\circ}{E}(\overline{\lambda},I) \tag{10.1}$$

so that $\overline{R}_\lambda$ exists for all non-real values of λ.

Let Z be a symmetric boundary condition represented by the maximal Q-nullspace $Z(a\cdot\cdot b)$ in $\overset{\circ}{E}(a\cdot\cdot b)$. The linear space

$$\begin{aligned} G(Z,\lambda) &= Z(a\cdot\cdot b) \cap E(\lambda, a\cdot\cdot b) \\ &= Z(a\cdot\cdot b) \cap \overset{\circ}{E}(\lambda, a\cdot\cdot b) \end{aligned} \tag{10.2}$$

will be called the Green's space belonging to Z and to the non-real value λ. Let x be a fixed point in I. According to a remark in Section 8 it can be assumed that the elements $U = u|\lambda u$ of $E(\lambda, a\cdot\cdot b)$ are defined in

$$\begin{aligned} a- &= \{y : a < y \leq x - 0\} , \\ -b &= \{y : x + 0 \leq y < b\} . \end{aligned} \tag{10.3}$$

On account of the existence theorem for the differential equation $Su - \lambda Tu = 0$ there are elements $U^k = u^k|\lambda u^k$ in $E(\lambda, a\cdot\cdot b)$ such that

$$\Delta_j(x)\, U^k = D^j u^k(x+0) - D^j u^k(x-0) = \delta_{jk} \tag{10.4}$$

for $j,k = 0, 1, \ldots, M-1$, and with Kronecker's δ_{jk}. Theorem 9.1 with $v = 0$ shows that there are elements $U_0^k \in E(\lambda, I)$ such that $W^k = U^k - U_0^k$ belongs to $Z(a\cdot\cdot b)$ i.e. to $G(Z,\lambda)$ for $k = 0, 1, \ldots, M-1$. Evidently

$$\Delta_j(x) W^k = \delta_{jk} \tag{10.5}$$

when $j,k = 0, 1, \ldots, M-1$, because of the continuity properties of the elements U_0^k in $E(\lambda, I)$. Thus all W^k are linearly independent. The conditions $\Delta_j(x)U = 0$, $j = 0, 1, \ldots, M-1$, cut out $Z(I) \cap E(\lambda, I) = \{0\}$ from $G(Z,\lambda)$. As a consequence $W^0, W^1, \ldots, W^{M-1}$ form a base for $G(Z,\lambda)$, the discontinuity base at x.

The functions w^k in $W^k = w^k|\lambda w^k$ belong to the hilbert space H_0 for $k = n, n+1, \ldots, M-1$.

<u>11</u>. Let $u = R_{\bar{\lambda}} v$ so that $U = u|(\bar{\lambda} u + v)$ belongs to Z and let $W = w|\lambda w$ be an element of $G(Z,\lambda)$. If $x - 0$ and $x + 0$ are taken as endpoints of L in the definition (8.3) of Q, it follows from $Q(U,W) = 0$ that

$$\Big[q(U,W) \Big]_{x-0}^{x+0} + i^{-1}((\dot{u}, w) - (u, \dot{w})) = 0 , \tag{11.1}$$

where $\dot{u} = \bar{\lambda} u + v$, $\dot{w} = \lambda w$. An immediate reduction gives

$$\left[q(U,W)\right]_{x-0}^{x+0} = i(v,w) . \tag{11.2}$$

Put $W = W^{M-1}$ i.e. $w = w^{M-1}$ in this formula. Because of (10.5), $\left[D^{M-1}w\right]_{x-0}^{x+0} = 1$ while $\left[D^j w\right]_{x-0}^{x+0} = 0$ for $j < M - 1$. Computations on account of the formula (2.4) for q lead to the result

$$s_M(x)u(x) = i(v,w^{M-1}) , \tag{11.3}$$

where s_M is the leading coefficient of S when this operator is written in the form $S = \sum_{j=0}^{M} s_j D^j$. With $w^{M-1} = w^{M-1}(x,y,\lambda)$ Green's function belonging to Z and λ and having its pole at x is defined by

$$g(x,y,\lambda) = \frac{1}{i\, s_M(x)}\, w^{M-1}(x,y,\lambda) . \tag{11.4}$$

By the help of $g(x,y,\lambda)$ the result (11.3) is written

$$u(x) = (v(\cdot),\ g(x,\cdot,\lambda)) . \tag{11.5}$$

This formula is equivalent to $u = R_{\bar{\lambda}} v$ and solves the equation

$$Su - \bar{\lambda}\, Tu = Tv \quad \text{in } I ,$$

for every v in $\overset{o}{C}{}^{(N)}(I)$ so that $u \mid (\bar{\lambda}u + v) \in Z$.

As a function of y Green's function satisfies the conditions

$$(S - \lambda T)_y\, g(x,y,\lambda) = 0 \quad \text{in } I \text{ for } y \neq x ,$$

$$g(x,y,\lambda) \mid \lambda\, g(x,y,\lambda) \in Z ,$$

$$\left[D_y^j\, g(x,y,\lambda)\right]_{y=x-0}^{x+0} = \begin{cases} 0 & \text{when } j < M - 1 , \\ \dfrac{1}{i\, s_M(x)} & \text{when } j = M - 1 . \end{cases}$$

As a function of λ it is analytic in both half-planes $\operatorname{Im}\lambda > 0$ and $\operatorname{Im}\lambda < 0$.

12. Let $U = u \mid \lambda u$ with $u = g(x_1,y,\lambda)$ and let $V = v \mid \mu v$ with $v = g(x_2,y,\mu)$. Here x_1 and x_2 are fixed points in I and $U = u \mid \lambda u$, $V = v \mid \mu v$ shall satisfy the same boundary condition Z. In generalization of (11.1) one obtains

$$\left[q(U,V) \right]_{x_1-0}^{x_1+0} + \left[q(U,V) \right]_{x_2-0}^{x_2+0} + i^{-1}((\dot{u},v) - (u,\dot{v})) = 0 ,$$

where $\dot{u} = \lambda u$, $\dot{v} = \mu v$. Similar reductions as in the preceeding section lead to the result

$$g(x_1,x_2,\lambda) - \overline{g(x_2,x_1,\mu)} = (\lambda - \bar{\mu})(g(x_1,\cdot,\lambda),\ g(x_2,\cdot,\mu)). \tag{12.1}$$

For $\mu = \bar{\lambda}$ this gives the symmetry relation

$$\overline{g(x_1,x_2,\lambda)} = g(x_2,x_1,\bar{\lambda}) . \tag{12.2}$$

If (12.2) is used in (12.1) and μ is replaced by $\bar{\mu}$ it follows that

$$g(x_1,x_2,\lambda) - g(x_1,x_2,\mu) = (\lambda - \mu)(g(x_1,\cdot,\lambda),\ \overline{g(\cdot,x_2,\mu)})$$

which corresponds to the resolvent relation $R_\lambda - R_\mu = (\lambda - \mu)R_\lambda R_\mu$. The symmetry relation gives information about $g(x,y,\lambda)$ as a function of x. For instance, $(\bar{S} - \lambda\bar{T})_x\, g(x,y,\lambda) = 0$ for $x \neq y$, where the bars indicate complex conjugation.

13. If $U = u|(\bar{\lambda}u + v) \in Z$ and W is replaced by $W^{M-1}, W^{M-2}, \ldots, W^0$ in (11.2), a system of equations is obtained which can be solved for $D^j u(x)$, $j = 0, 1, \ldots, M-1$, with a result of the form

$$D^j u(x) = (v(\cdot)\ ,\ g_j(x,\cdot,\lambda))\ ,\ 0 \leq j < M-n\ , \tag{13.1}$$

$$D^j u(x) = (v(\cdot),\ g_j(x,\cdot,\lambda)) + \sum_{k=M-j}^{j-M+N} \tau_{jk} D^k v(x),\ M-n \leq j < M. \tag{13.2}$$

Here the kernels are linear combinations of the functions w^k in the discontinuity base at x with coefficients continuously depending on x. Clearly $g_j(x,y,\lambda) = \overline{D_x^j}\, g(x,y,\lambda)$, $j = 0, 1, \ldots, M-1$. The quantities τ_{jk} are continuous in x .

Properties of the discontinuity base $W[x] = \{W^k[x]\}_0^{M-1}$ at x, hence of the kernels in (13.1), (13.2), can be obtained from the linear transformation $W[x] = T_c^x W[c]$ of a discontinuity base at a fixed point c to one at x . We observe for instance that the

Dirichlet integrals $(g_j(x,\cdot,\lambda), g_j(x,\cdot,\lambda))$, $j = 0, 1, \ldots, M-1$, are continuous functions of x, thus bounded when x varies in any compact subset of I. This is a consequence of the fact that the coefficients of the matrix T_c^x are continuous in x.

14. The equations (13.1), (13.2) are valid if $u = R_{\bar\lambda} v$. If $u = \bar R_{\bar\lambda} v$, let $\{v_\nu\}_1^\infty$ be a sequence of functions in $\overset{\circ}{C}{}^{(N)}(I)$ which tend to v in H. Since $\bar R_{\bar\lambda}$ is bounded, the elements in $\{u_\nu = R_{\bar\lambda} v_\nu\}_1^\infty$ tend to u in H. From the first set (13.1) of the system (13.1), (13.2) it follows, on account of the observation in the preceeding section, that $D^j u_\nu$ for $j < M - n$ converge pointwise and uniformly on compact subsets of I when $\nu \to \infty$. From this it is easy to see that $u = \bar R_{\bar\lambda} v$ has continuous derivatives of all orders $\leq M - n - 1$ which satisfy (13.1).

As an element of $H = \overset{\circ}{C}{}^{(n)}(I)$ the function v has $n-1$ continuous derivatives. The application of $\bar R_{\bar\lambda}$ gives a function with $M - n - 1$ continuous derivatives and $M - n - 1 > n - 1$ since $M > N$. If v has more than $n-1$ continuous derivatives a corresponding number of the first equations in (13.2) can be taken into account to show that the number of continuous derivatives of $u = \bar R_{\bar\lambda} v$ is accordingly increased above the value $M - n - 1$. An application of this idea gives the result stated in Section 7.

BIBLIOGRAPHY

[1] Bennewitz, Chr. and Pleijel, Åke. Selfadjoint extension of ordinary differential operators. To appear in the Proceedings of the Colloquium on Mathematical Analysis at Jyväskylä, Finland, August 17-21, 1970. (Mimeographed preprint available from the Department of Mathematics at Uppsala University.)

[2] Niessen, Heinz-Dieter. Singuläre S-hermitesche Rand-Eigenwertprobleme. Manuscripta Mathematica, 3, 35-68, 1970. (Contains references to earlier papers by F.W. Schäfke and A. Schneider.)

[3] Pleijel, Åke. Generalization of the spectral theory of ordinary formally self-adjoint differential operators. To appear in Journal of the Indian Mathematical Society, 34, 1972. (Mimeographed preprint available from the Department of Mathematics at Uppsala University.)

Recent Results in the Theory of Variational Inequalities

G. Stampacchia

This will be a report of some results which have been obtained in recent months in the theory of variational inequalities at the Scuola Normale Superiore in Pisa.

First of all I would like to recall some results already known. Among them a paper of H. Lewy and Stampacchia which appeared in 1971 in the Archive for Rational Mechanics and Analysis [1].

Let $a(p)$ be a continuous vector field on $\mathbb{R}^n$. The field is called monotone if

$$\big(a(p)-a(q)\big)(p-q) \geq 0$$

and is called strictly monotone if equality holds only for $p=q$.

Let Ω be an open bounded convex set of $\mathbb{R}^n$ and let $\psi(x)$ be a Lipschitz function defined on Ω with $\psi(x)<0$ on $\partial\Omega$. Denote by $H_o^{1,\infty}(\Omega)$ the space of all Lipschitz functions defined in Ω and vanishing on $\partial\Omega$. Let $\mathbb{K}$ be the convex set of those functions u of $H_o^{1,\infty}(\Omega)$ such that $u \geq \psi$ in Ω.

The following theorem holds [1].

Theorem 1. *For any monotone continuous field $a(p)$ there exists a function u such that:*

(1) $$u\in\mathbb{K} \;:\; \int_\Omega a_i(u_x)(v-u)_{x_i}\,dx \geq 0 \qquad \text{for all } v\in\mathbb{K}.$$

The Lipschitz coefficient of u is no greater than that of the obstacle ψ. Such a solution u is unique provided the field $a(p)$ is strictly monotone.

In the statement the summation convention is understood and u_x stands for the gradient of u.

Theorem 1 solves a problem of the theory of variational inequalities which contains many interesting special cases. For instance theorem 1 contains as a special case the problem of minimizing the integral

$$\int_\Omega f(v_x)\,dx$$

among all functions of $\mathbb{K}$ where f is a convex function over $\mathbb{R}^n$.

The case $f(p) = \sqrt{1+p^2}$ deals with the surface of minimal area among all surfaces $u(x)$ with given boundary values which stay above the obstacle represented by ψ.

The problem for linear $a(p)$ does not require the assumption of convexity on Ω and was treated by H. Lewy and Stampacchia in a previous paper [2].

The assumption of convexity of Ω can be omitted when $a(p)$ is non linear provided that, beside the monotonicity, $a(p)$ is coercive in a suitable sense.

Theorem 1 gave rise to many interesting questions. In two dimensions D. Kinderlehrer [3] showed that from theorem 1 can be derived a proof of the existence and smoothness of the solution to a variational inequality when the obstacle ψ is a function defined only on a segment of the domain Ω.

The idea of Kinderlehrer allowed Stampacchia and Vignoli to generalize theorem 1 in the following sense [4] : Let ψ be a smooth obstacle as in theorem 1 and C an uniformly convex open set (see e.g. [4]). Consider the problem (1) when the obstacle is obtained as the restriction of ψ to $\Omega - C$. The solution of the problem is still a Lipschitz function.

One of the most interesting questions arising from the problem considered in theorem 1 is to study the nature of the set where the solution u coincides with ψ. This problem was considered in [2] in the case where $a(p)=p$ for $n=2$. Recently, in a not yet published paper, D. Kinderlehrer has considered this problem for the case of $a(p) = \frac{p}{\sqrt{1+p^2}}$ i.e. the problem of minimal surfaces constrained to lie above the obstacle ψ.

Let me state the result.

Theorem 2 . Let ψ be a strictly convex domain with smooth boundary $\partial\Omega$ in the plane $z = x_1+ix_2$.

Let ψ be an analytic function in $\bar{\Omega}$, strictly concave and satisfying

$$\max_{\Omega} \psi > 0 \quad , \quad \psi < 0 \quad \text{on} \quad \partial\Omega .$$

Let $\mathbb{K} = \{v \in H_o^{1,\infty}(\Omega) \,:\, v \geq \psi \quad \text{in} \quad \Omega\}$ and

$$u \in \mathbb{K} \,:\, \int_{\Omega} \frac{u_{x_i}}{\sqrt{1+u_x^2}} (v-u)_{x_i} \, dx \geq 0 \qquad v \in \mathbb{K} .$$

The existence of such a function u follows from theorem 1.

Denote by

$$I = \{z \in \Omega \,:\, u(z)=\psi(z)\} .$$

Then, first,

∂I is a Jordan curve of class C^1

from which it follows that,

∂I is an analytic Jordan curve free of cusps.

In order to obtain such a result it is necessary to use the fact that $u \in H^{2,\infty}(\Omega)$ proven by himself in [5].

In a different direction Silvia Mazzone is considering the problem of existence and regularity of the solution when the variational inequality is of the type

$$(2) \qquad u \in \mathbb{K} : \int_\Omega a_i\,(u_x)(v-u)_{x_i}\,dx \geq \int_\Omega f(x)\,(v-u)\,dx \qquad v \in \mathbb{K}.$$

assuming that $f(x) \in L^\infty(\Omega)$. Provided that the field $a(p)$ is smooth with $\partial\Omega$ and satisfies suitable conditions of coercitivity of the type

$$\nu_o\big(1+\max(|p|^2,|q|^2)\big)^\tau |p-q|^2 \leq \big(a(p)-a(q)\big)(p-q) \quad , \quad \nu_o>0,\ \tau \geq -\frac{1}{2}$$

the solution is still $C^{1,\alpha}(\Omega)$.

In all these problems we look for solutions u where the values of u are given on all of the boundary of Ω.

In the applications it turns out that also the problems where u is free on a part of $\partial\Omega$ are interesting.

Such a problem is being considered by Murthy and Stampacchia.

Consider the case of an elliptic operator of the form

$$(3) \qquad Au = -\frac{\partial}{\partial x_j}(a_{ij}\,u_{x_i})$$

with smooth coefficients such that

$$a_{ij}(x)\,\xi_i\xi_j \geq \nu|\xi|^2 \qquad \text{for all } \xi \in \mathbb{R}^n .$$

Let Ω be a bounded domain and $\partial\Omega = \partial_1\Omega \cup \partial_2\Omega$ with $\partial_1\Omega$ "not too small". Let $H^1_\#(\Omega) \equiv \{v \in H^1(\Omega) : v=0 \text{ on } \partial_1\Omega\}$ and ψ be a smooth function such that

$$\psi \leq 0 \quad \text{on} \quad \partial_1\Omega .$$

Consider the convex set $\mathbb{K} = \{v \in H^1_{\#} : v \geq \psi \text{ in } \Omega\}$ and let f be in $L^\infty(\Omega)$. Then there exists

$$(4) \qquad u \in \mathbb{K} : \int_\Omega a_{ij}(x) u_{x_i} (v-u)_{x_j} dx \geq \int_\Omega f(v-u) dx \qquad \text{for all } v \in \mathbb{K} .$$

We shall make one of the following assumptions on the operator A and on the domain Ω.

Assumption I. (global)

The mixed problem

$$Au = F$$

$$u=0 \quad \text{on} \quad \partial_1\Omega , \quad \frac{\partial u}{\partial n} = g \quad \text{on} \quad \partial_2\Omega$$

admits a solution u for every $F \in L^p$, $g \in H^{1/p,p}(\partial_2\Omega)$ and the following a priori estimate holds

$$\| u \|_{H^{2,p}(\Omega)} \leq \text{const} \left(\| F \|_{L^p(\Omega)} + \| g \| \right) \quad \text{where} \quad p>n$$

Assumption I'. (local)

The estimate

$$\| u \|_{H^{2,p}(\Omega')} \leq \text{const}\left(\| F \|_{L^p(\Omega)} + \| u \|_{L^p(\Omega)} + \| g \| \right)$$

holds for a certain class of subdomains Ω' of Ω for $p>n$.
A sufficient condition on the domain Ω in order that an estimate

of this type holds can be deduced from the results of Shamir on the mixed boundary value problems [6] .

Assumption II.

The following estimate holds

$$\| u \|_{C^{0,\alpha}(\Omega)} \leq \text{const}(\| F \|_{L^p(\Omega)} + \| g \|_{L^p(\partial_2\Omega)}) \text{ with } p>n .$$

Then we have

Theorem 3. Under the assumption I, assuming that, on $\partial_2\Omega$ either "$u>\psi$" or "$u=\psi$ but $\frac{\partial\psi}{\partial n} \leq 0$" the solution of (4) is in $H^{2,p}(\Omega)$ (and via Sobolev in $C^{1,\alpha}(\bar{\Omega})$); moreover $\frac{\partial u}{\partial n} = 0$ on $\partial_2\Omega$.

Theorem 3'. Under the assumption I' for Ω' such that $\bar{\Omega}' \cap \{x\in\partial_2\Omega : \frac{\partial\psi}{\partial n} > 0\} = \emptyset$, and under the assumption II, the solution u is in $H^{2,p}(\Omega') \cap C^{0,\alpha}(\bar{\Omega})$.

In all these problems the inequality instead of an equation depends on the fact that a unilateral constraint was imposed on the solutions. But it is worthy to note that variational inequalities can arise also when no unilateral constraint is imposed, such as in a result of Vergara Caffarelli which appeared recently [7] .

I shall present his result in a simple situation. This problem is suggested by the physical problem of the equilibrium of two membranes pushed against each other by some given forces. It is natural to require that one membrane does not cross the other.

This can be formulated mathematically as follows; denote by $H^1_h(\Omega)$ the affine manifold of the function v in $H^1(\Omega)$ such that $v-h \in H^1_o(\Omega)$. Given $\rho>\sigma$ on $\partial\Omega$ consider the convex set of

$H^1_\rho(\Omega) \times H^1_\sigma(\Omega)$ defined by

$$\mathbb{K} \equiv \{(\xi,\eta) \; ; \; \xi \geq \eta \quad \text{in} \quad \Omega\} .$$

Consider the variational inequality

$$(5) \quad (u,v) \in \mathbb{K} : \int_\Omega a_{ij}(x) \left[u_{x_i} (\xi-u)_{x_j} + v_{x_i} (\eta-u)_{x_j} \right] dx \geq$$

$$\geq \int_\Omega \left[f(\xi-u) + g(\eta-v) \right] dx \quad \text{for all} \quad (\xi,\eta) \in \mathbb{K} .$$

If f and g are small enough, then

$Au = f$, $Av = g$ in Ω , and $u > v$;

otherwise there exists a set I of Ω where $u \equiv v$.

In the first case the functions u and v are sufficiently smooth provided that the data of the problem are smooth; in the latter case the solution (u,v) cannot be very regular even if the data are highly smooth. In fact if $f,g \in L^p(\Omega)$ with $p>n$ and $a_{ij}(x)$ and $\partial\Omega$ are smooth then $u,v \in C^{1,\alpha}(\Omega)$.

Vergara Caffarelli in a not yet published paper has extended this result also to the case of variational inequality defined, as at the beginning, for a monotone field $a(p)$.

REFERENCES

[1] Lewy H. and G.Stampacchia, On existence and smoothness of solutions of some non-coercive variational inequalities, Arch. Rational Mech.Anal. 41 (1971) 241-253.

[2] Lewy H. and G.Stampacchia, On the regularity of the solution of a variational inequality, Comm. Pure Appl.Math. 22 (1969), 153-188.

[3] Kinderlehrer D., Variational inequalities with lower dimensional obstacles, Israel Jr.Math. 10 (1971), 339-348.

[4] Stampacchia G. and A.Vignoli, A remark on variational inequalities for a second order non linear differential operator with non Lipschitz obstacles, to appear in Bull.Unione Matematica Italiana, 5 (1972).

[5] Kinderlehrer D., The regularity of the solution to a certain variational inequality, Proc.Symp.Pure Math. 23, Amer. Math. Soc.

[6] Shamir E., Elliptic systems of singular inetgral operators I, Trans.Amer.Math.Soc. 127 (1967), 107-124.

[7] Vergara Caffarelli Giorgio, Regolarità di un problema di disequazioni variazionali relativo a due membrane, Rendiconti Accad.Naz.Lincei, 50 (1971), 659-662.

Singular Perturbations of Elliptic Sesquilinear Forms

F. Stummel

This paper establishes a perturbation theory for sequences of strongly elliptic sesquilinear forms a_ι on Hilbert spaces E_ι, $\iota = 0,1,2,\cdots$, that are continuously embedded subspaces of a Hilbert space E. An essential tool for this perturbation theory is the theory of discrete convergence as developed in Stummel [15], [16]. By these means we prove in Section 2 under necessary and sufficient stability and consistency conditions the strong convergence $u_\iota \to u_0$ and the weak convergence $u_\iota \rightharpoonup u_0$ for $\iota \to \infty$ of the solutions u_ι of inhomogeneous, uniformly strongly elliptic equations $a_\iota(\varphi, u_\iota) = \ell_\iota(\varphi)$, $\varphi \in E_\iota$, $\iota = 0,1,2,\cdots$. The convergence of eigenvalues and associated algebraic eigenspaces is shown in Section 3 for general, not necessarily Hermitian eigenvalue problems $a_\iota(\varphi, w_\iota) = \overline{\lambda}_\iota b_\iota(\varphi, w_\iota)$, $\varphi \in E_\iota$, $\iota = 0,1,2,\cdots$. The first applications of our theory in Section 4 yield interesting results for elliptic problems in the range of the asymptotic perturbation theory (s. Kato [9], Chapter VIII) and of the Ritz-Galerkin methods and projection methods (s. Grigorieff [5], Stummel [17], Vainikko [18]). Then we apply the theory to singular perturbations of elliptic problems studied by Greenlee [4], Huet [7], [8] and others. Here we obtain new results by generalizing convergence theorems in [4], [7], [8] from elliptic problems in the sense of Lions to strongly elliptic problems in the sense of Gårding and from special Hermitian to general, not necessarily Hermitian eigenvalue problems. Moreover, our singular perturbation theory of strongly elliptic Dirichlet problems in Section 5 admits, additionally, very general perturbations of the domains of definition. Thus, Section 5 extends considerably results of Babuška [2], Huet [7], Greenlee [4], Nečas [12]. This perturbation theory for Dirichlet problems is a special example of the general perturbation theory for elliptic boundary value problems studied in a forthcoming paper of the author.

1. Discrete Convergence

Let E be a real resp. complex Hilbert space with scalar product $(.,.)_E$. For every $\iota = 0,1,2,\cdots$ let E_ι be subspaces of E that are Hilbert spaces with scalar products $(.,.)_{E_\iota}$. We assume that the related norms satisfy

(1) $$\|u\|_E = \|u\|_{E_0}, \quad u \in E_0, \qquad \|u\|_E \leq \mu \|u\|_{E_\iota}, \quad u \in E_\iota, \quad \iota = 1,2,\cdots,$$

with some positive constant μ. Then the scalar product $(.,.)_E$ defines bounded sesquilinear forms on $E_\iota \times E$ for every $\iota = 0,1,2,\cdots$, which determine uniquely linear operators $R_\iota \in \mathcal{B}(E,E_\iota)$ with the property

(2) $$(u,v)_E = (u,R_\iota v)_{E_\iota}, \quad u \in E_\iota, \quad v \in E, \quad \iota = 0,1,2,\cdots.$$

By (1) we have $(.,.)_E = (.,.)_{E_0}$ on E_0 such that R_0 is the orthogonal projection P_0 of E onto E_0 and thus

(3) $$R_0 u = P_0 u = u, \quad u \in E_0 .$$

In general, however, the operators R_ι are no projections. By (1) and the representation (2) for $u = R_\iota v$ one obtains the inequalities

(4) $$\frac{1}{\mu} \|R_\iota v\|_E \leq \|R_\iota v\|_{E_\iota} \leq \mu \|v\|_E, \quad v \in E, \quad \iota = 0,1,2,\cdots.$$

In this paper, we want to make the basic assumption (R): $R_\iota \xrightarrow[E]{} R_0$ $(\iota \to \infty)$, i.e.

(R) $$R_\iota v \xrightarrow[E]{} R_0 v \quad (\iota \to \infty), \quad v \in E .$$

An immediate consequence of this assumption is

(5) $$\|R_\iota v\|^2_{E_\iota} = (R_\iota v, v)_E \longrightarrow (R_0 v, v)_E = \|R_0 v\|^2_E \quad (\iota \to \infty), \; v \in E .$$

An essential tool for our perturbation theory is the notion of discrete convergence. For elements $u \in E_0$, for sequences $u_\iota \in E_\iota$, $\iota \in N'$, and infinite subsequences of natural numbers $N' \subseteq N = (1,2,3,\cdots)$, the <u>(strong) discrete convergence</u> $\longrightarrow$ is defined by

(6) $$u_\iota \longrightarrow u \iff u_\iota \xrightarrow[E]{} u, \qquad \|u_\iota\|_{E_\iota} \longrightarrow \|u\|_E \quad (\iota \to \infty, \iota \in N').$$

Under the assumption (R) we can prove the following basic Theorem.

(7) <u>The strong discrete convergence</u> (6) <u>is equivalent to the condition</u>

$$u_\iota \longrightarrow u \iff \|u_\iota - R_\iota u\|_{E_\iota} \longrightarrow 0 \quad (\iota \to \infty, \iota \in N').$$

<u>Proof.</u> (i) From (R) and (3) we have the convergence of $R_\iota u \xrightarrow[E]{} u$ $(\iota \to \infty)$ for every $u \in E_0$. Hence we obtain from (5), (6) the convergence of

$$\|u_\iota - R_\iota u\|^2_{E_\iota} = \|u_\iota\|^2_{E_\iota} + \|R_\iota u\|^2_{E_\iota} - 2\,\mathrm{Re}\,(u_\iota, u)_E \rightarrow 0 \quad (\iota\rightarrow\infty, \iota\in N').$$

(ii) Conversely, the convergence of $u_\iota \underset{E}{\rightarrow} u$ follows in view of $R_\iota u \underset{E}{\rightarrow} u$ by the estimate

$$\|u_\iota - R_\iota u\|_E \leq \mu \|u_\iota - R_\iota u\|_{E_\iota} \rightarrow 0 \quad (\iota\rightarrow\infty, \iota\in N').$$

From (5) we conclude the convergence of the expression

$$\big|\|u_\iota\|_{E_\iota} - \|u\|_E\big| \leq \|u_\iota - R_\iota u\|_{E_\iota} + \big|\|R_\iota u\|_{E_\iota} - \|u\|_E\big| \rightarrow 0 \ (\iota\rightarrow\infty, \iota\in N')$$

for $u = R_0 u$. This shows the discrete convergence of $u_\iota \rightarrow u$ $(\iota\rightarrow\infty, \iota\in N')$ defined by (6).

The sequence of Hilbert spaces E_ι, $\iota = 1, 2, \cdots$, constitutes a discrete approximation $\mathcal{A}(E_0, \Pi E_\iota, R)$ of the space E_0 in the sense of [15] with the mapping

$$R(u) = \{(u_\iota)\in \Pi E_\iota \mid \|u_\iota - R_\iota u\|_{E_\iota} \rightarrow 0 \quad (\iota\rightarrow\infty)\}, \quad u\in E_0.$$

The operators $R_\iota: E_0 \mapsto E_\iota$, $\iota\in N$, form a sequence of linear restriction operators (s. [15], §4.1) with the property

(8) $$R_\iota u \rightarrow u \quad (\iota\rightarrow\infty), \quad u\in E_0.$$

Next, we introduce the notion of weak discrete convergence. For elements $u\in E_0$, for sequences $u_\iota \in E_\iota$, $\iota\in N'$, and infinite subsequences of natural numbers $N'\subseteq N$, the <u>weak discrete convergence</u> $\rightharpoonup$ is defined by

(9) $$u_\iota \rightharpoonup u \iff u_\iota \underset{E}{\rightharpoonup} u, \quad \|u_\iota\|_{E_\iota} \text{ bounded } (\iota\rightarrow\infty, \iota\in N').$$

Strong and weak discrete convergence permit the following important characterization.

(10) <u>For every</u> $u\in E_0$ <u>and every sequence</u> $u_\iota \in E_\iota$, $\iota\in N'\subseteq N$, <u>the weak discrete convergence</u> $u_\iota \rightharpoonup u$ <u>is equivalent to</u>

(11) $$(u_\iota, w_\iota)_{E_\iota} \rightarrow (u, w)_E \quad (\iota\rightarrow\infty, \iota\in N')$$

<u>for every</u> $w\in E_0$ <u>and every sequence</u> $w_\iota \in E_\iota$, $w_\iota \rightarrow w$ $(\iota\rightarrow\infty, \iota\in N')$.

(12) <u>For every</u> $u\in E_0$ <u>and every sequence</u> $u_\iota\in E_\iota$, $\iota\in N'\subseteq N$, <u>the strong discrete convergence</u> $u_\iota \rightarrow u$ <u>is equivalent to</u>

(13) $$(u_\iota, w_\iota)_{E_\iota} \rightarrow (u, w)_E \quad (\iota\rightarrow\infty, \iota\in N')$$

<u>for every</u> $w\in E_0$ <u>and every sequence</u> $w_\iota \in E_\iota$, $w_\iota \rightharpoonup w$ $(\iota\rightarrow\infty, \iota\in N')$.

<u>Proof.</u> (i) The discrete convergence $w_\iota \rightarrow w$ implies $\|w_\iota - R_\iota w\|_{E_\iota} \rightarrow 0$. From $u_\iota \rightharpoonup u$

and definition (9) we infer the convergence of

$$(u_\iota, w_\iota)_{E_\iota} = (u_\iota, w_\iota - R_\iota w)_{E_\iota} + (u_\iota, w)_E \rightarrow (u, w)_E \quad (\iota\rightarrow\infty, \iota\in N').$$

Conversely, for every $v \in E$ and $w_\iota = R_\iota v \rightarrow P_o v = w$ we obtain from (11) the convergence of

$$(u_\iota, v)_E = (u_\iota, w_\iota)_{E_\iota} \rightarrow (u, w)_E = (u, v)_E, \quad v \in E,$$

i.e. the weak convergence of $u_\iota \rightharpoonup u$ $(\iota\rightarrow\infty, \iota\in N')$. The sequence of norms $\|u_\iota\|_{E_\iota}$, $\iota \in N'$, is necessarily bounded. Otherwise, there would exist a subsequence $N'' \subseteq N'$ with the property

$$\alpha_\iota = \|u_\iota\|_{E_\iota} \rightarrow \infty, \quad w_\iota = \frac{1}{\alpha_\iota^2} u_\iota \rightarrow 0 \quad (\iota\rightarrow\infty, \iota\in N'').$$

Hence, (11) leads to the contradiction $1 = (u_\iota, w_\iota)_{E_\iota} \rightarrow 0$ $(\iota\rightarrow\infty, \iota\in N'')$.

(ii) For every sequence $u_\iota \rightarrow u$ $(\iota\rightarrow\infty, \iota\in N')$, Theorem (10) shows the statement (13). Conversely, the discrete convergence of $w_\iota \rightarrow w$, obviously implies the weak discrete convergence $w_\iota \rightharpoonup w$ $(\iota\rightarrow\infty, \iota\in N')$. Hence, for every sequence $w_\iota \rightarrow w$ we obtain from (13) the statement (11) and thus the weak discrete convergence of $u_\iota \rightharpoonup u$ $(\iota\rightarrow\infty, \iota\in N')$. On setting $u = w, u_\iota = w_\iota, \iota\in N'$, (13) implies the convergence of $\|u_\iota\|^2_{E_\iota} \rightarrow \|u\|^2_E$ $(\iota\rightarrow\infty, \iota\in N')$. Now, using (8) and Theorem (7), the assertion $u_\iota \rightarrow u$ follows from (13) by

$$\|u_\iota - R_\iota u\|^2_{E_\iota} = \|u_\iota\|^2_{E_\iota} + \|R_\iota u\|^2_{E_\iota} - 2\,\mathrm{Re}\,(u_\iota, u)_E \rightarrow 0 \quad (\iota\rightarrow\infty, \iota\in N').$$

Bounded linear functionals $\ell \in E', \ell_\iota \in E'_\iota$ can be represented always by uniquely determined elements $v \in E, v_\iota \in E_\iota$ in the form

(14) $$\ell(\varphi) = (\varphi, v)_E, \quad \ell_\iota(\varphi) = (\varphi, v_\iota)_{E_\iota}, \quad \varphi \in E_\iota, \quad \iota = 0,1,2,\cdots.$$

This leads to the definition of the <u>(strong) discrete convergence</u> $\rightarrow$ for sequences of functionals $\ell_\iota \in E'_\iota, \iota\in N' \subseteq N$, by

(15) $$\ell_\iota \rightarrow \ell_o \iff v_\iota \rightarrow v_o \quad (\iota\rightarrow\infty, \iota\in N').$$

Any functional $\ell_o \in E'_o$ has extensions $\ell \in E'$ with $\ell_o = \ell | E_o$. This extension property permits the following convergence theorem.

(16) <u>The discrete convergence $\ell_\iota \rightarrow \ell_o$ is equivalent to the condition</u>

$$\sup_{0\neq\psi\in E_\iota} |\ell_\iota(\psi) - \ell(\psi)| / \|\psi\|_{E_\iota} \rightarrow 0 \quad (\iota\rightarrow\infty, \iota\in N')$$

<u>for an arbitrary extension $\ell \in E'$ of $\ell_o = \ell | E_o$.</u>

<u>Proof.</u> Let $\ell \in E'$ be an extension of ℓ_o and $\ell(\varphi) = (\varphi, v)_E, \varphi\in E$. Then ℓ_o

satisfies the equation $\ell_0(\varphi) = (\varphi, P_0 v)$, $\varphi \in E_0$, so that $v_0 = P_0 v$ and $P_0(v - v_0) = 0$. In view of assumption (R),

$$\| R_\iota (v - v_0) \|_{E_\iota} \longrightarrow \| P_0 (v - v_0) \|_E = 0 \quad (\iota \to \infty).$$

Hence the discrete convergence of $v_\iota \to v_0$ becomes equivalent to $\| v_\iota - R_\iota v \|_{E_\iota} \to 0 \ (\iota \to \infty)$. This is just the above assertion.

Analogously, the <u>weak discrete convergence</u> $\rightharpoonup$ for sequences of functionals $\ell_\iota \in E'_\iota$, $\iota \in N' \subseteq N$, is defined by

(17) $$\ell_\iota \rightharpoonup \ell_0 \iff v_\iota \rightharpoonup v_0 \quad (\iota \to \infty, \iota \in N').$$

By Theorem (10), this condition is equivalent to the statement

(18) $$\ell_\iota(w_\iota) \longrightarrow \ell_0(w_0) \quad (\iota \to \infty, \iota \in N')$$

for every $w_0 \in E_0$ and every sequence $w_\iota \in E_\iota$, $w_\iota \to w_0$ $(\iota \to \infty, \iota \in N')$. We only mention another equivalent condition for the weak discrete convergence (17) or (18), given by

(19) $$\ell_\iota(\varphi_\iota) \longrightarrow \ell_0(\varphi), \quad \| \ell_\iota \|_{E'_\iota} \text{ bounded} \quad (\iota \to \infty, \iota \in N'),$$

for every φ of a dense subspace $D_\ell \subseteq E_0$ with a suitable sequence $\varphi_\iota \to \varphi$ $(\iota \to \infty)$ (s. [15], 2.1.(3)).

Finally, we will state a very useful Theorem asserting the validity of the assumption (R).

(20) <u>The following conditions</u> (R1), (R2), (R3) <u>are necessary and sufficient for the convergence condition</u> (R): $R_\iota \underset{E}{\to} R_0$ $(\iota \to \infty)$.

(R1) <u>There exists a dense subspace</u> $D_0 \subseteq E_0$ <u>and for every</u> $\varphi \in D_0$ <u>there is a sequence</u> $\varphi_\iota \in E_\iota$ <u>with the properties</u>

$$\varphi_\iota \underset{E}{\to} \varphi, \quad \| \varphi_\iota \|_{E_\iota} \longrightarrow \| \varphi \|_E \quad (\iota \to \infty).$$

(R2) <u>For every weakly convergent subsequence</u> $w_\iota \underset{E}{\rightharpoonup} w$ $(\iota \to \infty, \iota \in N')$ <u>with</u> $w_\iota \in E_\iota$, $\| w_\iota \|_{E_\iota} \le \beta$, $\iota \in N'$, <u>the limit</u> w <u>belongs to</u> E_0.

(R3) $$\overline{\lim_{\iota \to \infty}} \| R_\iota v \|_E \le \| R_0 v \|_E, \quad v \in E.$$

<u>Proof.</u> (i) Obviously, $R_\iota \underset{E}{\to} R_0$ $(\iota \to \infty)$ implies (R3). From (3) it is seen that (R1) is satisfied by $D_0 = E_0$ and by $\varphi_\iota = R_\iota \varphi$, $\iota = 1, 2, \cdots$, for every $\varphi \in E_0$. For any

$h \in E \ominus E_0$ we have

$$\| R_\iota h \|^2_{E_\iota} = (R_\iota h, h)_E \longrightarrow (R_0 h, h)_E = 0 \quad (\iota \to \infty).$$

Hence, condition (R2) follows from the relation

$$0 = \lim_{\iota \in N'} (w, R_\iota h)_{E_\iota} = \lim_{\iota \in N'} (w_\iota, h)_E = (w, h)_E , \quad h \in E \ominus E_0 .$$

(ii) First, by the assumption (R3) and the representations (2) and (3), we obtain the estimate

(21) $$\overline{\lim_{\iota \to \infty}} \| R_\iota u \|^2_{E_\iota} = \overline{\lim_{\iota \to \infty}} (R_\iota u, u)_E \leq \| u \|^2_E , \quad u \in E_0 .$$

Then condition (R1) gives for every $\varphi \in D_0$ and $\varphi_\iota \to \varphi$,

$$\| \varphi_\iota - R_\iota \varphi \|^2_{E_\iota} = \| \varphi_\iota \|^2_{E_\iota} + \| R_\iota \varphi \|^2_{E_\iota} - 2 \operatorname{Re} (\varphi_\iota, \varphi)_E \to 0 \quad (\iota \to \infty).$$

By the inequality (4), the sequence $w_\iota = R_\iota v$, $\iota = 1, 2, \cdots$, is bounded in E for every $v \in E$ and thus weakly compact. We now prove the weak convergence of $R_\iota v \underset{E}{\rightharpoonup} R_0 v$ $(\iota \to \infty)$. For this it is sufficient to show that every weakly convergent subsequence $R_\iota v \underset{E}{\rightharpoonup} w$ $(\iota \to \infty, \ \iota \in N')$ has the limit $w = R_0 v$. From condition (R2) follows $w \in E_0$. For every $\varphi \in D_0$, $\varphi_\iota \to \varphi$ $(\iota \to \infty)$, the convergence of $w_\iota \underset{E}{\rightharpoonup} w$ implies

$$(\varphi_\iota, w_\iota)_{E_\iota} = (\varphi_\iota - R_\iota \varphi, w_\iota)_{E_\iota} + (\varphi, w_\iota)_E \longrightarrow (\varphi, w)_E \quad (\iota \to \infty, \iota \in N').$$

Furthermore,

$$(\varphi_\iota, w_\iota)_{E_\iota} = (\varphi_\iota, v)_E \longrightarrow (\varphi, v)_E \quad (\iota \to \infty), \ \varphi \in D_0 ,$$

so that $w = R_0 v$, which proves $R_\iota v \underset{E}{\rightharpoonup} R_0 v$ $(\iota \to \infty)$, $v \in E$.

(iii) Finally, the assertion $R_\iota \underset{E}{\to} R_0$ $(\iota \to \infty)$ follows, using (R3) and the weak convergence $R_\iota \underset{E}{\rightharpoonup} R_0$ $(\iota \to \infty)$, from the convergence of the expression

$$\| R_\iota v - R_0 v \|^2_E = \| R_0 v \|^2_E + \| R_\iota v \|^2_E - 2 \operatorname{Re} (R_0 v, R_\iota v)_E \to 0 \quad (\iota \to \infty).$$

The Theorem just proved, immediately implies the following Corollary.

(22) <u>The condition</u> (R): $R_\iota \underset{E}{\to} R_0$ $(\iota \to \infty)$ <u>is satisfied, if</u> (R1), (R2) <u>and the following condition</u> (R3') <u>are valid,</u>

(R3') $$\| u \|_E \leq \| u \|_{E_\iota} , \quad u \in E_\iota , \quad \iota = 1, 2, \cdots .$$

<u>Proof.</u> Condition (R3') implies $\mu = 1$ in the inequalities (1) and (4). From the estimate $\| R_\iota v \|_E \leq \| v \|_E$, $v \in E$, we infer the inequality (21). Consequently, as in the proof of Theorem (20), one obtains the weak convergence $R_\iota v \underset{E}{\rightharpoonup} R_0 v$ $(\iota \to \infty)$ for

every $v \in E$. Since $R_0 = P_0$ and

$$\|R_\iota v\|^2_{E_\iota} = (R_\iota v, v)_E \longrightarrow (R_0 v, v)_E = \|R_0 v\|^2_E \quad (\iota \to \infty),$$

we conclude from condition (R3'),

$$\|R_\iota v - R_0 v\|^2_E \le \|R_0 v\|^2_E + \|R_\iota v\|^2_{E_\iota} - 2\,\mathrm{Re}(R_0 v, R_\iota v)_E \to 0 \quad (\iota \to \infty).$$

2. Inhomogeneous Equations

Bounded sesquilinear forms a_ι on E_ι and bounded linear functionals ℓ_ι on E_ι define a sequence of inhomogeneous equations

$$(1) \qquad a_\iota(\varphi, u_\iota) = \ell_\iota(\varphi), \quad \varphi \in E_\iota, \quad \iota = 0,1,2,\cdots .$$

In this Section we will show the strong discrete convergence $u_\iota \to u_0$ resp. the weak discrete convergence $u_\iota \rightharpoonup u_0$ of the solutions of (1) for $\ell_\iota \to \ell_0$ resp. $\ell_\iota \rightharpoonup \ell_0$ for $\iota \to \infty$. It is a well-known fact that bounded sesquilinear forms a_ι on E_ι and bounded linear functionals ℓ_ι on E_ι may be represented uniquely by bounded linear operators $A_\iota \in \mathcal{B}(E_\iota)$ and elements $v_\iota \in E_\iota$ in the form

$$(2) \qquad a_\iota(\varphi, \psi) = (\varphi, A_\iota \psi)_{E_\iota}, \quad \ell_\iota(\varphi) = (\varphi, v_\iota)_{E_\iota}, \quad \varphi, \psi \in E_\iota, \quad \iota = 0,1,2,\cdots.$$

Consequently, the equations (1) are equivalent to the inhomogeneous equations

$$(3) \qquad A_\iota u_\iota = v_\iota, \quad \iota = 0,1,2,\cdots.$$

To begin the study, some basic definitions and notations will be given. A sequence of sesquilinear forms a_ι on E_ι, $\iota = 0,1,2,\cdots$, is called <u>stable</u> if it is uniformly bounded by some positive constant α_1,

$$(b) \qquad |a_\iota(\varphi, \psi)| \le \alpha_1 \|\varphi\|_{E_\iota} \|\psi\|_{E_\iota}, \quad \varphi, \psi \in E_\iota, \quad \iota = 1,2,\cdots .$$

For a stable sequence a_ι on E_ι, $\iota = 0,1,2,\cdots$, the corresponding sequence of operators $A_\iota \in \mathcal{B}(E_\iota)$ and their adjoints $A^*_\iota \in \mathcal{B}(E_\iota)$ satisfy the <u>stability condition</u> $\|A_\iota\| = \|A^*_\iota\| \le \alpha_1$, $\iota = 0,1,2,\cdots$. The <u>inverse stability condition</u> requires

$$(b_{-1}) \qquad \alpha_0 \|\varphi\|_{E_\iota} \le \sup_{0 \ne \psi \in E_\iota} |a_\iota(\psi, \varphi)| / \|\psi\|_{E_\iota}, \quad \varphi \in E_\iota, \quad \iota = 1,2,\cdots,$$

for some positive constant α_0 . This means that the corresponding operators $A_\iota \in \mathcal{B}(E_\iota)$ are invertible and have uniformly bounded inverses, according to $\alpha_0 \|\varphi\|_{E_\iota} \le \|A_\iota \varphi\|_{E_\iota}$, $\iota = 0,1,2,\cdots$.

A sequence of sesquilinear forms a_ι on E_ι, $\iota = 0,1,2,\cdots$, is called <u>consistent</u> if there exists a dense subset $\mathcal{D}_a \subseteq E_0$ and to every $\varphi \in \mathcal{D}_a$ some sequence $\varphi_\iota \in E_\iota, \varphi_\iota \to \varphi$, with the property

(4) $$A_\iota \varphi_\iota \to A_0 \varphi \iff \| A_\iota \varphi_\iota - R_\iota A_0 \varphi \|_{E_\iota} \to 0 \quad (\iota \to \infty).$$

However, there is still another form of this consistency condition which will be more convenient for applications to sesquilinear forms. First we note that bounded sesquilinear forms a_0 on E_0 can always be extended to sesquilinear forms a on E such that the restriction of a to E_0 equals a_0. For example, one obtains such an extension by the orthogonal projection $P_0 : E \mapsto E_0$, on setting $a(\varphi, \psi) = a_0(P_0\varphi, P_0\psi), \varphi, \psi \in E$. The bounded linear operators $A_0 \in \mathcal{B}(E_0)$, $A \in \mathcal{B}(E)$ representing a_0 on E_0, a on E satisfy

(5) $$a_0 = a | E_0 \iff A_0 \varphi = P_0 A \varphi, \quad \varphi \in E_0.$$

Hence, using condition (R),

(6) $$\| R_\iota A_0 \varphi - R_\iota A \varphi \|_{E_\iota} \to \| A_0 \varphi - P_0 A \varphi \|_E = 0 \quad (\iota \to \infty), \; \varphi \in E_0.$$

Consequently, the sequence a_ι on E_ι, $\iota = 0,1,2,\cdots$, is consistent if and only if it satisfies the <u>consistency condition</u> $\| A_\iota \varphi_\iota - R_\iota A \varphi \|_{E_\iota} \to 0$ or

(c) $$\sup_{0 \neq \psi \in E_\iota} | a_\iota(\psi, \varphi_\iota) - a(\psi, \varphi) | / \| \psi \|_{E_\iota} \to 0 \quad (\iota \to \infty)$$

for every $\varphi \in \mathcal{D}_a$ with an associated sequence $\varphi_\iota \in E_\iota$, $\varphi_\iota \to \varphi$ $(\iota \to \infty)$.

Stability and consistency of a sequence a_ι, $\iota = 0,1,2,\cdots$, are equivalent to the <u>convergence</u> $a_\iota \to a_0$ resp. $A_\iota \to A_0$ defined by the relation

(7) $$u_\iota \to u_0 \implies v_\iota = A_\iota u_\iota \to v_0 = A_0 u_0 \quad (\iota \to \infty)$$

for every sequence $u_\iota \in E_\iota$, $\iota = 0,1,2,\cdots$ (s. [15], Theorem 1.2.(6)).

Having made these basic definitions, we are in a position to prove the first Theorem which establishes the strong discrete convergence of the solutions of (1) resp. (3).

(8) <u>Let</u> A_0 <u>be surjective in</u> E_0 <u>and let</u> A_ι <u>be bijective in</u> E_ι <u>for each</u> $\iota = 1,2,\cdots$. <u>Then the inverse stability</u> (b_{-1}) <u>and the consistency</u> (c) <u>of the sequence</u> a_ι <u>on</u> E_ι, $\iota = 0,1,2,\cdots$, <u>are necessary and sufficient for the existence of</u> A_0^{-1} <u>on</u> E_0 <u>and</u>

the discrete convergence of the solutions

(9) $$u_\iota = A_\iota^{-1} v_\iota \longrightarrow u_o = A_o^{-1} v_o \quad (\iota\to\infty)$$

for every sequence $v_\iota \in E_\iota,\ v_\iota \to v_o$ resp. $\ell_\iota \in E_\iota',\ \ell_\iota \to \ell_o$ $(\iota\to\infty)$.

Proof. (i) It follows by a well-known Theorem of Banach that the bijective operators A_ι have continuous inverses A_ι^{-1} in E_ι, $\iota = 1,2,\cdots$. Consequently, the sufficiency part of the assertion follows from [15], Theorem 1.3.(3).

(ii) Conversely, (9) implies the stability of the sequence A_ι^{-1}, $\iota=1,2,\cdots$, i.e. the inverse stability condition (b_{-1}) (s. [15], Theorem 1.2.(6)). Finally, the consistency condition (c) follows for every $\varphi \in D_a = E_o$ and for $\psi = A_o\varphi$ from (9) by

$$\varphi_\iota = A_\iota^{-1} R_\iota \psi \to A_o^{-1}\psi = \varphi, \quad A_\iota\varphi_\iota = R_\iota\psi \to \psi = A_o\varphi \quad (\iota\to\infty).$$

This Theorem may be applied to the adjoint sesquilinear forms a_ι^* defined by

(10) $$a_\iota^*(\varphi,\psi) = \overline{a_\iota(\psi,\varphi)}, \quad \varphi,\psi \in E_\iota,\ \iota = 0,1,2,\cdots.$$

Then the inverse stability condition for the sequence a_ι^* on E_ι, $\iota=1,2,\cdots$, requires

(b_{-1}^*) $$\alpha_o \|\varphi\|_{E_\iota} \le \sup_{0\neq\psi\in E_\iota} |a_\iota(\varphi,\psi)| / \|\psi\|_{E_\iota}, \quad \varphi \in E_\iota, \quad \iota = 1,2,\cdots,$$

with some positive constant α_o. The sequence a_ι^*, $\iota=0,1,2,\cdots$, is consistent if and only if it satisfies the adjoint consistency condition

(c^*) $$\sup_{0\neq\psi\in E_\iota} |a_\iota(\varphi_\iota,\psi) - a(\varphi,\psi)| / \|\psi\|_{E_\iota} \to 0 \quad (\iota\to\infty)$$

with some sequence $\varphi_\iota \in E_\iota$, $\varphi_\iota \to \varphi$ $(\iota\to\infty)$ for every φ in a dense subset $D_{a^*} \subseteq E_o$ and for an extension a on E of $a_o = a|E_o$.

Our next Theorem shows the weak discrete convergence of the solutions of the inhomogeneous equations (1) resp. (3).

(11) Let A_o be injective in E_o and let A_ι be bijective in E_ι for $\iota=1,2,\cdots$. Then the inverse stability (b_{-1}^*) and the consistency (c^*) of the sequence a_ι, $\iota=0,1,2,\cdots$, are necessary and sufficient for the existence of A_o^{-1} on E_o and the weak discrete convergence

(12) $$u_\iota = A_\iota^{-1} v_\iota \rightharpoonup u = A_o^{-1} v_o \quad (\iota\to\infty)$$

for every sequence $v_\iota \in E_\iota,\ v_\iota \to v_o$ resp. $\ell_\iota \in E_\iota',\ \ell_\iota \rightharpoonup \ell_o$ $(\iota\to\infty)$.

Proof. Note that the sesquilinear forms a_ι^* on E_ι, $\iota=0,1,2,\cdots$, satisfy the hypotheses of

Theorem (8). Hence the conditions (b^*_{-1}), (c^*) are necessary and sufficient for the existence of $(A_0^*)^{-1}=(A_0^{-1})^*$ and, in view of $(A_\iota^*)^{-1}=(A_\iota^{-1})^*$, for the discrete convergence

$$(13)\qquad (A_\iota^{-1})^* w_\iota \longrightarrow (A_0^{-1})^* w_0 \quad (\iota\to\infty)$$

for every sequence $w_\iota \in E_\iota, w_\iota \to w\ (\iota\to\infty)$. From this statement and Theorem 1.(10) we infer (12) by the convergence of

$$(u_\iota, w_\iota)_{E_\iota} = (v_\iota, (A_\iota^{-1})^* w_\iota)_{E_\iota} \to (v_0, (A_0^{-1})^* w_0)_E = (u_0, w_0)_E \quad (\iota\to\infty)$$

for every sequence $w_\iota \in E_\iota, w_\iota \to w_0\ (\iota\to\infty)$. Conversely, (12) implies (13) since

$$(v_\iota, (A_\iota^{-1})^* w_\iota)_{E_\iota} = (u_\iota, w_\iota)_{E_\iota} \to (u_0, w_0)_E = (v_0, (A_0^{-1})^* w_0)_E$$

for every sequence $v_\iota \in E_\iota,\ v_\iota \to v_0\ (\iota\to\infty)$.

Now we turn to an analysis of the important class of elliptic sesquilinear forms. Here we call a sesquilinear form a_ι on E_ι <u>strongly elliptic</u> if there exist a positive constant γ_ι and a compact strictly positive sesquilinear form k_ι on E_ι with the property

$$(14)\qquad \gamma_\iota \|\varphi\|^2_{E_\iota} \le \operatorname{Re} a_\iota(\varphi) + k_\iota(\varphi), \quad \varphi \in E_\iota .$$

Correspondingly, the sequence a_ι on $E_\iota, \iota=0,1,2,\cdots$, is called <u>uniformly strongly elliptic</u> if there exist a positive constant γ and a compact strictly positive sesquilinear form k on E with the property

$$(15)\qquad \gamma \|\varphi\|^2_{E_\iota} \le \operatorname{Re} a_\iota(\varphi) + k(\varphi), \quad \varphi \in E_\iota,\ \iota=0,1,2,\cdots .$$

The restriction $k_\iota = k|E_\iota$ of k to E_ι is compact and strictly positive so that condition (15) implies the strong ellipticity of a_ι for each $\iota=0,1,2,\cdots$.

We now show strong and weak discrete convergence of the solutions of the uniformly strongly elliptic inhomogeneous equations (1) resp. (3).

(16) <u>Let</u> E_0 <u>be separable, let</u> A_0 <u>be injective and let the sequence</u> a_ι <u>on</u> $E_\iota, \iota=0,1,2,\cdots$, <u>be uniformly strongly elliptic. Then</u> A_0^{-1} <u>exists in</u> $\mathcal{B}(E_0)$, <u>the consistency condition</u> (c) <u>is necessary and sufficient for the existence of</u> A_ι^{-1} <u>in</u> $\mathcal{B}(E_\iota)$ <u>for almost all</u> ι <u>and for the discrete convergence of the solutions</u>

$$(17)\qquad u_\iota = A_\iota^{-1} v_\iota \longrightarrow u_0 = A_0^{-1} v_0 \quad (\iota\to\infty)$$

<u>for every sequence</u> $v_\iota \in E_\iota, v_\iota \to v_0$ <u>resp.</u> $\ell_\iota \in E'_\iota, \ell_\iota \to \ell_0 (\iota\to\infty)$. <u>The adjoint consistency</u>

condition (c^*) is necessary and sufficient for the existence of A_ι^{-1} in $\mathcal{B}(E_\iota)$ for almost all ι and for the weak discrete convergence of the solutions

(18) $$u_\iota = A_\iota^{-1} v_\iota \rightharpoonup u_o = A_o^{-1} v_o \quad (\iota \to \infty)$$

for every sequence $v_\iota \in E_\iota, v_\iota \to v_o$ resp. $\ell_\iota \in E'_\iota, \ell_\iota \rightharpoonup \ell_o \quad (\iota \to \infty)$.

Proof. Strongly elliptic sesquilinear forms a_ι on E_ι and the corresponding operators A_ι in E_ι, $\iota = 0,1,2,\cdots$, are Fredholm with index 0 (s.[14], p.48). Then A_o injective implies A_o surjective and the existence of A_o^{-1} and $(A_o^*)^{-1}$ in $\mathcal{B}(E_o)$. Now the assertions follow from [15], Theorem 3.2.(5). Note that there exist compact strictly positive linear operators K in E and $K_\iota = R_\iota K | E_\iota$ in E_ι with the property

(19) $$k(u,v) = (u, Kv)_E = (u, K_\iota v)_{E_\iota} = k_\iota(u,v), \quad u,v \in E_\iota, \quad \iota = 0,1,2,\cdots.$$

The sequence $(K_\iota) = (K_\iota^*)$ is weakly discretely compact since for every weakly convergent null sequence $w_\iota \rightharpoonup 0, w_\iota \in E_\iota$, one has $w_\iota \underset{E}{\rightharpoonup} 0$ and thus $Kw_\iota \underset{E}{\to} 0$ or

(20) $$\|K_\iota w_\iota\|_{E_\iota} = \|R_\iota K w_\iota\|_{E_\iota} \le \mu \|K w_\iota\|_E \to 0 .$$

The uniform ellipticity condition (15) implies the inequalities

$$\gamma \|\varphi\|_{E_\iota} \le \|A_\iota^* \varphi\|_{E_\iota} + \|K_\iota \varphi\|_{E_\iota}, \quad \gamma \|\varphi\|_{E_\iota} \le \|A_\iota \varphi\|_{E_\iota} + \|K_\iota \varphi\|_{E_\iota}, \quad \varphi \in E_\iota,$$

for $\iota = 0,1,2,\cdots$. Then, by [15], Theorem 3.2.(5), the consistency condition (c) is sufficient for the existence of A_ι^{-1} for almost all ι and for the discrete convergence $A_\iota^{-1} \to A_o^{-1} (\iota \to \infty)$, i.e. for the convergence statement (17). Correspondingly, the adjoint consistency condition (c^*) establishes the existence of $(A_\iota^*)^{-1} = (A_\iota^{-1})^*$ for almost all ι and the discrete convergence $(A_\iota^{-1})^* \to (A_o^{-1})^* (\iota \to \infty)$ so that the convergence statement (13) and thus (12) resp. (18) are demonstrated. Finally, one obtains the necessity of the consistency condition (c) resp. (c^*) just as in Theorem (8) resp. Theorem (11).

3. Eigenvalue Problems

This section deals with general, not necessarily Hermitian eigenvalue problems, defined by sequences of bounded sesquilinear forms a_ι, b_ι on E_ι in the form

(1) $$a_\iota(\varphi, w_\iota) = \bar{\lambda}_\iota \, b_\iota(\varphi, w_\iota), \quad \varphi \in E_\iota, \quad \iota = 0,1,2,\cdots.$$

We now assume $E, E_\iota, \iota = 0,1,2,\cdots$, to be complex Hilbert spaces. Together with the eigenvalue problems (1), we consider the sequence of inhomogeneous equations

(2) $$a_\iota(\varphi, u_\iota) - \bar{z}\, b_\iota(\varphi, u_\iota) = \ell_\iota(\varphi), \quad \varphi \in E_\iota, \quad \iota = 0,1,2,\cdots,$$

with complex numbers $z \in \mathbb{C}$ and bounded linear functionals $\ell_\iota \in E'_\iota$. According to the representation 2.(2), there are bounded linear operators $A_\iota, B_\iota \in \mathcal{B}(E_\iota)$ and elements $v_\iota \in E_\iota$ such that the eigenvalue problem (1) can be written

(3) $$A_\iota w_\iota = \lambda_\iota B_\iota w_\iota, \quad \iota = 0,1,2,\cdots,$$

analogously, the inhomogeneous equations (2) can be expressed in the form

(4) $$A_\iota u_\iota - z\, B_\iota u_\iota = v_\iota, \quad \iota = 0,1,2,\cdots.$$

Then the <u>resolvent set</u> $P(a_\iota, b_\iota) = P(A_\iota, B_\iota)$ is the set of all complex numbers $z \in \mathbb{C}$ such that the inverse operator $A_\iota(z)^{-1} = (A_\iota - zB_\iota)^{-1}$ exists in $\mathcal{B}(E_\iota)$. Hence z is in $P(a_\iota, b_\iota)$ if and only if the inhomogeneous equations (2) resp. (4) are uniquely and continuously solvable for every $\ell_\iota \in E'_\iota$ resp. $v_\iota \in E_\iota$. The complementary set in $\mathbb{C}$ of the resolvent set is called the <u>spectrum</u> $\Sigma(a_\iota, b_\iota) = \Sigma(A_\iota, B_\iota)$ of a_ι, b_ι resp. A_ι, B_ι.

In the following, a pair a_ι, b_ι of sesquilinear forms on E_ι is said to be <u>strongly definite</u> if

(5) $$|\mathrm{Re}\, a_\iota(\varphi)| + |\mathrm{Re}\, b_\iota(\varphi)| > 0, \quad 0 \neq \varphi \in E_\iota.$$

Our first Theorem now is basic in the study of strongly elliptic eigenvalue problems.

(6) <u>Let a_ι, b_ι be a pair of bounded sesquilinear forms on E_ι, let a_ι be strongly elliptic, b_ι compact and let the pair a_ι, b_ι be strongly definite. Then $a_\iota - \bar{z} b_\iota$ resp. $A_\iota - zB_\iota$ is Fredholm with index 0 for every $z \in \mathbb{C}$ and the spectrum $\Sigma(a_\iota, b_\iota)$ consists of a countable set, having no finite accumulation point in $\mathbb{C}$, of eigenvalues with finite multiplicities.</u>

<u>Proof.</u> According to the proof of [14], Theorem II-2.2.(14) (p.151), one concludes from condition (5) the existence of a strictly positive sesquilinear form q_ι on E_ι,

(7) $$q_\iota(\varphi) = \alpha_\iota \mathrm{Re}\, a_\iota(\varphi) + \beta_\iota \mathrm{Re}\, b_\iota(\varphi) > 0, \quad 0 \neq \varphi \in E_\iota,$$

with some real constants $\alpha_\iota, \beta_\iota$. In view of [14], Theorem I-2.3.(9)(p.36), the compact sesquilinear form k_ι on E_ι in the ellipticity condition 2.(14) satisfies, for every $\varepsilon > 0$, the inequality

$$k_\iota(\varphi) \leq \varepsilon \|\varphi\|^2_{E_\iota} + \eta_\iota(\varepsilon)\, q_\iota(\varphi), \quad \varphi \in E_\iota,$$

with some constant $\eta_\iota(\varepsilon) > 0$. On setting $\varepsilon = \gamma_\iota/2$ and $\eta_\iota > \eta_\iota(\gamma/2)$, we get from the ellipticity condition 2.(14),

$$\frac{1}{2}\gamma_\iota \|\varphi\|^2_{E_\iota} \le (1+\alpha_\iota\eta_\iota) \operatorname{Re} a_\iota(\varphi) + \beta_\iota\eta_\iota \operatorname{Re} b_\iota(\varphi), \quad \varphi \in E_\iota .$$

For a suitable η_ι we have $1+\alpha_\iota\eta_\iota \neq 0$ and hence the inequality

(8) $$\gamma'_\iota \|\varphi\|^2_{E_\iota} \le |a_\iota(\varphi) - \mu_\iota b_\iota(\varphi)|, \quad \varphi \in E_\iota ,$$

with the constants $2\gamma'_\iota = \gamma_\iota/|1+\alpha_\iota\eta_\iota|$, $\mu_\iota = -\beta_\iota\eta_\iota/(1+\alpha_\iota\eta_\iota)$. Consequently, the inverse operator $(A_\iota - \mu_\iota B_\iota)^{-1}$ exists in $\mathcal{B}(E_\iota)$ which means that $\mu_\iota \in P(A_\iota, B_\iota)$ so that the resolvent sets of A_ι, B_ι resp. a_ι, b_ι are not empty. Hence the assertion follows from [14], Theorem I-3.1.(10) (p.54) or from [16], Theorem 1.1.(8).

The next Theorems need the notion of weakly discretely compact sequences of adjoint sesquilinear forms b^*_ι resp. of adjoint operators B^*_ι, $\iota = 1,2,\cdots$. This notion is defined by the property

(9) $$\sup_{0\neq\psi\in E_\iota} |b_\iota(w_\iota,\psi)|/\|\psi\|_{E_\iota} \to 0 \iff \|B^*_\iota w_\iota\|_{E_\iota} \to 0 \quad (\iota\to\infty)$$

for every weakly discretely convergent null sequence $w_\iota \in E_\iota$, $w_\iota \to 0$ $(\iota\to\infty)$.

In order to simplify the formulations to follow, let us now state the following assumptions.

(A) The space E_0 is separable and $E, E_\iota, \iota = 0,1,2,\cdots$, are complex Hilbert spaces satisfying (R): $R_\iota \xrightarrow[E]{} R_0$ $(\iota \to \infty)$.

(B) The sequences of bounded sesquilinear forms a_ι on $E_\iota, \iota = 0,1,2,\cdots$, and b_ι on $E_\iota, \iota = 0,1,2,\cdots$, are consistent.

(C) The sesquilinear forms b_ι are compact for each $\iota = 0,1,2,\cdots$ and the sequence $b^*_\iota, \iota = 1,2,\cdots$, is weakly discretely compact.

(D) The pairs a_ι, b_ι are strongly definite on E_ι for each $\iota = 0,1,2,\cdots$.

(E) The sequence a_ι on $E_\iota, \iota = 0,1,2,\cdots$, is uniformly strongly elliptic.

Under the assumptions (A),...,(E) we then obtain the following Theorem which shows the convergence of the resolvent sets $P(a_\iota, b_\iota)$ and the discrete convergence of the resolvents $A_\iota(z)^{-1} \to A_0(z)^{-1}$ and thus of the solutions $u_\iota \to u_0 (\iota\to\infty)$ of the equations (2), (4).

(10) For every compact subset $\Gamma \subseteq P(a_0, b_0)$, it follows that $\Gamma \subseteq P(a_\iota, b_\iota)$ for almost all $\iota = 1,2,\cdots$. The condition $z \in P(a_0, b_0)$ is necessary and sufficient for the existence of

$A_\iota(z)^{-1}$ in $\mathcal{B}(E_\iota)$ for almost all ι and for the discrete convergence of the solutions

(11) $$u_\iota = A_\iota(z)^{-1} v_\iota \to u_o = A_o(z)^{-1} v_o \quad (\iota \to \infty)$$

for every sequence $v_\iota \in E_\iota, v_\iota \to v_o$ resp. $\ell_\iota \in E'_\iota, \ell_\iota \to \ell_o \ (\iota \to \infty)$.

Proof. The assertion will be proved by using [16], Theorem 2.2.(4). First we conclude from Theorem (6) that the resolvent sets $P(a_\iota, b_\iota)$ are not empty for $\iota = 0,1,2,\cdots$. By assumption (C), the sesquilinear forms b_ι on E_ι and thus the associated operators $B_\iota \in \mathcal{B}(E_\iota)$ are compact. By [15], Theorem 3.1.(2), the sequence $b_\iota, \iota = 1,2,\cdots$, is stable. For every complex number $z \in \mathbb{C}$ we have the representation

$$a_\iota(\varphi) = (\varphi, A_\iota \varphi)_{E_\iota} = (\varphi, A_\iota(z)\varphi)_{E_\iota} + (\varphi, z B_\iota)_{E_\iota}, \quad \varphi \in E_\iota, \ \iota = 0,1,2,\cdots .$$

Then the ellipticity condition 2.(15) and the representation 2.(19) of k give the inequalities

$$\gamma \, \|\varphi\|_{E_\iota} \leq \|A_\iota(z)^* \varphi\|_{E_\iota} + \|(K_\iota + \bar{z} B_\iota^*)\varphi\|_{E_\iota}, \quad \varphi \in E_\iota, \ \iota = 1,2,\cdots .$$

According to hypothesis (B), the sequences of operators $A_\iota, \iota = 0,1,2,\cdots$, and $B_\iota, \iota = 0,1,2,\cdots$, are consistent. By assumption (C) respectively by 2.(20) the sequences $(B_\iota^*), (K_\iota)$ are weakly discretely compact. Consequently, by [16], Theorem 2.2.(4), we obtain

(12) $$P(A_o, B_o) = \Delta_s = \Delta_b \neq \emptyset$$

which means, that for every $z \in P(a_o, b_o) = P(A_o, B_o)$ the inverse operators $A_\iota(z)^{-1}$ exist for almost all $\iota = 1,2,\cdots$ and converge discretely,

(13) $$A_\iota(z)^{-1} \to A_o(z)^{-1} \quad (\iota \to \infty), \quad z \in P(A_o, B_o).$$

This proves the convergence statement (11). Finally, the first assertion concerning the convergence of the resolvent sets follows from [16], Theorem 3.2.(1).

In view of Theorem (6), the eigenvalue problems (1) resp. (3) have discrete spectra of eigenvalues of finite multiplicities, with no finite accumulation point. Our next Theorem establishes the convergence of spectra and algebraic eigenspaces under the assumptions (A),...,(E).

(14) Let λ_o be an eigenvalue of a_o, b_o with the algebraic multiplicity m and let $w_o^{(1)}, \cdots, w_o^{(m)}$ be a basis of the associated algebraic eigenspace. Let $\mathcal{U}$ be any compact neighbourhood of λ in $\mathbb{C}$ and $\Sigma(a_o, b_o) \cap \mathcal{U} = \{\lambda\}$. Then there are for almost all $\iota = 1,2,\cdots$ exactly m eigen-

values $\lambda_\iota^{(1)},\dots,\lambda_\iota^{(m)}$, counted repeatedly according to their algebraic multiplicities, and linearly independent vectors $w_\iota^{(1)},\dots,w_\iota^{(m)}$ in the sum of the algebraic eigenspaces of $\lambda_\iota^{(1)},\dots,\lambda_\iota^{(m)}$ with the properties

$$\Sigma(a_\iota,b_\iota)\cap\mathcal{U}=\{\lambda_\iota^{(1)},\dots,\lambda_\iota^{(m)}\}$$

and

$$\lambda_\iota^{(k)}\to\lambda_0\,,\qquad w_\iota^{(k)}\to w_0^{(k)}\qquad(\iota\to\infty),\qquad k=1,\dots,m.$$

Proof. The assertion is an immediate consequence of [16], Theorem 3.2.(8). By assumption (B), the sequences of operators $A_\iota,\ \iota=0,1,2,\cdots,$ and $\mathcal{B}_\iota,\ \iota=0,1,2,\cdots,$ are consistent. In view of (12) we have $P(a_0,b_0)=\Delta_b\neq\emptyset$. From assumption (C) we will conclude in part (ii) below that the sequence of compact operators $\mathcal{B}_\iota,\iota=1,2,\cdots,$ is discretely compact. Hence the assumptions of [16], Theorem 3.2.(8) are satisfied so that the above assertion follows.
(ii) We now prove that the sequence of operators $\mathcal{B}_\iota,\iota=1,2,\cdots$, is discretely compact (s.[15], p.66). First we note that the sequence $(\mathcal{B}_\iota)$ is stable. Thus for every bounded sequence $u_\iota\in E_\iota, \|u_\iota\|_{E_\iota}\le\beta,\iota=1,2,\cdots,$ the sequence $w_\iota=\mathcal{B}_\iota u_\iota$ is bounded in E by $\|w_\iota\|_E\le$ $\le\mu\|w_\iota\|_{E_\iota}\le\mu'$. This sequence has a weakly convergent subsequence $w_\iota\underset{E}{\rightharpoonup} w\ (\iota\to\infty,\iota\in N'\subseteq N)$ in the Hilbert space E. The weak limit w lies in E_0 because assumption (R) implies by Theorem 1.(20) the condition (R2). Since $R_\iota w\to w$, it follows that $w_\iota-R_\iota w\rightharpoonup 0\ (\iota\to\infty,\iota\in N')$. The weak discrete compactness of (b_ι^*) resp. $(\mathcal{B}_\iota^*)$ then implies the convergence of

$$(w_\iota-R_\iota w,\,w_\iota)_{E_\iota}=(\mathcal{B}_\iota^*(w_\iota-R_\iota w),\,u_\iota)_{E_\iota}\to 0\qquad(\iota\to\infty,\iota\in N').$$

Consequently, we obtain

$$\|w_\iota-R_\iota w\|^2_{E_\iota}=(w_\iota-R_\iota w,\,w_\iota)_{E_\iota}-(w_\iota-R_\iota w,\,w)_E\to 0\quad(\iota\to\infty,\iota\in N').$$

In view of Theorem 1.(7), this shows the discrete convergence of $w_\iota=\mathcal{B}_\iota u_\iota\to w$ $(\iota\to\infty,\iota\in N')$ thus proving the discrete compactness of the sequence $\mathcal{B}_\iota$ on $E_\iota,\ \iota=1,2,\cdots$.

Let us now consider Hermitian sesquilinear forms $a_\iota,\ b_\iota$ on $E_\iota,\ \iota=0,1,2,\cdots$. In this case, the associated quadratic forms are real. A strongly definite pair a_ι,b_ι satisfies the inequality (7), i.e.

(15) $$\alpha_\iota a_\iota(\varphi)+\beta_\iota b_\iota(\varphi)>0\,,\qquad 0\neq\varphi\in E_\iota,\quad \iota=0,1,2,\cdots,$$

with some real constants α_ι,β_ι. By means of the natural isometric isomorphisms $\mathfrak{J}_\iota$ of E_ι onto E_ι', we obtain $\mathfrak{J}_\iota$-definite pairs a_ι,b_ι resp. $A_\iota,\mathcal{B}_\iota$ in the sense of [16], 1.3.

Consequently, all eigenvalues λ_ι of a_ι, b_ι are real and the associated algebraic eigenspaces coincide with the geometric eigenspaces $N(A_\iota - \lambda_\iota B_\iota)$. By Theorem (10), each number $\alpha \in P(a_0, b_0)$ lies in $P(a_\iota, b_\iota)$ for almost all $\iota = 1,2,\cdots$. In this case, the eigenvalues of a_ι, b_ι can be arranged in ascending order,

(16) $$\cdots \leq \lambda_\iota^{(-2)} \leq \lambda_\iota^{(-1)} < \alpha \leq \lambda_\iota^{(0)} \leq \lambda_\iota^{(1)} \leq \lambda_\iota^{(2)} \leq \cdots, \quad \iota = 0,1,2,\cdots,$$

where each eigenvalue is counted repeatedly according to its multiplicity. Under the assumptions (A),$\cdots$,(E), one obtains from Theorem (14) the convergence of the ordered sequence of eigenvalues

(17) $$\lambda_\iota^{(k)} \to \lambda_0^{(k)} \quad (\iota \to \infty)$$

for $k = 0, \pm 1, \pm 2, \cdots$ as far as eigenvalues of a_0, b_0 exist (s. [16], Theorem 3.2.(13)).

Finally, we state a very convenient criterion for the weak discrete compactness (9) of sequences of adjoint sesquilinear forms. Here we assume only the validity of the condition (R).

(18) <u>Let b_ι on E_ι, $\iota = 0,1,2,\cdots$, be a consistent sequence of bounded sesquilinear forms and let k on E be a compact strictly positive sesquilinear form. Then the sequence of adjoint forms b_ι^*, $\iota = 1,2,\cdots$, is weakly discretely compact if and only if for every $\varepsilon > 0$ there exist positive numbers $\eta(\varepsilon)$ and $\nu(\varepsilon)$ with the property</u>

(19) $$|b_\iota(\varphi)| \leq \varepsilon \|\varphi\|_{E_\iota}^2 + \eta(\varepsilon) k(\varphi), \quad \varphi \in E_\iota, \quad \iota \geq \nu(\varepsilon).$$

<u>Proof.</u> (i) Note that (19) implies the stability of the sequence b_ι on E_ι, $\iota = 0,1,2,\cdots$. Let (w_ι) be any weakly discretely convergent null sequence $w_\iota \in E_\iota$, $w_\iota \rightharpoonup 0$ $(\iota \to \infty)$. By 1.(9), then $w_\iota \underset{E}{\rightharpoonup} 0$ and thus $k(w_\iota) \to 0$ $(\iota \to \infty)$. In particular, the sequence (w_ι) is bounded, $\|w_\iota\|_{E_\iota} \leq \beta$, $\iota = 1,2,\cdots$. Hence, from (19) we obtain the estimate $\varlimsup_{\iota \to \infty} |b_\iota(w_\iota)| \leq \varepsilon\beta$ for every $\varepsilon > 0$ so that $b_\iota(w_\iota) \to 0$ $(\iota \to \infty)$. For every pair of null sequences $w_\iota \rightharpoonup 0$, $v_\iota \rightharpoonup 0$ one has the convergence of $w_\iota \pm v_\iota \rightharpoonup 0$, $w_\iota \pm i v_\iota \rightharpoonup 0$ and thus, by a well-known formula, the convergence of

(20) $$b_\iota(w_\iota, v_\iota) = \frac{1}{4}\{b_\iota(w_\iota + v_\iota) - b_\iota(w_\iota - v_\iota) + i b_\iota(w_\iota + i v_\iota) - i b_\iota(w_\iota - i v_\iota)\} \to 0 \quad (\iota \to \infty).$$

The associated operators $B_\iota \in \mathcal{B}(E_\iota)$, $\iota = 0,1,2,\cdots$, of a stable and consistent sequence (b_ι) converge discretely, according to 2.(7), $B_\iota \to B_0$ $(\iota \to \infty)$, so that the adjoints converge weakly discretely, $B_\iota^* \rightharpoonup B_0^*$ $(\iota \to \infty)$ (s. [15], 2.2.(3)). Hence $w_\iota \rightharpoonup 0$ implies $v_\iota = B_\iota^* w_\iota \rightharpoonup 0$ and, using (20),

$$\| \mathcal{B}_\iota^* w_\iota \|_{E_\iota}^2 = (\mathcal{B}_\iota^* w_\iota, v_\iota)_{E_\iota} = b_\iota(w_\iota, v_\iota) \to 0 \qquad (\iota \to \infty).$$

This verifies the weak discrete compactness of the sequence b_ι^* on E_ι, $\iota = 1,2,\cdots$.

(ii) Conversely, the weak discrete compactness of (b_ι^*) leads to the inequalities (19). If this is false, then for some $\varepsilon_0 > 0$, for a subsequence $N' \subseteq N = (1,2,\cdots)$, for a sequence $\eta_\iota \to \infty$ $(\iota \to \infty, \iota \in N')$ and elements $w_\iota \in E_\iota$, $\|w_\iota\|_{E_\iota} = 1$, it follows

(21) $\qquad \|w_\iota\|_{E_\iota} = 1, \qquad |b_\iota(w_\iota)| > \varepsilon_0 + \eta_\iota k(w_\iota), \qquad \iota \in N'.$

The sequence (w_ι) is bounded and hence weakly compact in E so that $w \in E$ and $N'' \subseteq N'$ exist with $w_\iota \underset{E}{\rightharpoonup} w$ $(\iota \to \infty, \iota \in N'')$. Thus we have $k(w_\iota) \to k(w)$ $(\iota \to \infty, \iota \in N'')$. Using the inequalities (21), we obtain $k(w_\iota) \to 0$ $(\iota \to \infty, \iota \in N')$ because the sequence $b_\iota(w_\iota)$, $\iota \in N$, is bounded and $\eta_\iota \to \infty$ $(\iota \to \infty, \iota \in N')$. Consequently, $k(w) = 0$ which means $w = 0$ and $w_\iota \rightharpoonup 0$ $(\iota \to \infty, \iota \in N'')$. Since the weak discrete compactness of (b_ι^*) gives $|b_\iota(w_\iota)| \le \|\mathcal{B}_\iota^* w_\iota\|_{E_\iota} \to 0$ $(\iota \to \infty, \iota \in N'')$, the inequality (21) leads to the contradiction $|b_\iota(w_\iota)| > \varepsilon_0$, $\iota \in N''$.

It is easily seen, using Theorem [14], I-2.3.(6), that the above Theorem has the following Corollary.

(22) <u>If the inqualities</u> (19) <u>are valid with</u> $\nu(\varepsilon) = 1$ <u>for every</u> $\varepsilon > 0$ <u>then the sesquilinear forms</u> b_ι <u>on</u> E_ι <u>are compact for each</u> $\iota = 0,1,2,\cdots$ <u>and the sequence</u> b_ι^*, $\iota = 1,2,\cdots$, <u>is weakly discretely compact.</u>

4. Applications

This Section gives some applications of the theory established in this paper to perturbation theory in the sense of strong convergence, to Ritz-Galerkin and projection methods and to the so-called singular perturbations of elliptic sesquilinear forms.

4.1. Perturbation theory in the strong convergence.

Let E be a Hilbert space and let E_ι be equal to E with the scalar products $(.,.)_{E_\iota} = (.,.)_E$ for $\iota = 0,1,2,\cdots$. This simple special case leads to a perturbation theory for sequences of elliptic sesquilinear forms on E which can be considered as an interesting supplement to the perturbation theory in the sense of generalized strong convergence of linear operators in Stummel [17], Section 1. The results of our theory, obtained in this case, exceed essentially those of Kato [9], Chapter VIII, as far as our subject is concerned.

Here, the discrete convergence is the ordinary convergence in E. The condition (R) is trivially satisfied by $R_\iota = 1_E, \iota = 0,1,2,\cdots; 1_E$ being the identity mapping in E. For a stable sequence of operators $A_\iota \in \mathcal{B}(E), \iota = 0,1,2,\cdots$, the consistency condition (c) is equivalent as well to the strong convergence $A_\iota \to A_0 (\iota \to \infty)$ as to the conditions

$$(1) \qquad \|A_\iota \varphi - A_0 \varphi\|_E \longrightarrow 0 \iff \sup_{0 \neq \psi \in E} |a_\iota(\psi,\varphi) - a_0(\psi,\varphi)| / \|\psi\|_E \longrightarrow 0 \quad (\iota \to \infty)$$

for a dense subspace $\mathcal{D}_a \subseteq E$ and the constant sequence $\varphi_\iota = \varphi, \iota = 1,2,\cdots$. For a strongly convergent sequence of operators $\mathcal{B}_\iota \to \mathcal{B}_0 (\iota \to \infty)$, the associated sesquilinear forms b_ι on E_ι are compact for each $\iota = 0,1,2,\cdots$ and the sequence (b_ι^*) is weakly discretely compact if there is a compact strictly positive sesquilinear form k on E and, for every $\varepsilon > 0$, a positive number $\eta(\varepsilon) > 0$ with the property

$$(2) \qquad |b_\iota(\varphi)| \le \varepsilon \|\varphi\|_E^2 + \eta(\varepsilon) k(\varphi), \qquad \varphi \in E, \quad \iota = 0,1,2,\cdots.$$

4.2. <u>Projection methods.</u> Let E be a Hilbert space and let $E_\iota \subseteq E$ be closed linear subspaces with the scalar products $(.,.)_{E_\iota} = (.,.)_E$ for $\iota = 0,1,2,\cdots$. Thus we obtain a perturbation theory of elliptic sesquilinear forms in the framework of projection methods. Important examples are the well-known Ritz-Galerkin methods (s. Grigorieff [5], Stummel [17], Section 2, Vainikko [18]). Another example of projection methods will be given in Section 5 below by the perturbation theory of the Dirichlet problems in case $m = \ell$.

Here again, the discrete convergence is the ordinary convergence in E. The operators R_ι are the orthogonal projections $P_\iota : E \mapsto E_\iota, \iota = 0,1,2,\cdots$, defined by 1.(2), and the condition (R) requires the strong convergence $P_\iota \xrightarrow[E]{} P_0 (\iota \to \infty)$. For operators $A, \mathcal{B} \in \mathcal{B}(E)$ one easily obtains approximating sequences of operators $A_\iota, \mathcal{B}_\iota \in \mathcal{B}(E_\iota)$ on setting $A_\iota = P_\iota A | E_\iota$, $\mathcal{B}_\iota = P_\iota \mathcal{B} | E_\iota, \iota = 0,1,2,\cdots$. Under the assumption $P_\iota \to P_0 (\iota \to \infty)$, the sequences $A_\iota, \iota = 0,1,2,\cdots$, and $\mathcal{B}_\iota$, $\iota = 0,1,2,\cdots$, are stable and consistent. If $\mathcal{B}$ is compact then the operators $\mathcal{B}_\iota$ are compact for each $\iota = 0,1,2,\cdots$, the sequence $(\mathcal{B}_\iota^*)$ is weakly discretely compact and the sequence $(\mathcal{B}_\iota)$ is discretely compact (s. [15], p.86, p.66). An operator $A \in \mathcal{B}(E)$ or the associated sesquilinear form a on E is strongly elliptic on E, if there are a positive constant γ and a compact strictly positive sesquilinear form k on E with the property

(3) $$\gamma\|\varphi\|_E^2 \leq \operatorname{Re} a(\varphi) + k(\varphi), \quad \varphi \in E.$$

Obviously, then the sequence $a_\iota = a|E_\iota$ on E_ι and the sequence of associated operators $A_\iota = P_\iota A|E_\iota, \iota = 0,1,2,\cdots$, are uniformly strongly elliptic.

4.3. Singular perturbation theory. Friedman, Greenlee, Huet, Moser, O'Malley and others have studied so-called singular perturbations of strongly elliptic sesquilinear forms which easily can be treated by the perturbation theory developed in this paper. Moreover, we get some important new results, in particular, we obtain the extension of convergence theorems from elliptic sesquilinear forms in the sense of Lions to strongly elliptic forms in the sense of Gårding and from special Hermitian to general not necessarily Hermitian eigenvalue problems. For special examples of singular perturbations in elliptic differential equations we refer to the literature mentioned and to Section 5 below.

Let $E = E_0$ be a Hilbert space with the scalar product $(.,.)_E = (.,.)_{E_0}$ and let D be a dense subspace of E which is a Hilbert space with scalar product $(.,.)_D$. We assume $\|u\|_E \leq \beta\|u\|_D, u \in D$, so that every null sequence of positive numbers $x_\iota \to 0$ $(\iota\to\infty)$ defines a sequence of Hilbert spaces E_ι by

(4) $$E_\iota = D, \quad (u,v)_{E_\iota} = (u,v)_E + x_\iota^2 (u,v)_D, \quad u,v \in D, \iota = 1,2,\cdots.$$

The discrete convergence 1.(6) of a sequence $u_\iota \in E_\iota = D, \iota = 1,2,\cdots$, to an element $u \in E_0 = E$ then becomes equivalent to

(5) $$u_\iota \to u \ (\iota\to\infty) \iff u_\iota \underset{E}{\to} u, \quad x_\iota u_\iota \underset{D}{\to} 0 \quad (\iota\to\infty),$$

and the weak discrete convergence 1.(9) is equivalent to

(6) $$u_\iota \rightharpoonup u \ (\iota\to\infty) \iff u_\iota \underset{E}{\rightharpoonup} u \ (\iota\to\infty), \ \|x_\iota u_\iota\|_D, \ \iota = 1,2,\cdots, \text{ bounded.}$$

In particular, one has the discrete convergence $\varphi_\iota \to \varphi \ (\iota\to\infty)$ for the constant sequence $\varphi_\iota = \varphi$ for every $\varphi \in D$. Next we show that the assumption (R) is satisfied with $R_0 = 1_E$.

(7) $R_\iota v \underset{E}{\to} v \ (\iota\to\infty)$ for every $v \in E$.

Proof. Obviously, $\|u\|_E \leq \|u\|_{E_\iota}, u \in D, \iota = 1,2,\cdots$, so that the assumption 1.(1) is true with the constant $\mu = 1$. In view of the inequalities 1.(4), the sequence $R_\iota v_\iota, \iota = 1,2,\cdots$, is bounded in E by $\|R_\iota v\|_E \le \|R_\iota v\|_{E_\iota} \le \|v\|_E$. Hence we obtain from 1.(2), (4) for every $\varphi \in D$ the convergence of

$$|(\varphi, v - R_\iota v)_E| = \varkappa_\iota^2 |(\varphi, R_\iota v)_D| \le \varkappa_\iota \|\varphi\|_D \|v\|_E \to 0 \quad (\iota \to \infty).$$

This shows the weak convergence of $R_\iota v \underset{E}{\rightharpoonup} v \ (\iota \to \infty)$. Thus $\|R_\iota v\|^2_{E_\iota} = (R_\iota v, v)_E \to \|v\|_E^2 \ (\iota \to \infty)$ and finally

$$\|v - R_\iota v\|_E^2 \le \|v\|_E^2 + \|R_\iota v\|^2_{E_\iota} - 2\,\mathrm{Re}(v, R_\iota v)_E \to 0 \qquad (\iota \to \infty).$$

Bounded sesquilinear forms $a = a_0$ on E and c on D define sequences of special sesquilinear forms a_ι on E_ι in the form

(8) $$a_\iota(\varphi, \psi) = a(\varphi, \psi) + \varkappa_\iota^2 c(\varphi, \psi), \quad \varphi, \psi \in E_\iota, \ \iota = 0,1,2,\cdots, \quad \varkappa_0 = 0.$$

These sequences a_ι on $E_\iota, \iota = 0,1,2,\cdots$, satisfy the stability condition (b) with the constant $\alpha_1 = \|a\|_E + \|c\|_D$. Furthermore, we can state the consistency condition (c) in the form

(9) $$\sup_{0 \ne \psi \in D} |a_\iota(\psi, \varphi_\iota) - a(\psi, \varphi)| / \|\psi\|_{E_\iota} \le \varkappa_\iota \|c\|_D \|\varphi\|_D \to 0 \quad (\iota \to \infty)$$

for every $\varphi \in D_a = D$ with the constant sequence $\varphi_\iota = \varphi, \iota = 1,2,\cdots$.

(10) <u>If $a = a_0$ on E and c on D are strongly elliptic then the sequence</u> $a_\iota = a + \varkappa_\iota^2 c$ <u>on $E_\iota, \iota = 0,1,2,\cdots$, is uniformly strongly elliptic.</u>

<u>Proof.</u> From our definition 2.(14) of strong ellipticity we obtain a positive constant γ_0 and compact strictly positive sesquilinear forms k_E on E, k_D on D with the properties

$$\gamma_0 \|\varphi\|_E^2 \le \mathrm{Re}\, a(\varphi) + k_E(\varphi), \quad \varphi \in E, \quad \gamma_0 \|\varphi\|_D^2 \le \mathrm{Re}\, c(\varphi) + k_D(\varphi), \quad \varphi \in D.$$

Making use of Theorem I - 2.3.(9) in [14] (p.36) we have the following inequality

$$k_D(\varphi) \le \varepsilon \|\varphi\|_D^2 + \eta(\varepsilon) k_E(\varphi), \quad \varphi \in D,$$

for every $\varepsilon > 0$ with a positive constant $\eta(\varepsilon)$. On setting $\varepsilon = \gamma = \gamma_0/2$ and $k = (1 + \varkappa^2 \eta(\gamma)) k_E$, $\varkappa = \max \varkappa_\iota$, we infer the uniform strong ellipticity condition

$$\gamma \|\varphi\|^2_{E_\iota} \le \mathrm{Re}\, a_\iota(\varphi) + k(\varphi), \quad \varphi \in E_\iota, \quad \iota = 0,1,2,\cdots.$$

Now, let $b = b_0$ on E and d on D be compact sesquilinear forms and denote by b_ι on E_ι the sequence

(11) $$b_\iota(\varphi, \psi) = b(\varphi, \psi) + \varkappa_\iota^2 d(\varphi, \psi), \quad \varphi, \psi \in E_\iota, \ \iota = 0,1,2,\cdots.$$

Like a_ι, the sequence b_ι on $E_\iota, \iota=0,1,2,\cdots$, is consistent. Additionally, we state the following properties.

(12) <u>The sesquilinear forms</u> $b_\iota = b + x_\iota^2 d$ <u>on</u> E_ι <u>are compact for each</u> $\iota = 0,1,2,\cdots$ <u>and the sequence</u> $b_\iota^*, \iota=1,2,\cdots$, <u>is weakly discretely compact.</u>

<u>Proof.</u> Let k_E be a compact strictly positive sesquilinear form on E. Then the following inequalities are valid

$$|b(\varphi)| \le \varepsilon \|\varphi\|_E^2 + \eta(\varepsilon) k_E(\varphi),\ \varphi \in E, \quad |d(\varphi)| \le \varepsilon \|\varphi\|_D^2 + \eta(\varepsilon) k_E(\varphi),\ \varphi \in D,$$

for every $\varepsilon > 0$ with a positive constant $\eta(\varepsilon)$ (s.Theorem I-2.3.(6) in [14],p.36). On setting $k = (1+x^2)k_E$, we thus obtain

$$|b_\iota(\varphi)| \le \varepsilon \|\varphi\|_{E_\iota}^2 + \eta(\varepsilon) k(\varphi), \quad \varphi \in E_\iota, \quad \iota = 0,1,2,\cdots.$$

Hence the assertion follows from Theorem 3.(18) or Corollary 3.(22).

Finally, let E be separable, let E, E_ι be complex Hilbert spaces and let $a_\iota = a + x_\iota^2 c$, $b_\iota = b + x_\iota^2 d$ on E_ι be pairs of strongly definite sesquilinear forms, i.e.

$$|\mathrm{Re}\, a_\iota(\varphi)| + |\mathrm{Re}\, b_\iota(\varphi)| > 0, \quad 0 \neq \varphi \in E_\iota, \quad \iota = 0,1,2,\cdots.$$

Then the assumptions of this Section 4.3 imply the assumptions (A),$\cdots$,(E) of Section 3. Thus the assertions of the Theorems 3.(10) and 3.(14) are valid such that the convergence of the resolvents and resolvent sets and the convergence of eigenvalues and associated algebraic eigenspaces of a_ι, b_ι is established.

5. Perturbation Theory of Dirichlet Problems

Our perturbation theory of strongly elliptic Dirichlet problems of order $2m$ is concerned with perturbations of the coefficients and the inhomogeneous terms, the order of the equations and the domains of definition. Our results exceed considerably for example those of Babuška [2], Greenlee [4], Huet [7], Nečas [12], by the treatment of very general perturbations of the domain of definition, by the treatment of singular perturbations of strongly $\mathring{H}^{m_\iota}(\Omega_\iota)$-elliptic sesquilinear forms in the sense of Gårding instead of the $\mathring{H}^{m_\iota}(\Omega_\iota)$-elliptic forms in the sense of Lions and by the treatement of general, not necessarily Hermitian eigenvalue problems. In this Section, we state sufficient conditions (A'),$\cdots$,(E') for the validity of the conditions (A),$\cdots$,(E) of

Section 3. Hence, under the assumptions (A'),⋯,(E') the assertions of the general convergence theorems 2.(16), 3.(10) and 3.(14) are valid.

Let $M \subseteq \mathbb{R}^n$ be a bounded open interval of $\mathbb{R}^n$ and let $H^0(M)=L^2(M)$ be the Hilbert space of all complex-valued functions on M that are square integrable in the sense of Lebesgue, with the well-known scalar product $(.,.)_0$ and the norm $\|\cdot\|_0$. Let $H^s = H^s(M), s \geq 1$, be the associated Sobolew spaces (s. AGMON [1]) with the scalar products

(1) $$(u,v)_s = \sum_{|\sigma| \leq s} (\partial^\sigma u, \partial^\sigma v)_0, \quad u,v \in H^s(M).$$

For any open subset $\Omega \subseteq M$ we denote by $H_0^s(\Omega)$ the closure in $H^s(M)$ of the subspace $C_0^\infty(\Omega)$ of test functions with compact support in Ω.

We consider in the following an infinite sequence of open subsets $\Omega_\iota \subseteq M$, a sequence m_ι of natural numbers, $m \leq m_\iota \leq \ell$, and the associated spaces

(2) $$E = H^m(M), \quad E_\iota = H_0^{m_\iota}(\Omega_\iota), \quad \iota = 0,1,2,\cdots, \quad m_0 = m.$$

Thus E, E_0 are separable complex Hilbert spaces with the scalar product $(.,.)_E = (.,.)_m$. Let $\varkappa_\iota \to 0 \ (\iota \to \infty)$ be any sequence of nonnegative numbers such that the spaces $E_\iota = H_0^{m_\iota}(\Omega_\iota)$ become Hilbert spaces with the scalar products

(3) $$(u,v)_{E_\iota} = (u,v)_m + \varkappa_\iota^2 (u,v)_{m_\iota}, \quad u,v \in H_0^{m_\iota}(\Omega_\iota), \ \iota = 1,2,\cdots.$$

Now we formulate the assumption (A') concerning the behaviour of the sequence of open subsets $\Omega_\iota \subseteq M$.

(A') <u>The open set</u> $\Omega_0 \subseteq M$ <u>has the segment property. For every compact subset</u> $\Gamma \subseteq \Omega_0$ <u>it follows</u> $\Gamma \subseteq \Omega_\iota$ <u>for almost all</u> $\iota = 1,2,\cdots$ <u>and the Lebesgue measure of the following sets tends to zero,</u>

(4) $$\text{mes}\,(\Omega_\iota \cap \complement\, \Omega_0) \to 0 \quad (\iota \to \infty).$$

Then we can show the following Theorem.

(5) <u>Under the assumption</u> (A'), <u>the convergence condition</u> (R) <u>is valid,</u> $R_\iota v \to R_0 v \ (\iota \to \infty)$ <u>for all</u> v <u>in</u> $H^m(M)$.

<u>Proof.</u> The assertion will be proved by means of Corollary 1.(22). Obviously, we have $\|u\|_E \leq \|u\|_{E_\iota}$ for all $u \in H_0^{m_\iota}(\Omega_\iota), \iota = 1,2,\cdots$, such that condition (R3') is satisfied. Every test function $\varphi \in C_0^\infty(\Omega_0)$ has compact support Γ in Ω_0. Then, by assumption (A'), we get $\Gamma \subseteq \Omega_\iota$

for almost all natural numbers $\iota \in N_1 \subseteq N = (1,2,\cdots)$. Hence, a sequence $\varphi_\iota \in C_0^\infty(\Omega_\iota) \subseteq H_0^{m_\iota}(\Omega_\iota)$ is obtained on setting $\varphi_\iota = \varphi, \iota \in N_1$, and $\varphi_\iota = 0, \iota \in N - N_1$. Thus $\varphi_\iota \longrightarrow \varphi$ in H^m and

$$\|\varphi_\iota\|^2_{E_\iota} = \|\varphi\|^2_m + x_\iota^2 \|\varphi\|^2_{m_\iota} \longrightarrow \|\varphi\|^2_m \qquad (\iota \to \infty, \iota \in N_1)$$

so that condition (R1) is valid. Further, let $w_\iota \rightharpoonup w$ in H^m $(\iota \to \infty, \iota \in N')$ be any weakly convergent subsequence of functions $w_\iota \in H_0^{m_\iota}(\Omega_\iota)$, bounded by $\|w_\iota\|_{E_\iota} \le \beta, \iota \in N'$. From the compactness of the natural embedding of $H^m(M)$ into $L^2(M)$ follows the strong convergence $w_\iota \to w$ in $L^2(M)$ $(\iota \to \infty, \iota \in N')$. For every function $\psi \in L^2(M), \psi = 0$ in Ω_0, one has the equation

$$(\psi, w_\iota)_0 = \int_{\Omega_\iota \cap \complement\Omega_0} \psi(x)\overline{w_\iota(x)}\,dx = \int_{\Omega_\iota \cap \complement\Omega_0} \psi(x)\overline{w(x)}\,dx + \int_{\Omega_\iota \cap \complement\Omega_0} \psi(x)\overline{(w_\iota(x) - w(x))}\,dx.$$

Here, using (4), the first term on the right side tends to zero and, using $w_\iota \to w$ in $L^2(M)$, the second term tends to zero for $\iota \to \infty, \iota \in N'$. Hence $(\psi, w)_0 = 0$ for all $\psi \in L^2(M)$, $\psi = 0$ in Ω_0, so that $w \in H^m(M)$ and $w = 0$ in $\complement\Omega_0$. In view of the segment property of Ω_0, this means $w \in H_0^m(\Omega_0)$. Consequently, the condition (R2) is verified.

Under the assumption (A') the sequence of spaces $H_0^{m_\iota}(\Omega_\iota), \iota = 1,2,\cdots$, constitutes a discrete approximation of the space $H_0^m(\Omega_0)$. A sequence of functions $u_\iota \in H_0^{m_\iota}(\Omega_\iota)$ converges discretely to $u_0 \in H_0^m(\Omega_0)$ in case

$$u_\iota \longrightarrow u_0 \Leftrightarrow u_\iota \longrightarrow u_0 \text{ in } H^m, \quad \|x_\iota u_\iota\|_{m_\iota} \longrightarrow 0 \quad (\iota \to \infty),$$

it converges weakly discretely to u_0 in case

$$u_\iota \rightharpoonup u_0 \Leftrightarrow u_\iota \rightharpoonup u_0 \text{ in } H^m, \quad \|x_\iota u_\iota\|_{m_\iota} \text{ bounded} \quad (\iota \to \infty).$$

Let us now introduce sequences of bounded sesquilinear forms a_ι, b_ι on E_ι by

(6) $$a_\iota(u,v) = \sum_{\substack{|\sigma| \le m_\iota \\ |\tau| \le m_\iota}} (\partial^\sigma u, a_{\iota\sigma\tau} \partial^\tau v)_0, \quad b_\iota(u,v) = \sum_{\substack{|\sigma| \le m_\iota \\ |\tau| \le m_\iota}} (\partial^\sigma u, b_{\iota\sigma\tau} \partial^\tau v)_0$$

for $u, v \in H_0^{m_\iota}(\Omega_\iota)$, where $a_{\iota\sigma\tau}, b_{\iota\sigma\tau}$ denote bounded measurable coefficients on Ω_ι of the form

(7) $$a_{\iota\sigma\tau} = a'_{\iota\sigma\tau} + x_\iota^2 a''_{\iota\sigma\tau}, \qquad b_{\iota\sigma\tau} = b'_{\iota\sigma\tau} + x_\iota^2 b''_{\iota\sigma\tau}, \qquad |\sigma|, |\tau| \le m_\iota,$$

with $a'_{\iota\sigma\tau} = b'_{\iota\sigma\tau} = 0$ for $m < |\sigma|, |\tau| \le m_\iota$ and $x_0 = 0$, $m_0 = m$, $\iota = 0,1,2,\cdots$.

(B') <u>The coefficients</u> $a'_{\iota\sigma\tau}, b'_{\iota\sigma\tau}, a''_{\iota\sigma\tau}, b''_{\iota\sigma\tau} \in L^\infty(\Omega)$ <u>are bounded by some constant</u> α <u>for all</u> $\iota = 0,1,2,\cdots$ <u>and almost all</u> $x \in \Omega_\iota$,

(8) $$|a'_{\iota\sigma\tau}(x)| + |b'_{\iota\sigma\tau}(x)| \le \alpha, \ |\sigma|,|\tau| \le m, \quad |a''_{\iota\sigma\tau}(x)| + |b''_{\iota\sigma\tau}(x)| \le \alpha, \ |\sigma|,|\tau| \le m_\iota .$$

<u>For every compact subset</u> $\Gamma \subseteq \Omega_0$ <u>the coefficients</u> $a'_{\iota\sigma\tau}, b'_{\iota\sigma\tau}$ <u>converge in measure to</u> $a_{0\sigma\tau}, b_{0\sigma\tau}$,

(9) $a'_{\iota\sigma\tau} \xrightarrow[\text{mes}]{} a_{o\sigma\tau}$ on Γ, $\quad b'_{\iota\sigma\tau} \xrightarrow[\text{mes}]{} b_{o\sigma\tau}$ on Γ $\quad (\iota\to\infty)$, $|\sigma|,|\tau| \le m$.

Under these assumptions and with the notions of Section 2 we will show the following Theorem.

(10) The sequences of sesquilinear forms $a_\iota, b_\iota, a_\iota^*, b_\iota^*$ on $H_o^{m_\iota}(\Omega_\iota)$, $\iota=0,1,2,\cdots$, are stable and consistent.

Proof. (i) The sesquilinear forms a_ι can be written in the form $a_\iota = a'_\iota + \varkappa_\iota^2 a''_\iota$ with the coefficients $a'_{\iota\sigma\tau}$ of a'_ι and $a''_{\iota\sigma\tau}$ of a''_ι, $\iota=0,1,2,\cdots$. From (8) we obtain the estimates

$$|a'_\iota(\varphi,\psi)| \le \nu\alpha \|\varphi\|_m \|\psi\|_m, \quad \|a''_\iota(\varphi,\psi)\| \le \nu\alpha \|\varphi\|_{m_\iota} \|\psi\|_{m_\iota}, \quad \varphi,\psi \in H_o^{m_\iota}(\Omega_\iota).$$

The constant ν depends only on n, m, ℓ. This proves the stability condition (b) with $\alpha_1 = \nu\alpha$.

(ii) The coefficients $a_{o\sigma\tau} = a'_{o\sigma\tau}$ of the sesquilinear forms $a_o = a'_o$ can be extended to functions $a_{\sigma\tau}$ in $L^\infty(M)$. The associated sesquilinear form a on $H^m(M)$ satisfies $a_o = a|H_o^m(\Omega_o)$. Now we set $D_a = C_o^\infty(\Omega_o)$ and for every $\varphi \in C_o^\infty(\Omega_o)$ we set $\varphi_\iota = \varphi$, if the support of φ is in Ω_ι, $\varphi_\iota = 0$ else, for $\iota = 1,2,\cdots$. By assumption (A'), then $\varphi_\iota = \varphi$ for almost all ι and $\varphi_\iota \to \varphi$ $(\iota\to\infty)$. Finally,

$$|a_\iota(\psi,\varphi_\iota) - a(\psi,\varphi)| \le |a'_\iota(\psi,\varphi_\iota) - a(\psi,\varphi)| + \varkappa_\iota^2 |a''_\iota(\psi,\varphi_\iota)|$$

for all $\psi,\varphi_\iota \in H_o^{m_\iota}(\Omega_\iota)$, $\varphi \in D_a = C_o^\infty(\Omega_o)$. Hence the consistency condition (c) follows from (9) by the estimate

$$\sup_{0\ne\psi\in E_\iota} |a_\iota(\psi,\varphi_\iota) - a(\psi,\varphi)| / \|\psi\|_{E_\iota} \le \nu \max_{\substack{|\sigma|\le m \\ |\tau|\le m}} \|(a'_{\iota\sigma\tau} - a_{o\sigma\tau})\partial^\tau\varphi\| + \nu\alpha\varkappa_\iota \|\varphi\|_\ell \to 0 \ (\iota\to\infty).$$

(iii) The sesquilinear forms b_ι are of the same type as the forms a_ι. The adjoint sesquilinear forms a_ι^*, b_ι^*, defined by 2.(10), have the coefficients $\overline{a}_{\iota\tau\sigma}$, $\overline{b}_{\iota\tau\sigma}$, $|\sigma|,|\tau| \le m_\iota$, $\iota=0,1,2,\cdots$. Hence the coefficients of a_ι^*, b_ι^* satisfy also the assumption (B'). Then part (i) of the proof shows further the stability and consistency of the sequences $b_\iota, a_\iota^*, b_\iota^*$ on $H_o^{m_\iota}(\Omega_\iota)$, $\iota = 0,1,2,\cdots$.

Condition (C) in Section 3 can easily be proved under the following assumption.

(C') The coefficients $b_{\iota\sigma\tau} = b'_{\iota\sigma\tau} + \varkappa_\iota^2 b''_{\iota\sigma\tau}$ of the sesquilinear forms b_ι have the property

$$b'_{\iota\sigma\tau} = 0, \quad |\sigma| = |\tau| = m, \qquad b''_{\iota\sigma\tau} = 0, \quad |\sigma| = |\tau| = m_\iota, \quad \iota = 0,1,2,\cdots.$$

Then, from the assumptions (B'),(C'), we get the Theorem

(11) <u>The sesquilinear forms</u> b_ι <u>on</u> E_ι <u>are compact for each</u> $\iota = 0,1,2,\cdots$, <u>the sequence</u> b_ι^* <u>on</u> $E_\iota, \iota=1,2,\cdots$, <u>is weakly discretely compact.</u>

<u>Proof.</u> We have $b_o = b_o'$ and $b_\iota = b_\iota' + \varkappa_\iota^2 b_\iota''$, $\iota = 1,2,\cdots$. From (B'),(C') follow the estimates

$$|b_\iota'(\varphi)| \leq \nu\alpha \|\varphi\|_m \|\varphi\|_{m-1}, \quad |b_\iota''(\varphi)| \leq \nu\alpha \|\varphi\|_{m_\iota} \|\varphi\|_{m_\iota-1}, \quad \varphi \in H_o^{m_\iota}(\Omega_\iota),$$

for $\iota = 0,1,2,\cdots$. For every $\varepsilon > 0$ there exists $\eta(\varepsilon) > 0$ such that

$$\nu\alpha \|\varphi\|_s \|\varphi\|_{s-1} \leq \varepsilon \|\varphi\|_s^2 + \eta(\varepsilon)\|\varphi\|_o^2, \quad \varphi \in H_o^s(M),$$

for all s in $m \leq s \leq \ell$. Hence we obtain

$$|b_\iota(\varphi)| \leq |b_\iota'(\varphi)| + \varkappa_\iota^2 |b_\iota''(\varphi)| \leq \varepsilon \|\varphi\|_{E_\iota}^2 + \eta(\varepsilon) k(\varphi), \quad \varphi \in H_o^{m_\iota}(\Omega_\iota),$$

for all $\iota = 0,1,2,\cdots$, with $k(\varphi) = (1+\varkappa^2)\|\varphi\|_o^2, \varkappa = \max \varkappa_\iota$. Thus the assertion follows from Corollary 2.(22).

The pairs of sesquilinear forms a_ι, b_ι on $H_o^{m_\iota}(\Omega_\iota)$ are strongly definite if

(D') $$|\mathrm{Re}\, a_\iota(\varphi)| + |\mathrm{Re}\, b_\iota(\varphi)| > 0, \quad 0 \neq \varphi \in H_o^{m_\iota}(\Omega_\iota), \quad \iota = 0,1,2,\cdots,$$

i.e. if $\mathrm{Re}\, a_\iota(\varphi) \neq 0$ or $\mathrm{Re}\, b_\iota(\varphi) \neq 0$. Evidently, this condition is satisfied if $\mathrm{Re}\, b_\iota(\varphi) \neq 0$, $0 \neq \varphi \in H_o^{m_\iota}(\Omega_\iota)$, for all $\iota = 0,1,2,\cdots$. For special eigenvalue problems one has $b_\iota = (.,.)_o$ so that $b_\iota(\varphi) = \|\varphi\|_o^2 > 0, \ 0 \neq \varphi \in H_o^{m_\iota}(\Omega_\iota), \ \iota = 0,1,2,\cdots$.

Finally, we formulate a sufficient condition for the uniform strong ellipticity of the sequence a_ι on $H_o^{m_\iota}(\Omega_\iota), \iota = 0,1,2,\cdots$. Let a', a'' be bounded strongly elliptic sesquilinear forms on $H_o^m(M), H_o^\ell(M)$ such that a', a'' satisfy the Gårding inequalities

(12) $$\gamma \|\varphi\|_m^2 \leq \mathrm{Re}\, a'(\varphi) + \mu \|\varphi\|_o^2, \quad \gamma \|\varphi\|_\ell^2 \leq \mathrm{Re}\, a''(\varphi) + \mu \|\varphi\|_o^2, \ \varphi \in C_o^\infty(M),$$

with some constants $\gamma > 0, \mu \geq 0$ (s. Agmon [1], p.78). Then the sequence $a_\iota = a' + \varkappa_\iota^2 a''$ on $H_o^{m_\iota}(\Omega_\iota)$ is uniformly strongly elliptic according to

(E') $$\gamma \|\varphi\|_{E_\iota}^2 \leq \mathrm{Re}\, a_\iota(\varphi) + k(\varphi), \quad \varphi \in H_o^{m_\iota}(\Omega_\iota), \quad \iota = 0,1,2,\cdots,$$

with $k(\varphi) = \mu(1+\varkappa^2)\|\varphi\|_o^2, \quad \varkappa = \max \varkappa_\iota, \quad m_o = m, \quad m_\iota = \ell, \ \iota = 1,2,\cdots$.

<u>References</u>

1. Agmon, S.: Lectures on elliptic boundary value problems. Princeton: Van Nostrand 1965.
2. Babuška, I.: Continuous dependence of eigenvalues on the domain. Czechoslovak Math. J. <u>15</u>, 169-178 (1965).

3. Friedman, A.: Singular perturbations for partial differential equations. Arch. Rational Mech. Anal. 29, 289-303 (1968).

4. Greenlee, W.M.: Singular perturbation of eigenvalues. Arch. Rat. Mech. Anal. 34, 143-164 (1969).

5. Grigorieff, R.D.: Über die Lösung regulärer koerzitiver Rand- und Eigenwertaufgaben mit dem Galerkinverfahren. Manuscripta math. 1, 385-411 (1969).

6. —— Approximation von Eigenwertproblemen und Gleichungen zweiter Art. Math.Ann. 183, 45-77 (1969).

7. Huet, D.: Phénomènes de perturbation singulière dans les problèmes aux limites. Ann. Inst. Fourier (Grenoble) 10, 61-150 (1960).

8. —— Singular perturbations of elliptic variational inequalities. Ann.Math.Pura Appl., to appear.

9. Kato, T.: Perturbation theory for linear operators. Berlin - Heidelberg - New York: Springer 1966.

10. Morgenstern, D.: Singuläre Störungstheorie partieller Differentialgleichungen. J. Rational Mech. Anal. 5, 204-216 (1956).

11. Moser, J.: Singular perturbation of eigenvalue problems for linear differential equations of even order. Comm. Pure Appl. Math. 8, 251-278 (1955).

12. Nečas, J.: Les méthodes directes en théorie des équations elliptiques. Paris: Masson 1967.

13. O'Malley, R.E.: Topics in singular perturbations. Advances in Math. 2, 365-470 (1968).

14. Stummel, F.: Rand- und Eigenwertaufgaben in Sobolewschen Räumen. Lecture Notes in Mathematics 102. Berlin - Heidelberg - New York: Springer 1969.

15. —— Diskrete Konvergenz linearer Operatoren I. Math.Ann. 190, 45-92 (1970).

16. —— Diskrete Konvergenz linearer Operatoren II. Math.Z. 120, 231-264 (1971).

17. —— Diskrete Konvergenz linearer Operatoren III. To appear, Proceedings of the Oberwolfach Conference on Linear Operators and Approximation 1971, Vol.20, Int. Series of Numerical Mathematics. Basel: Birkhäuser-Verlag.

18. Vainikko, G.M.: The compact approximation principle in the theory of approximation methods. USSR Comp. Math. and Math. Phys. 9, 1-32 (1969).

Some New Aspects of the Line Method for Parabolic Differential Equations

W. Walter

This lecture deals with three subjects and correlations between them, namely (a) parabolic differential equations, (b) ordinary differential equations in Banach space, and (c) flow-invariant sets.

I. Parabolic differential equations. Let us consider the parabolic differential equations in one space variable

$$(1) \qquad Pu := u_t - f(t,x,u,u_x,u_{xx}) = 0 \ ,$$

where $f(t,x,z,p,r)$ is assumed to be increasing in r. For simplicity, let $\overline{G}$ be a rectangle $J \times [a,b]$, $J = [0,T]$, and $G_p = J_o \times (a,b)$, $J_o = (0,T]$ the parabolic interior, $R_p = \overline{G} - G_p$ the parabolic boundary of $\overline{G}$. Then Nagumo's lemma (1939) states that

(A) $Pu < Pv$ in G_p, $u < v$ on R_p implies $u < v$ in G_p.

If, in addition, f satisfies a uniqueness condition (e.g., a Lipschitz condition with respect to z), then

(B) $Pu \leq Pv$ in G_p, $u \leq v$ on R_p implies $u \leq v$ in G_p.

A statement like (A) or (B) is a monotonicity theorem for the first boundary value problem, the latter being a "Aufgabe von monotoner Art" (Collatz). As for the many applications with respect to uniqueness, maximum principle, asymptotic behavior and stability, we refer to Chapter IV of the book by Walter (1970).

II. Ordinary differential equations. A similar monotonicity theorem holds for ordinary differential equations. Let $Py := y'(t) - f(t,y,(t))$. Then

(A) $Pu < Pv$ in J_o, $u(0) < v(0)$ implies $u < v$ in J.

If f(t,y) satisfies a uniqueness condition, then the corresponding theorem (B) (with equality permitted) is also valid. These theorems are classical for one equations. In the case of a system of differential equations, i.e. if y and f are n-vectors and if inequalities are interpreted component-wise, theorems (A) (B) are not true any longer. Yet they hold, if the right hand side f(t,y) is quasimonotone increasing in y, i.e. if f_i is increasing in y_j for $i \neq j$, or, more exactly, if

(Q) $x \leq y$, $x_i = y_i$ implies $f_i(t,x) \leq f_i(t,y)$.

This was discovered by M. Müller (1926).

III. The line method for parabolic differential equations will now be described. Let $x_i = a + ih$, $h = (b-a)/n$, $i = 0,\dots,n$. We replace in (1) the derivatives with respect to x by finite differences, $u(t,x_i) \approx u_i(t)$,

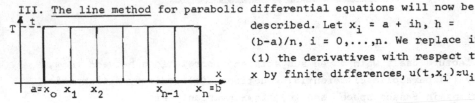

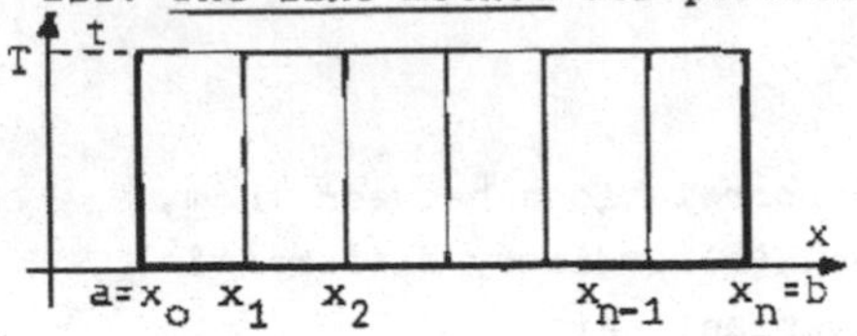

$$u_x(t,x_i) \approx \delta u_i := \frac{u_{i+1}-u_i}{h}, \quad u_{xx}(t,x_i) \approx \delta^2 u_i := \frac{u_{i+1}+u_{i-1}-2u_i}{h^2}$$

(one could also take central or backward differences for δu_i). In this way a system of ordinary differential equations for $u_1(t),\dots,u_{n-1}(t)$

$$(2) \qquad u_i' = f(t,x_i,u_i,\delta u_i,\delta^2 u_i) \qquad (i=1,\dots,n-1)$$

is obtained, where u_o, u_n and $u_i(0)$ are given by the prescribed boundary data.

It is easily proved that the right hand side of the system (2) has the quasimonotonicity property (Q). This means that for (2) the theorems (A) (B) of II hold. They are the discretized versions of the corresponding theorems for the parabolic equation, given in I.

Using these theorems, one can prove estimation and convergence theorems for the line method. More important, one can give constructive existence proofs for boundary value problems with respect to the nonlinear equation (1). A composition of this theory and a bibliography are given in Sections 35 and 36 of Walter (1970).

If we deal with the Cauchy problem (or any other boundary value problem with an infinite region) for equation (1), then the corresponding system (2) is infinite. In this way one is led to

IV. Ordinary differential equations in ordered Banach spaces. Let f be a map from $J \times B$ to B, where B is a real Banach space. Then it is well known that there exists one and only one solution (in the strong topology) of the initial value problem

$$y' = f(t,y) \text{ in } J, \quad y(0) = y_o,$$

if f is (strongly) continuous and satisfies a Lipschitz condition

$$\|f(t,y)-f(t,z)\| \le L\|y-z\| \quad \text{for } y,z \in B .$$

Let B_+ be a positive cone in B, i.e., a closed subset of B such that

$x,y \in B_+$, $\lambda \geq 0$ implies $\lambda x \in B_+$, $x + y \in B_+$. The cone B+ defines an order relation in B in the usual way,

$$x \leq y \iff y - x \in B_+ \; ; \qquad x < y \iff y - x \in \operatorname{int} B_+ \; ,$$

when int B_+ is the interior of B_+ (which may be empty).

We are dealing with the problem of defining quasimonotonicity in an appropriate way and of proving theorems (A) (B) as given in II. Let us note that (B) is well-defined, but that (A) makes sense only if int $B_+ \neq \emptyset$.

If B is a "sequence space" of elements $y = (y_i)_{i \in A}$ (A arbitrary index set) with finite norm $||y|| := \sup\{p_i |y_i| : i \in A\}$, ($p_i > 0$ given) and with the natural positive cone $B_+ = \{y \in B : y_i \geq 0$ for all $i \in A\}$, then quasimonotonicity is defined in a natural way, namely by (Q) as given in II. Theorems (A) (B) are true in this case; see Walter (1969). Similar theorems for countably infinite A, but in the weak topology, have been given by Mlak and Olech (1963).

In an arbitrary ordered Banach space, monotonicity is defined as usual: f(t,y) is increasing in y, if $x \leq y$ implies $f(t,x) \leq f(t,y)$. A possibility to define quasimonotonicity is given by

(Q_1) $f(t,y) + Cy$ <u>is increasing in</u> y <u>for some</u> $C > 0$.

In the case $B = \mathbb{R}^n$ considered in II, (Q_1) implies (Q); but if f is Lipschitz-continuous in y, then (Q) and (Q_1) are equivalent.

If f(t,y) is Lipschitz-continuous in y and quasimonotone according to (Q_1), then (B) holds. This result of Walter (1971) is true in arbitrary ordered Banach spaces.

A more general concept of quasimonotonicity in an arbitrary ordered Banach space B was given by Volkmann (1971). The function f(t,y) is said to be quasimonotone increasing in y according to Volkmann, if

(Q_2) $x \leq y$, $\phi(x) = \phi(y)$ <u>implies</u> $\phi(f(t,x)) \leq \phi(f(t,y))$ <u>for every</u> $\phi \in B_+^*$,

when B_+^* is the positive cone in the dual space B^* induced by B_+, i.e. the set of all continuous linear functionals $\phi \in B^*$ sucht that $\phi(B_+) \geq 0$.

Volkmann's definition is the most general one in the sense that if f(t,y) is quasimonotone according to one of the definitions given by various authors in connection with monotonicity theorems, then it is also quasimonotone according to (Q_2); see Volkmann (1971). In the case $B = \mathbb{R}^n$, $B_+ = \mathbb{R}^n_+ = \{x \in \mathbb{R}^n : x_i \geq 0$ for $i = 1,\dots,n\}$ the coordinate functionals $\phi_i(x) = x_i$ are in B_+^*, and each $\phi \in B_+^*$ is a linear combination of the ϕ_i with nonnegative coefficients. If we take $\phi = \phi_i$ in (Q_2), then

(Q_2) becomes (Q).

The following example shows that, even in finite-dimensional spaces and for linear (hence Lipschitz-continuous) functions, (Q_2) is more general then (Q_1).

<u>Example.</u> Let $B = \mathbb{R}^3$, $B_+ = \{x \in \mathbb{R}^3 : x_1^2 + x_2^2 \leq x_3^2, x_3 \geq 0\}$,

$$f(x) = (-x_2, x_1, 0) .$$

For $x = (1,0,1) \in B_+$ we get

$$f(x) + Cx = (C,1,C) \notin B_+ \quad \text{for} \quad C > 0 ,$$

which shows that f does not satisfy (Q_1). But f satisfies (Q_2). For, B_+^* is equal to B_+, if we identify $\mathbb{R}^n$ with its dual space in the usual way. Since f is linear, one has to prove that $\phi, x \in B_+$, $\phi(x) = \phi_1 x_1 + \phi_2 x_2 + \phi_3 x_3 = 0$ implies $\phi(f(x)) \geq 0$. Now, if $x \in B_+$ and $\phi(x) = 0$, then x is necessarily on the boundary of B_+ and $\phi = (-x_1, -x_2, +x_3)$; whence

$$\phi(f(x)) = (-x_1)(-x_2) + (-x_2)x_1 + x_3 \cdot 0 = 0$$

follows.

The main result of Volkmann's paper can be stated as follows. If int $B_+ \neq \emptyset$ and if f satisfies (Q_2), then (A) holds (without any regularity assumption on f); if, in addition, f satisfies a uniqueness condition, then (B) holds. These theorems are even valid in locally convex spaces.

V. <u>The line method in infinite regions.</u> As we have already stated, a parabolic boundary value problem for an infinite region gives rise to an infinte system of ordinary differential equations of the form (2). Let us take, as a simple prototype, the Cauchy problems for the heat equation

(3) $$u_t = u_{xx} \text{ in } J_o \times \mathbb{R}, \quad u(0,x) = \phi(x) \quad \text{for } x \in \mathbb{R} .$$

This problem is transformed by the line method into

(4) $$u_i' = \delta^2 u_i = \frac{1}{h^2}(u_{i+1} + u_{i-1} - 2u_i), \quad u_i(0) = \phi(x_i) \quad \text{for } i \in \mathbb{Z},$$

when $\mathbb{Z}$ is the set of integers, $x_i = ih$ and $h > 0$ a constant.

Using the results quoted in IV one can prove approximation and convergence theorems for the line method, if the class of admissible solutions u is restricted by an inequality

(5) $$|u(t,x)| \leq Ke^{K|x|}$$

for some $K > 0$. Now it is known that the classical theorems for the heat equation (and also for more general linear equations) hold under a milder growth condition

(6) $$|u(t,x)| \le Ke^{Kx^2}.$$

If the line method is considered under the assumption (6), there arise two difficulties. First, the difference $\delta^2 u_i$ is no longer a good approximation for u_{xx}, even for such simple functions as $u = e^{x^2}$. Second, if we take as Banach space B the set of all $y = (y_i)_{i\in\mathbb{Z}}$ such that

$$||y|| := \sup\{|y_i|e^{-Kx_i^2} : i\in\mathbb{Z}\}$$

is finite, then the function f corresponding to the right hand side of (4) is not continuous. In fact, the shift operator $(y_i) \to (y_{i+1})$ is not continuous.

In order to overcome these difficulties, one has to work with a non-uniform grid and to consider solutions of (4) in the weak topology. In this framework, Voigt (1971) has investigated the line method for parabolic equations under the growth condition (6). He has proved, among others, that the discretized problem has one and only one solution, that a monocity theorem holds for the discretized problem, and that the solution of the discretized problem converges to the solution of the original parabolic problem, if the mesh size tends to zero.

VI. <u>Flow-invariant sets.</u> We consider an ordinary differential equation

(7) $$y' = f(t,y)$$

in a Banach space B. The set $M \subset B$ is said to be flow-invariant with respect to equation (7) (or to f), if for each solution y(t) of (7)

$$y(t_o)\in M \text{ implies } y(t)\in M \text{ for } t \ge t_o$$

(as long as the solution exists). In theorems on flow-invariante there are usually two conditions on f, a uniqueness condition, and an assumption (T) which states that at a point x on the boundary ∂M of M the vector f(t,x) is tangent to M or points into the interior of M. In the case $B = \mathbb{R}^n$, theorems of this kind are well-known under additional hypotheses on the smoothness of ∂M. Recently, Bony (1969) and Brezis (1970) have given such theorems which do without a smoothness condition on ∂M. Their results are as follows.

Let M be a closed subset of $B = \mathbb{R}^n$, and let f = f(x) be defined in a neighborhood of M, independent of t, and (locally) Lipschitz-continuous. If f satisfies the "tangent condition" (T_1) or (T_2), respectively, then M is flow-invariant. Here, (T_1) is Brezis' condition

(T_1) $\quad \mathrm{dist}(x+hf(x),M) = o(h)$ <u>as</u> $h \to +0$ <u>for every</u> $x\in\partial M$

where dist(a,M) is the distance from a to M.

Before giving Bony's condition, we define a normal in the sense of Bony. Let $x \in \partial M$ and let B be a closed ball with center a (and positive radius) such that $B \cap M = \{x\}$. Then $n_x := a - x$ is said to be an outer normal to M at x. Bony's condition reads

(T_2) $n_x \cdot f(x) \leq 0$, if n_x is an outer normal to M at $x \in \partial M$,

where the dot denotes the inner product in $\mathbb{R}^n$.

The following theorem is a generalization of Bony's theorem. Its proof is, in our opinion, of a remarkable simplicity. Let us note that a subset M of a normed space is said to be a distance set, if to every $x \notin M$ there exists a $y \in M$ such that $\text{dist}(x,M) = ||y-x||$. A distance set is necessarily closed.

VII. <u>Theorem.</u> <u>Let</u> H <u>be</u> <u>a</u> <u>real</u> <u>Hilbert</u> <u>space</u> <u>with</u> <u>inner</u> <u>product</u> (.,.). <u>Let</u> $M \subset H$ <u>be</u> <u>a</u> <u>distance</u> <u>set</u>. <u>If</u> f <u>satisfies</u> <u>the</u> <u>tangent</u> <u>condition</u>

(T_2') $(n_x, f(t,x)) \leq 0$ for all $x \in \partial M$ such that an outer normal n_x exists,

<u>and</u> <u>an</u> <u>Osgood</u> <u>condition</u>

$$(x-y, f(t,x)-f(t,y)) \leq ||x-y||\,g(||x-y||) \,, \tag{8}$$

<u>then</u> M <u>is</u> <u>flow-invariant</u>.

Here it is assumed that g(s) is an Osgood function, i.e. that g is continuous for $s \geq 0$, positive for $s > 0$, $g(0) = 0$ and $\int_0^1 [g(s)]^{-1} ds = \infty$. An example is $g(s) = Ls$ (Lipschitz-condition). We remark that, instead of (8), more general conditions with $g = g(t,s)$ are permitted. What is needed is that g is in the class $\mathfrak{L}_5$ defined in Walter (1970; p. 81).

<u>Proof.</u> Let us assume that y is a solution of (7) and that, contrary to our assertion, $y(t_0) \in M$, $y(t) \notin M$ for $t_0 < t < t_1$. Let $t_2 \in (t_0, t_1)$, let x be a point in M such that $\text{dist}(u(t_2),M) = ||u(t_2)-x||$ and let v(t), w(t) be scalar functions defined by

$$v(t) := \text{dist}(u(t),M) \,, \qquad w(t) := ||u(t)-x|| \,.$$

Now,

$$(w^2)' = 2(u-x,u') = 2(u-x,f(t,u)) \,.$$

Since $n_x := u(t_2) - x$ is an outer normal at x, we get from (T_2') and (8) for $t = t_2$

$$(w^2)' \leq 2(u(t_2)-x, f(t_2,u)-f(t_2,x)) \leq 2w(t_2)g(w(t_2)) \,,$$

i.e., $w'(t_2) \leq g(w(t_2))$. According to definition, $v(t) \leq w(t)$,

$v(t_2) = w(t_2)$. Therefore $D^+v(t_2) \leq w'(t_2)$, whence

(9) $$D^+v(t) \leq g(v(t))$$

at $t = t_2$ follows. But since t_2 is arbitrary, this inequality holds for $t_o < t < t_1$. Now, it follows from a classical theorem on differential inequalities that $v(t_o) = 0$ and (9) implies $v(t) = 0$ for $t_o < t < t_1$; see, e.g., Walter (1970; p.70 and p.81). Thus we have arrived at a contradiction.

Let us note that (8) is implied by a condition

(8') $$||f(t,x)-f(t,y)|| \leq g(||x-y||) .$$

Similarly, Brezis' theorem can be generalized as follows.

VIII. Theorem. A distance set M in a Banach space B is flow-invariant with respect to (7), if the Osgood condition (8') and

(T_1') $\operatorname{dist}(x+hf(t,x),M) = o(h)$ as $h \to +0$ for every $x \in \partial M$ holds.

Proof. Let $t_o, t_1, t_2, x, v(t)$ be defined as in the proof of VII. We write t instead of t_2 for simplicity. From

$$v(t+h) \leq ||u(t+h)-(x+hf(t,x))|| + \operatorname{dist}(x+hf(t,x),M)$$

the inequalities ($h > 0$)

$$\begin{aligned}\frac{v(t+h)-v(t)}{h} &\leq \frac{1}{h}\{||u(t+h)-(x+hf(t,x))||-||u(t)-x||\} + o(1)\\ &\leq \frac{1}{h}\ ||u(t+h)-u(t)-hf(t,x)|| + o(1)\\ &\leq \frac{1}{h}\ ||hf(t,u)+o(h)-hf(t,x)|| + o(1)\\ &\leq g(||u-x||) + o(1) = g(v(t)) + o(1)\end{aligned}$$

follow. As before, we have (9) and $v = 0$.

We remark that the original proof of Brezis, which relies on existence theory, cannot be used.

IX. Connection with quasimonotonicity. We consider equation (7) in an ordered Banach space B with positive cone B_+. Let $Py = y' - f(t,y)$. Theorem

(B) $Pu \leq Pv$ in $J = [0,T]$, $u(0) \leq v(0)$ implies $u \leq v$ in J

can be reformulated as follows. Let $y(t) := v(t) - u(t)$ and

(10) $$\tilde{f}(t,x) := f(t,u+x) - f(t,u) + p(t), \quad p(t) := Pv - Pu \in B_+ .$$

Then y is a solution of

$$(11) \qquad y' = \tilde{f}(t,y) \text{ in } J, \quad y(0) \in B_+ .$$

This shows that (B) is a statement on flow-invariance of B_+ with respect to $\tilde{f}$: $u \leq v$ in J is equivalent with $y(t) \in B_+$ for $0 \leq t \leq T$. Since

$$\text{dist}(a+p,B_+) \leq \text{dist}(a,B_+) \quad \text{for } p \in B_+ ,$$

the condition

(Q_3) $\text{dist}(x+h(f(t,u+x)-f(t,u)),B_+) = o(h)$ as $h \to +0$ for every $x \in \partial B_+$

is sufficient to guarantee that $\tilde{f}$ and B_+ satisfy (T_1'). It follows from Theorem VIII that (B) holds if f satisfies (Q_3) and if B_+ is a distance set. It can be shown that (Q_3) is equivalent with Volkmann's condition (Q_2). In this way a new proof of Theorem (B) under Volkmann's quasimonotonicity condition is obtained. This proof works whenever B_+ is a distance set, in particular in all reflexive Banach spaces. Summarizing, we have the following result.

The monotonicity theorem (B) holds under Volkmann's quasimonotonicity condition (Q_2) or (Q_3) if f satisfies a uniqueness condition (such as a Lipschitz or Osgood condition) and if

(a) int $B_+ \neq \emptyset$; or

(b) B_+ is a distance set.

It is an open problem, if the theorem is true in all ordered Banach spaces. Finally, let us remark that condition (Q_1) implies that $x + h(f(t,u+x)-f(t,u)) \in B_+$ for $x \in \partial B_+$ and small $h > 0$. Hence, (Q_3) is trivially satisfied in this case.

Theorems VII and VIII and the first part of IX were given by Professor Redheffer in his lectures at the University of Karlsruhe, where he was Visiting Professor during the Winter Semester 1971/72. The proof of VII given here is different from Redheffer's and is due to Walter, the proof of VIII is due to Redheffer; see also Redheffer (1972).

Literature

Bony, J.M.: Principe du maximum, inégalité de Harnack et unicité du problème de Cauchy pour les operateurs elliptiques dégénerés. Ann. Inst. Fourier 19, 277-309 (1969).

Brezis, H.: On the characterization of flow-invariant sets. Communications Pure Appl. Math. 23, 261-263 (1970).

Mlak, W. and C. Olech: Integration of infinite systems of differential inequalities. Ann. Polon. Math. 13, 105-112 (1963).

Müller, M.: Über das Fundamentaltheorem in der Theorie der gewöhnlichen Differentialgleichungen. Math. Zeitschr. 26, 619-645 (1926).

Redheffer, R.: The theorems of Bony and Brezis on flow-invariant sets. Amer. Math. Monthly (1972) (in print)

Voigt, A.: Die Linienmethode für das Cauchy-Problem bei nichtlinearen parabolischen Differentialgleichungen. Dissertation Karlsruhe 1971.

Volkmann, P.: Gewöhnliche Differentialungleichungen mit quasimonoton wachsenden Funktionen in topologischen Vektorräumen. Math. Zeitschr. (submitted for publication) (1971)

Walter, W.: Gewöhnliche Differential-Ungleichungen im Banachraum. Arch. Math. 20, 36-47 (1969).

--: Differential and integral inequalities. Ergebnisse der Mathematik und ihrer Grenzgebiete, Band 55. Springer Verlag 1970.

--: Ordinary differential inequalities in ordered Banach spaces. J. Diff. Equations 9, 253-261 (1971).

On Boundary Value Problems of Generalized Analytic Functions

W.Wendland

This paper deals with Riemann-Hilbert problems of generalized analytic functions :

(1) $$w_{\bar{z}} = Aw + B\bar{w} + C ,$$

(2) $$\mathrm{Re}\,(\bar{\lambda}\, w)\big|_{\dot{\Gamma}} = \varphi \; .$$

For this problem Hilbert used Green's functions . But he made an error , which W.Haack and G.Hellwig discovered in the fifties . They , together with the brothers J. and J.C.C.Nitsche , with J.Jaenicke and G.Bruhn developed the theory for a simply connected and small domain , again using Green's functions.[1)] One gets Fredholm integral equations with compact operators . If one uses the regularity-proposition of the similarity-principle which was proved by L.Bers and Vekua, then the Green's function method works for simply connected domains in the large, too.

Another method was developed by Vekua using Noether's and Muschelischwili's theory of singular integral equations. This method works in the case of multiply connected domains, too . (See [2])

The Green's function method is very easy and leads straight to the Noethernean properties of boundary value problems for a simply connected domain. This method is presented by W.Haack and W.Wendland in their book on Partial and Pfaffian Differential-Equations [1] . For multiply connected domains the same method using modified Green's functions again leads to a system of Fredholm integral equations with compact operators. From this system every solution of the boundary value problems can be calculated . It also leads to the

---

1) For a great many special references see [1] and [2] .

Noethernean property of the first boundary value problem- which we can take to be our standard problem. If all single indices are zero we get a canonical form and we can reduce this problem to a first boundary value problem . In the general case we get a problem on which in the case of analytic functions B.Bojarski based his classification of the so called special cases of the Riemann-Hilbert problem presented in Vekua's book ([2] , appendix to chap. IV). It seems possible to develop the whole solvability theory - especially the Noethernean properties - from these Fredholm equations without using singular integral equations . But at the present time I know the exact connections only in the cases presented here .

§1. Green's functions and integral equations for the 1-st boundary value problem

First we consider the 1-st boundary value problem (1-st BVP) for the elliptic system (1) , which means divided into real and imaginary part

$$(3)\qquad \begin{aligned} u_x - v_y &= au + bv + c\,, \\ u_y + v_x &= \tilde{a}u + \tilde{b}v + \tilde{c} \end{aligned}$$

with the 1-st boundary condition

$$(4)\qquad \operatorname{Re} w\big|_{\dot{\Gamma}} = u\big|_{\dot{\Gamma}} = \varphi\,.$$

Let the domain $\Gamma \subset R^2$ (or $\Gamma \subset C$) be bounded by (m+1) closed curves $\dot{\Gamma}_o, \dot{\Gamma}_1, \ldots, \dot{\Gamma}_m$ belonging to a Hölder class $C^{1+\alpha}$ $(\alpha > 0)$; $\dot{\Gamma} = \dot{\Gamma}_o \cup \ldots \cup \dot{\Gamma}_m$. Let $G^I(\,x,y\,;\xi,\eta\,)$

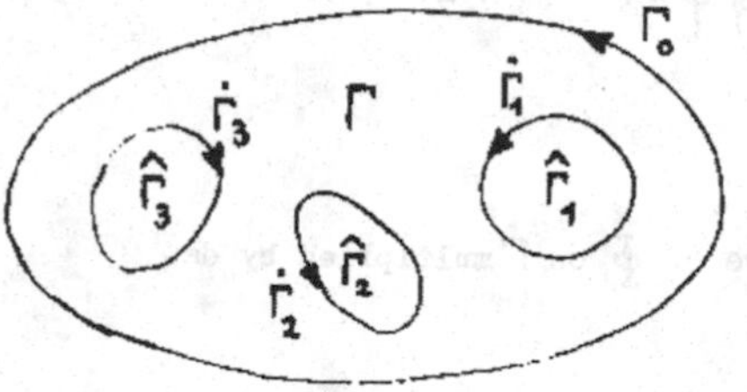

be the Green's function of the Laplacian in Γ . Because the harmonic conjugated function of G^I in the multiply connected domain ($m \geq 1$) is multivalent we need a modified Greens function using the canonical harmonic functions for Γ :

(5) $$\Delta e_k = 0 \text{ in } \Gamma \;, \; e_k\big|_{\dot{\Gamma}_j} = \delta_{jk} \;, \; j=0,1,\ldots,m \;;\; k=1,\ldots,m \;.$$

Then $\oint_{\dot{\Gamma}} e_j \, d_n e_k$ [2] is a symmetric positive definite matrix and its inverse

(6) $$c_{jk} := \Big(\Big(\oint_{\dot{\Gamma}} e_j \, d_n e_k \Big) \Big)^{-1}$$

is well defined . Now we define the modified Green's function by

(7) $$\Gamma^{I}(x,y\,;\xi,\eta) := G^{I}(x,y\,;\xi,\eta) - \sum_{j,k=1}^{m} c_{jk}\, e_j(x,y)\, e_k(\xi,\eta).$$

This modified Green's function has a univalent conjugate harmonic function Γ^{II}, the Green's function of 2-nd kind , for which we can demand the boundary properties

(8) $$d_n \Gamma^{II}\big|_{\dot{\Gamma}} = -\,\sigma\, ds \quad , \quad \oint \Gamma^{II} \sigma\, ds = 0$$

with any weight function $\sigma \geq 0$, $\sigma\big|_{\dot{\Gamma}_j} = 0$ for $j=1,\ldots,m$, $\oint_{\dot{\Gamma}} \sigma\, ds = 1$.

(These properties and the proof of the following theorem 1 will be presented in [3] .) Now we can prove that each solution w=u+iv of the 1-st BVP satisfies the following system of Fredholm integral equations :

(9) $$u(\xi,\eta) = \iint_{\Gamma} [\, -(au+bv)dy + (\tilde{a}u+\tilde{b}v)dx \;,\; d\Gamma^{I}] + \iint_{\Gamma} [\, -c\,dy + \tilde{c}\,dx \;,\; d\Gamma^{I}] - \oint_{\dot{\Gamma}} \varphi\, d_n \Gamma^{I} \;,$$

(10) $$v(\xi,\eta) = -\iint_{\Gamma} [\,(au+bv)dx + (\tilde{a}u+\tilde{b}v)dy \;,\; d\Gamma^{II}] - \iint_{\Gamma} [\, c\,dx + \tilde{c}\,dy \;,\; d\Gamma^{II}] + \oint_{\dot{\Gamma}} \varphi\, d\Gamma^{II} + K \;,$$

(11) $$K = \oint_{\dot{\Gamma}} v\,\sigma\, ds \;.$$

---

2) $d_n \Phi\big|_{\dot{\Gamma}}$ means the exterior normal derivative of Φ on $\dot{\Gamma}$ multiplied by ds where s means the naturally length of $\dot{\Gamma}$.

Theorem 1 : (See [3])

i) Every solution w (sufficiently smooth) of the 1-st BVP satisfies the integral equations (9) (10) with (11).

ii) Every solution w of the integral equations (9) (10) satisfies (11) and the differential equations (3) . The solution assumes the boundary values

(12) $$\operatorname{Re} w\Big|_{\dot\Gamma_j} = u\Big|_{\dot\Gamma_j} = \varphi + \alpha_j\ ,\ \alpha_o = 0\ ,\ \alpha_1,\dots,\alpha_m \in \mathbb{R}\ .$$

The constants α_j are determined by the equations

(13) $$\alpha_j = \sum_{k=1}^{m} c_{jk}\left\{-\oint_{\dot\Gamma}\varphi\, d_n e_k + \iint_{\Gamma}[(au+bv+c)dx - (au+bv+c)dy\ ,\ de_k]\right\},$$
$$= \sum_{k=1}^{m} c_{jk}\oint_{\dot\Gamma}(u-\varphi)\, d_n e_k\ .$$

From this theorem 1 we see that the real parts of the solutions of the 1-st BVP cannot have arbitrary boundary values .

Using theorem 1 with the special case of Cauchy-Riemann's equations we get the regularity propositions of the similarity-principle of L.Bers and Vekua as follows.

Theorem 2 :

If $w \in C^{\alpha}(\bar\Gamma)$ is a solution of the homogeneous system

(14) $$w_{\bar z} = Aw + B\bar w \qquad \text{in } \Gamma$$

then there exists a function $\tilde w \in C^{\alpha}(\bar\Gamma)$ such that

(15) $$f := w\, e^{-\tilde w}$$

We can now prove :

Theorem 4 :

For every given functions c , $\tilde{c}$, $\varphi \in C^{\alpha}(\dot{\Gamma})$ and every given constant $K \in \mathbb{R}$ the integral equations (9)(10) have one and only one continuous solution w .

Proof : Because the integral operators in (9)(10) have weak singularities they are both compact. So Fredholm's theorem holds . It suffices to look only at the homogeneous equations . Let $\mathring{w} = \mathring{u} + i\mathring{v}$ be the solution of the homogeneous equations (9)(10) ($c = \tilde{c} = 0$, $\varphi = 0$, $K = 0$) . Then $\mathring{w}$ satisfies - by theorem 1 - the homogeneous differential equations (14) and the boundary conditions

$$(20) \qquad \mathring{u}\big|_{\dot{\Gamma}_j} = \alpha_j \, , \quad j=0,\dots,m, \quad \alpha_0 = 0 \, , \quad \alpha_1,\dots,\alpha_m \ \text{const.}, \quad \oint_{\dot{\Gamma}} \mathring{v}\,\sigma\, ds = 0.$$

These boundary conditions together with the boundary-vector

$$\lambda = \lambda_1 + i\lambda_2 = \begin{cases} 1 & \text{if } \alpha_j = 0 \\ -\mathring{v}\big|_{\dot{\Gamma}_j} + i\alpha_j & \text{if } \alpha_j \neq 0 \end{cases}$$

lead to the homogeneous boundary condition

$$(21) \qquad \mathrm{Re}\,(\mathring{w}\bar{\lambda})\big|_{\dot{\Gamma}} = 0 \quad , \quad \lambda \neq 0 \quad .$$

Using theorem 3 and the function f which was defined in that theorem and which is analytic in $\tilde{\Gamma}$ we get for the boundary condition (20)

$$(22) \qquad \mathrm{Re}\; e^{-\mathring{\tilde{w}}} f\big|_{\dot{\Gamma}_j} = \alpha_j$$

and therefore

$$(23) \qquad \oint_{\dot{\Gamma}_j} d \arg f = 0 \qquad \text{for every } j=0,1,\dots,m.$$

is an analytic function in Γ. For the function $\tilde{w}$ we can demand that

$$(16) \qquad \operatorname{Im} \tilde{w}\big|_{\dot{\Gamma}_j} = \psi + c_j \;, \quad c_o = 0 \;, \quad \oint_{\dot{\Gamma}} \tilde{u}\,\sigma\, ds = \varkappa$$

with an arbitrary given function ψ and an arbitrary given real constant $\varkappa$.
But the real constants $c_1, \dots, c_m$ are determined.

Proof : Use the function $\tilde{w} = \tilde{u} + i\tilde{v}$ with

$$(17) \qquad \tilde{u}(\xi,\eta) = \varkappa - \oint_{\dot{\Gamma}} \psi\, d\Gamma^{II} - 2\operatorname{Re} \iint_{\Gamma} \Big(A + \frac{\overline{w}}{w} B \Big)\Big(\frac{\partial}{\partial x} - i\frac{\partial}{\partial y}\Big) \Gamma^{II} \,[dx,dy],$$

$$(18) \qquad \tilde{v}(\xi,\eta) = -\oint_{\dot{\Gamma}} \psi\, d_n \Gamma^{I} + 2\operatorname{Im} \iint_{\Gamma} \Big(A + \frac{\overline{w}}{w} B \Big)\Big(\frac{\partial}{\partial x} - i\frac{\partial}{\partial y}\Big) \Gamma^{I} \,[dx,dy],$$

where $\frac{\overline{w}}{w} = 0$ if $w = 0$.

Theorem 3 :

If $w \in C^{\alpha}(\overline{\Gamma})$ is a solution of the homogeneous system (14) and also satisfies a homogeneous boundary condition

$$(19) \qquad \operatorname{Re} (\overline{\lambda} w)\big|_{\dot{\Gamma}} = 0 \qquad \text{with } \lambda \neq 0 \quad ,$$

then we can find a function $\tilde{w} \in C^{\alpha}(\overline{\Gamma})$ such that

$$f := w\, e^{-\tilde{w}}$$

can be analytically continued into some domain $\tilde{\Gamma} \supset \overline{\Gamma}$.

For the proof see Vekua [2] p.241 and 178. It uses an analytic continuation of $\frac{\lambda}{|\lambda|}$, which we shall mention later.

On the other hand from $\sigma \geq 0$ and (20) it follows that there is at least one point $z_0 \in \dot{\Gamma}_0$ with $\overset{\circ}{w}(z_0) = 0$ and $f(z_0) = 0$. Using (23) and the argument principle this gives $f \equiv 0$ and $\overset{\circ}{w} \equiv 0$.

§2. The solution of the 1-st boundary value problem using the integral equations

With the properties of the integral equations (9)(10) we can now very easily prove the following well-known theorem.

Theorem 5 :

The 1-st boundary value problem

(24) $$w_{\bar{z}} = Aw + B\bar{w} \quad , \quad \mathrm{Re}\, w\big|_{\dot{\Gamma}} = u\big|_{\dot{\Gamma}} = \varphi$$

has the deficiency index either $(1,m)$ or $(0,m-1)$.

Proof : i) Let U_0+iV_0 be the solution of the integral equations (9)(10) with $C=0$ and with

(25) $$\varphi = 0 \;,\quad K = 1 \quad .$$

ii) Let $U(\varphi) + iV(\varphi)$ be the solution of (9)(10) with $C=0$ for any given

(26) $$\varphi \quad , \quad K = 0 \quad .$$

Then this solution is a bounded linear mapping $C^\alpha(\dot{\Gamma}) \to C^\alpha(\bar{\Gamma})$ with $\varphi \longmapsto U(\varphi) + iV(\varphi)$.
($C^\alpha(\bar{\Gamma})$ is here the linear real space of complex valued functions in $\bar{\Gamma}$ with the α - Hölder-norm .)

Now every solution of (9)(10) with given φ, K has the form

(27) $$w = K(U_o + iV_o) + U(\varphi) + iV(\varphi) \quad .$$

This solution satisfies the boundary condition

(28) $$u \Big|_{\dot\Gamma} = KU_o \Big|_{\dot\Gamma} + U(\varphi)\Big|_{\dot\Gamma} = \varphi$$

if and only if φ satisfies the conditions $\alpha_j = 0$ which we call the closing conditions :

(29) $$K \oint_{\dot\Gamma} U_o \, d_n e_j + \oint (U(\varphi) - \varphi) \, d_n e_j = 0 \quad \text{for every } j=1,\dots,m.$$

The $L_j(\varphi) := \oint_{\Gamma} U(\varphi) \, d_n e_j$ are bounded linear functionals. Now we have to distinguish two cases :

<u>Case i)</u> $$\oint_{\dot\Gamma} U_o \, d_n e_j = 0 \quad \text{for every } j=1,\dots,m.$$

Then $U_o + iV_o$ is an eigensolution of the 1-st BVP . Using theorem 3 we can easily show that every eigensolution has the form $\varkappa (U_o + iV_o)$ with $\varkappa \in R$.

On the other hand the necessary and sufficient closing or solvability conditions (29) are reduced to

(30) $$\oint_{\dot\Gamma} (U(\varphi) - \varphi) \, d_n e_j = 0 \quad \text{for every } j=1,\dots,m.$$

From $$\oint_{\dot\Gamma} e_k \, d_n \Gamma^{I} = 0 \; , \; \oint_{\Gamma} e_k \, d\Gamma^{II} = 0 \qquad ([3])$$

we get that $U(e_k) + iV(e_k)$ satisfies the homogeneous system (9)(10) . This yields

(31) $$U(e_k) + iV(e_k) = 0 \quad \text{for every } j=1,\dots,m.$$

So we get

(32) $$\operatorname{rank} \Big(\Big(\oint_{\dot\Gamma} (U(e_k) - e_k) \, d_n e_j \Big)\Big) = \operatorname{rank} \Big(\Big(- \oint_{\dot\Gamma} e_k \, d_n e_j \Big)\Big) = m \; .$$

Therefore the solvability conditions (30) are linearly independent and we have the deficiency index (1,m) .

<u>Case ii)</u> $\oint_{\dot\Gamma} U_o \, d_n e_{j_o} = \nu \neq 0$ for one index j_o .

Now we can eliminate K in (29) and get the (m-1) conditions

$$(33) \qquad \frac{1}{\nu}\oint_{\dot\Gamma} U_o \, d_n e_j \cdot \oint_{\dot\Gamma} (U(\varphi) - \varphi) \, d_n e_{j_o} + \oint_{\dot\Gamma} (U(\varphi) - \varphi) \, d_n e_j = 0$$

for every $j \neq j_o$, $j = 1,\dots,m$.

From $\nu \neq 0$ and (32) it follows that (33) are (m-1) linear independent solvability conditions . They are necessary and sufficient . In this case there are no eigensolutions because for each eigensolution $\overset{o}{w}$ would follow from (27)

$$(34) \qquad \overset{o}{w} = K (\overset{o}{\vartheta} + i\overset{o}{\vartheta})$$

but from (29) with $U(0) = 0$ and $j=j_o$

$$(35) \qquad K \cdot \nu = 0$$

so that $\overset{o}{w} = 0$. We have the deficiency index (0,m-1) . Q E D .

Vekua's theory shows that the conditions (30) or (33) must be equivalent to the solvability conditions

$$(36) \qquad \oint_{\dot\Gamma} \varphi \frac{1}{|\lambda|^2} \mathrm{Im} (iz'\lambda h) \, ds = 0$$

for every eigensolution h of the homogeneous adjoint BVP

$$(37) \qquad h_{\bar z} = - Ah + \overline{Bh} \quad , \quad \mathrm{Re} (iz'\lambda h)\big|_{\dot\Gamma} = 0 .$$

§3. Some remarks on the case of an arbitrary index

We first deal with the case where all the indices for the single curves are zero:

(38) $$n_j := \frac{1}{2\pi} \oint_{\dot{\Gamma}_j} d \arg \lambda = 0 \quad \text{for every } j=0,1,\ldots,m.$$

From (38) the function τ in

(39) $$e^{i\tau} := \frac{\overline{\lambda}}{|\lambda|}$$

is continuous on the boundary $\dot{\Gamma}$. Using the analytic continuation

(40) $$\Lambda_1 := e^{i \oint_{\dot{\Gamma}} \tau \{ - d_n \Gamma^I + i d \Gamma^{II} \}}$$

we get for the new unknown function

(41) $$w_1 := \Lambda_1 w$$

the so called <u>canonical form</u> of the boundary conditions:

(42) $$\operatorname{Re} w_1 e^{-i\alpha_j} \Big|_{\dot{\Gamma}_j} = \varphi_1$$

with $\alpha_0 = 0$ and real constants $\alpha_1,\ldots,\alpha_m$; and also the differential equations

(43) $$w_{1\overline{z}} = A w_1 + B \frac{\Lambda_1}{\overline{\Lambda_1}} \overline{w_1} \quad .$$

In the general case we cannot find any analytic continuation of the boundary vector for which all the α_j-s vanish. But with a <u>non analytic</u> continuation, e.g. with

(44) $$\Lambda_2 := e^{-i \sum_{j=0}^{m} \chi_j(x,y)\alpha_j} ,$$

where $\{\chi_j\}$ is a suitable C^∞- decomposition of 1 with

(45) $$0 \le \chi_j \le 1 \;, \; \sum_{j=0}^{m} \chi_j \equiv 1 \;, \; \chi_j \Big|_{\dot{\Gamma}_k} = \delta_{jk} \;, \quad j,k=0,\ldots,m$$

we get for the new unknown function

$$(46)\qquad \hat{w} := \Lambda_1 \Lambda_2 w$$

the 1-st BVP :

$$(47)\qquad \hat{w}_{\bar{z}} = \left(A + \frac{\Lambda_{2\bar{z}}}{\Lambda_2}\right) \hat{w} + B \frac{\Lambda_1 \Lambda_2}{\overline{\Lambda_1}\,\overline{\Lambda_2}} \overline{\hat{w}} \;, \quad \operatorname{Re} \hat{w}\Big|_{\dot{\Gamma}} = \hat{\varphi} \;.$$

In the general case with an arbitrary index

$$(48)\qquad n = \sum_{j=0}^{m} n_j$$

the well-known transformation

$$(49)\qquad w = f(z)\,\hat{w} \;,$$

$$(50)\qquad f(z) := \left(\prod_{k=1}^{|n|} (z-z_k)^{\operatorname{sign} n} \right) \prod_{j=1}^{m} \prod_{k=1}^{|n_j|} (z-z_k^{(j)})^{\operatorname{sign} n_j}$$

with the zeros or poles

$$(51)\qquad z_1,\dots,z_{|n|} \in \Gamma \;; \quad z_k^{(j)} \in \hat{\Gamma}_j \;, \quad j=1,\dots,m$$ [3)]

reduces to a BVP for $\hat{w}$ with all the single indices zero :

$$(52)\qquad \operatorname{Re} (\hat{w}\bar{\lambda} f)\Big|_{\dot{\Gamma}} = \varphi \;, \quad n_j(\lambda \bar{f}) = 0 \quad \text{for all } j=0,\dots,m.$$

If we transform this BVP to the canonical form (42), then the α_j-s are functions of the zeros or poles of f :

$$(53)\qquad \alpha_j = \alpha_j(x_1+iy_1, x_2+iy_2, \dots, x_n+iy_n \;;\; z_k^{(j)}) \;.$$

B.Bojarski has shown that the α_j-s do not depend on $z_k^{(j)}$ because the $z_k^{(j)}$ are in the exterior of $\bar{\Gamma}$. Using the α_j-s he has made a classification of the so called special classes of BVPs for analytic functions.(A = B = C = 0)([2],IV)

---

3) $\hat{\Gamma}_j$ is the component of the complementa $\bar{\Gamma}^c$ bounded by $\dot{\Gamma}_j$.

It seems that this classification can be used to develop the solvability theory of BVPs only from the integral equations (9)(10).

In the special case of a simply connected domain m=0 the closing conditions (30) or (33) drop out and the whole theory follows very easily and directly from the integral equations (9)(10) as presented in [1].

References

[1] Haack, W. u. W. Wendland : Vorlesungen über Partielle und Pfaffsche Differentialgleichungen. Basel, Stuttgart: Birkhäuser Verlag 1970.

- " " : Lectures on Partial and Pfaffian Differential Equations. Oxford: Pergamon Press 1972.

[2] Vekua, I.N.: Generalized Analytic Functions. Oxford: Pergamon Press 1962.

[3] Wendland, W.: Greensche Funktionen erster und zweiter Art und eine Anwendung auf Riemann-Hilbert Probleme. Meth. u. Verf. d. Mathematischen Physik (BI Mannheim) 1972 to appear.

Polynomial Solutions of Differential Equations with Bi - Orthogonal Properties

F.M. Arscott

1. Context of the Problem

"Ordinary" orthogonal polynomials on an interval (a,b) with weight function $\underline{w}$ form a singly-infinite set $\{p_n(x)\}$, n = 0 to ∞, with the property that

$$\int_a^b p_n(x)\, p_{n'}(x)\, w(x)\, dx = 0\,, \quad n \neq n', \qquad \neq 0\,, \quad n = n'. \tag{1}$$

Some of the properties of orthogonal polynomials are general in the sense that they follow from the orthogonality relation (1) alone; these include approximation properties, the recurrence relation and the Christoffel-Darboux formula. Other properties, however - differential equation, generating function, Rodrigues-type formula - are possessed only by the so-called "classical" orthogonal polynomials. These comprise, in effect, the polynomials of Jacobi (with many special cases), Laguerre and Hermite. ([5] , Chap. X).

Bi-orthogonal polynomials, by contrast, form a "triangularly-infinte" set $\{p_n^m(x)\}$, n = 0 to ∞, m = 0 to n, and satisfy an orthogonality relation involving a double integral of the form

$$\iint_R p_n^m(x)\, p_n^m(\xi)\, p_{n'}^{m'}(x)\, p_{n'}^{m'}(\xi)\, W(x,\xi)\, dx\, d\xi = 0 \tag{2}$$

unless n = n' and m = m', but is non-zero when n = n', m = m'.

Here $\underline{R}$ is a region (finite or infinite) of the x,ξ plane and $W(x,\xi) \geqslant 0$ a weight function such that $\iint_R x^M \xi^N W(x,\xi)\, dx\, d\xi$ exists for all $M,N \geqslant 0$.

From one point of view, this theory is embedded in that of orthogonal polynomials in two variables, i.e. polynomials such that ([2] ,[5] ,Chap XII.)

$$\iint_R P_n^m(x,\xi)\, P_{n'}^{m'}(x,\xi)\, W(x,\xi)\, dx\, d\xi = 0\,, \tag{3}$$

for $n \neq n'$ or $m \neq m'$; here $P(x,\xi)$ is not, in general, separable into the product of polynomials in x and in ξ. From the relation (3) one can obtain approximation properties, a recurrence relation and a Christoffel-Darboux type formula but none analogous to the properties possessed by the classical orthogonal polynomials. Indeed, until recently only one set of polynomials with a bi-orthogonality property were known; these were the Lamé polynomials ([1] , Chap.IX), for which the property (2) was equivalent to orthogonality over the surface of an ellipsoid. In 1962 it was noted that the Ince polynomials also possessed such a property ([4]), related to orthogonality over a paraboloid, and soon after the Heun polynomials ([7]) were observed to have the same; these are a generalisation of the Lamé polynomials. ([6])

The object of this article is to construct a number of families of polynomials satisfying a relation of the form (2), as solutions of differential equations. Some of these are simpler than any known bi-orthogonal polynomials and so may play a part in this theory similar to that of the classical polynomials in the older theory.

2. Bi-orthogonality derived from a differential equation

Consider the differential equation

$$A(x)\frac{d^2y}{dx^2} + B(x)\frac{dy}{dx} + C(x)\,y = 0, \tag{4}$$

where A,B,C are polynomials and for the moment we take C(x) in the special form

$$C(x) = \lambda f(x) + \mu g(x). \tag{5}$$

We also set

$$R(x) = \exp\left\{\int B(x)/A(x)\, dx\right\}. \tag{6}$$

Then a simple adaptation of the method used in [3] gives the following result: if $\underline{y}$ and $\underline{\hat{y}}$ are polynomial solutions of (4) corresponding to different values of λ and/or μ, then

$$\int_{\xi_1}^{\xi_2}\int_{x_1}^{x_2} y(x)\, y(\xi)\, \hat{y}(x)\, \hat{y}(\xi)\, W(x,\xi)\, dx\, d\xi = 0, \tag{7}$$

where

$$W(x,\xi) = \frac{R(x)\,R(\xi)}{A(x)\,A(\xi)}\left\{f(x)\,g(\xi) - f(\xi)\,g(x)\right\},$$

provided

$$\left[R(x)\left\{y(x)\,\hat{y}'(x) - y'(x)\,\hat{y}(x)\right\}\right]_{x_1}^{x_2} = 0,$$
$$\left[R(\xi)\left\{y(\xi)\,\hat{y}'(\xi) - y'(\xi)\,\hat{y}(\xi)\right\}\right]_{\xi_1}^{\xi_2} = 0 \tag{8}$$

and the paths of integration do not pass through a zero of $\underline{A}$. A zero of $\underline{A}$ may, however, occur at one end of a range provided the resulting integral converges.

3. The search for polynomial solutions of (4)

It would be an impossible task to obtain all polynomial solutions of an equation such as (4); instead we seek only those satisfying a reasonably simple differential equation and yielding a reasonably simple orthogonality relationship (2).

In (4), we write $A(x) = \sum_{i=0} \alpha_i x^i$, $B(x) = \sum_{i=0} \beta_i x^i$, $C(x) = \sum_{i=0} \gamma_i x^i$ and then assume a formal solution $y = x^c \sum_{r=0}^{\infty} a_r x^r$. This gives the recurrence relations

$$c(c-1)\alpha_0 a_1 = 0, \quad (c+1)c\alpha_0 a_1 + c(c-1)\alpha_1 a_0 + c\beta_0 a_0 = 0, \qquad (10a,b)$$

$$\sum_{s=0}^{r+2} (c+r+2-s)(c+r+1-s)\alpha_s a_{r+2-s} + \sum_{s=0}^{r+1} (c+r+1-s)\beta_s a_{r+1-s} + \sum_{s=0}^{r} \gamma_s a_{r-s} = 0. \qquad (10c)$$

One easily discovers that we cannot obtain polynomials of the required type by assuming a two-term recurrence relation between the coefficients a_r, as is the case for the classical polynomials. We therefore try to build up polynomial solutions on the basis of a three-term recurrence relationship.

Now, the complexity of equation (4) depends principally on the complexity of A(x), since its singularities occur, of course, at the zeros of A. So a natural classification of the resulting polynomials hangs on the number and location of the zeros of A(x), regarding x as a complex variable.

4. The simplest family; analogue of the Hermite polynomials

The obvious first choice is to take A = constant, but this is found to yield no polynomials of the type required. We therefore suppose A to have one simple zero, which for simplicity we take at the origin, thus A(x) = x; thus $\alpha_0 = 0$, $\alpha_1 = 1$, $\alpha_i = 0$, $i \geq 2$. Then (10b) gives the exponents c = 0, $c = 1 - \beta_0$, but since we require a polynomial we take c = 0 and assume $\beta_0 \neq 0, -1, -2, \ldots$. We next see that for a 3-term recurrence relation we need $\beta_i = 0$, $i \geq 3$, $\gamma_i = 0$, $i \geq 2$, and the relation is

$$(r+2)(r+3+\beta_0)\, a_{r+2} + \{(r+1)\beta_1 + \gamma_0\}\, a_{r+1} + (r\beta_2 + \gamma_1)\, a_r = 0. \qquad (11)$$

Now, to obtain a polynomial of degree n, we must (i) take $n\beta_2 + \gamma_1 = 0$ and (ii) choose γ_0 so that $a_{n+1} = 0$. This gives $a_r = 0$, $r \geq n+2$ also, so that y is a polynomial of precise degree n. The differential equation is then

$$x \frac{d^2y}{dx^2} + (\beta_0 + \beta_1 x + \beta_2 x^2)\frac{dy}{dx} + (\gamma_0 - n\beta_2 x)\, y = 0. \qquad (12)$$

Now we examine the orthogonality relation. We have

$R(x) = x^{\beta_0} \exp(\beta_1 x + \frac{1}{2}\beta_2 x^2)$, $f(x) = 1$, $g(x) = \beta_2 x$, $\lambda = \gamma_0$, $\mu = n$,

and
$$W(x,\xi) = (x\xi)^{\beta_0 - 1} e^{\beta_1(x+\xi)} e^{\frac{1}{2}\beta_2(x^2+\xi^2)} \beta_2(\xi - x). \tag{13}$$

It remains to choose the paths of integration (x_1, x_2) and (ξ_1, ξ_2). If we impose the restrictions $\beta_0 > 0$, $\beta_2 < 0$ we can conveniently take $(x_1, x_2) = (-\infty, 0)$ and $(\xi_1, \xi_2) = (0, \infty)$.

A particularly simple case is obtained by taking $\beta_0 = 1$, $\beta_1 = 0$, $\beta_2 = -2$, when the equation becomes

$$x \frac{d^2y}{dx^2} + (1 - 2x^2)\frac{dy}{dx} + (\lambda + 2nx)\, y = 0, \tag{14}$$

the recurrence relation is

$$(r+2)^2 a_{r+2} + \lambda a_{r+1} + 2(n-r)\, a_r = 0, \tag{15}$$

and the orthogonality property is that if y, $\hat{y}$ are solutions of this with different λ and/or n , then

$$\int_0^{\infty} d\xi \int_{-\infty}^{0} y(x)\, y(\xi)\, \hat{y}(x)\, \hat{y}(\xi)\, e^{-x^2-\xi^2} (\xi - x)\, dx = 0. \tag{16}$$

Our choice $\beta_2 < 0$ has the useful consequence that for given n, the n+1 values of λ which make $a_{n+1} = 0$ are all real and different, so that we do indeed obtain precisely n+1 distinct polynomials of degree n and thus the bi-orthogonal set $\{p_n^m(x)\}$; the form of (16) makes it reasonable to regard these as somewhat analogous to the Hermite polynomials.

The first few polynomials are: $n = 0 : p(x) = 1$,

$n = 1: p(x) = x \pm 1/\sqrt{2}$,

$n = 2,\ p(x) = x^2 - 1,\ x^2 \pm \sqrt{3}\, x + \frac{1}{2}$.

5. Further families

(a) Let $A(x)$ have two zeros; take them at 0,1 i.e. $A(x) = x(1-x)$, then by similar reasoning we get the equation

$$x(1-x)\frac{d^2y}{dx^2} + (\beta_0 + \beta_1 x + \beta_2 x^2)\frac{dy}{dx} + (\lambda - n\beta_2 x)\, y = 0,$$

and
$$W(x,\xi) = (x\xi)^{\beta_0 - 1}\{(1-x)(1-\xi)\}^{\delta - 1} e^{-\beta_2(x+\xi)} (\xi - x),$$

where $\delta = -\beta_0 - \beta_1 - \beta_2$. For the ranges of integration there are two convenient possibilities

(i) $\beta_0 > 0$, $\delta > 0$, $\beta_2 > 0$, $(x_1, x_2) = (0,1)$, $(\xi_1, \xi_2) = (1, \infty)$.

To make $W(x,\xi)$ real, we replace the factor $(1-\xi)$ by $(\xi - 1)$.

(ii) $\beta_0 > 0$, $\delta > 0$, $\beta_2 < 0$, $(x_1,x_2) = (-\infty,0)$, $(\xi_1,\xi_2) = (0,1)$.

For convenience, write $-x$ for x in $W(x,\xi)$.

The Ince polynomials arise from the case $\beta_0 = \delta = \frac{1}{2}$, but a simpler case is given by $\beta_0 = 1$, $\beta_1 = 0$, $\beta_2 = -2$, giving the orthogonality relation

$$\int_0^1 d\xi \int_{-\infty}^0 y(x)\, y(\xi)\, \hat{y}(x)\, \hat{y}(\xi)\, e^{2(x+\xi)}(\xi - x)\, dx = 0. \qquad (18)$$

The weight function indicates an analogy with the Laguerre polynomials. It is again found that there are n+1 polynomials of degree n.

(b) Finally, let A have three finite zeros, at 0,1,b where for simplicity we take b real, $b > 1$. Then the differential equation is found to be

$$x(x-1)(x-b)\frac{d^2y}{dx^2} + (\beta_0 + \beta_1 x + \beta_2 x^2)\frac{dy}{dx} + \left\{\lambda - n(n-1+\beta_2)\, x\right\} y = 0, \qquad (19)$$

which is seen to be, in fact, Heun's equation. The weight function is

$$(x\,\xi)^{\gamma-1}\left\{(1-x)(\xi-1)\right\}^{\delta-1}\left\{(b-x)(b-\xi)\right\}^{\epsilon-1}(\xi - x) \qquad (20)$$

and $(x\ ,x_2) = (0,1)$, $(\xi_1, \xi_2) = (1,b)$. Here γ, δ, ϵ are linear combinations of $\beta_0, \beta_1, \beta_2$ and we must specify $\gamma > 0$, $\delta > 0$, $\epsilon > 0$.

Lamé polynomials arise from $\gamma = \delta = \epsilon = \frac{1}{2}$, but the simplest case arises from $\gamma = \delta = \epsilon = 1$, when the equation takes the self-adjoint form

$$\frac{d}{dx}\left\{x(x-1)(x-b)\frac{dy}{dx}\right\} + \left\{\lambda - n(n+2)\, x\right\} y = 0. \qquad (21)$$

Both the ranges of integration now being finite, the Heun polynomials can be thought of as analogous to the classical Jacobi polynomials.

REFERENCES

1. F.M.Arscott, Periodic Differential Equations, Pergamon Press, 1964
2. F.M.Arscott, Orthogonal Bipolynomials, Z.A.M.M. , Vol.48, 1968
3. F.M.Arscott, Two-parameter eigenvalue problems in differential equations, Proc.Lond.Math.Soc. 14, 1964, 459-70
4. F.M.Arscott, The Whittaker-Hill equation and the wave equation in paraboloidal coordinates, Proc.Roy.Soc.Edin. LXVII, 1965.
5. A.Erdélyi et al: Higher Transcendental Functions, Vol.2, Mc-Graw-Hill 1953.
6. A.Erdélyi et al: ------ ------------ --------, Vol.3, -------,1955
7. B.D.Sleeman, Thesis, University of London, 1966.

Champs de Vecteurs et Régularité des Solutions de Certains Problèmes de Dirichlet Variationnels

M. Authier

1- Introduction : L'objet de ce travail est essentiellement de cerner les limites de la méthode Nirenberg (ou méthode des quotients différentiels, Cf.[6]) dans l'étude de la régularité des solutions de problèmes de Dirichlet variationnels non nécessairement elliptiques.

Soit Ω un ouvert de R^n tel que $\bar{\Omega}$ soit une variété C^∞ à bord; soit A(D), $D = -i\frac{\partial}{\partial x}$, un opérateur différentiel linéaire à coefficients constants, positif dans Ω, i.e. vérifiant:

$$\langle A(D)\varphi, \bar{\varphi}\rangle \geq 0 \quad , \quad \forall \varphi \in \mathcal{D}(\Omega).$$

Soit V' le sous-espace hilbertien de $\mathcal{D}'(\Omega)$ de noyau A (Cf. [7]); si V' est normal (par exemple si Ω est borné), son dual V est un sous-espace hilbertien de $\mathcal{D}'(\Omega)$ normal, et A(D) est un isomorphisme de V sur V'. Nous appellerons problème de Dirichlet associé à A le problème aux limites ainsi posé. On peut évidemment poser ce même problème par la méthode variationnelle. On note A^{-1} l'isomorphisme réciproque de A.

Soit maintenant X un champs de vecteurs sur $\bar{\Omega}$, autrement dit un opérateur différentiel du premier ordre à coefficients C^∞ réels, tangent au bord. On note D(X,V) l'ensemble des $u \in V$ tels que $Xu \in V$. Soit d'autre part

$$V_1' = \left\{ f \in V' \; ; \; D^\alpha f \in V' \; , \; |\alpha| \leq 1 \right\}$$

on s'intéresse au problème suivant:

(R) | caractériser les champs de vecteurs X sur $\bar{\Omega}$ vérifiant: si $f \in V_1'$ alors $A^{-1}f \in D(X,V)$.

Les champs de vecteurs vérifiant (R) seront dits réguliers pour A dans Ω.

L'intérêt de tels champs de vecteurs est le suivant: d'une part (R) est en lui même un résultat de régularité des solutions du problème de Dirichlet dans Ω associé à A; d'autre part, lorsqu'on analyse la méthode Nirenberg dans le cas elliptique, on constate qu'une étape de cette méthode (régularité à l'intérieur et tangentielle au bord) revient à montrer que certaines dérivées (i.e. champs de vecteurs à coefficients constants) constituent des champs de vecteurs réguliers dans un demi-espace.

La considération de champs de vecteurs à coefficients non constants dans des problèmes de régularité a déja été employée, par exemple pour des problèmes elliptiques dans des ouverts "à coins" [3].

On va s'intéresser ici essentiellement au cas où A a des caracteristiques réelles, le cas elliptique étant bien connu; et, plus précisément au cas où cet opérateur est hyperbolique. On sait (Cf. [1]) qu'alors sa partie principale est le carré d'un opèrateur hyperbolique: $A_{2m}(\xi) = [B(\xi)]^2$; on étudiera plus spécialement le cas où B est strictement hyperbolique. D'autre part on ne donnera ici que des conditions nécessaires pour que X soit régulier pour A dans Ω. La question de la suffisance de ces conditions fait l'objet d'un autre travail: [2]. La condition obtenue (proposition 2) est purement géométrique, et d'ailleurs assez naturelle (Cf. remarque).

<u>2- Une condition nécessaire</u>: Soit donc A(D) un opérateur positif d'ordre 2m, A_{2m} sa partie principale, on désigne par C(x,D) la partie principale de X^*AX, où X^* est l'adjoint formel de X. On a alors la proposition:

<u>Proposition 1</u> : <u>Une condition nécessaire pour que le champs de vecteurs X sur $\overline{\Omega}$ soit régulier pour A dans Ω est que, pour tout $x \in \overline{\Omega}$, si $A_{2m}(\xi) = 0$, $\xi \in R^n$, alors $C(x,\xi) = 0$.</u>

<u>Démonstration</u>: On vérifie facilement que D(X,V) muni de la norme du graphe et V_1' muni de la norme naturelle sont des espaces de Banach. D'autre part A^{-1} est linéaire et de graphe fermé de V_1' dans D(X,V), donc continue. On a donc

(1) $$\|XA^{-1}f\|_V + \|A^{-1}f\|_V \leq C\left[\sum_{i=1}^{n} \|D_i f\|_{V'} + \|f\|_{V'}\right], \quad \forall f \in V_1'.$$

D'où, en utilisant le fait que V_1' contient $A(D)\mathcal{D}(\Omega)$ et que, pour $\varphi \in \mathcal{D}(\Omega)$, $A^{-1}(A\varphi)$ n'est autre que φ, l'inégalité

(2) $$\|X\varphi\|_V \leq C\left[\sum_{i=1}^{n} \|D_i A\varphi\|_{V'} + \|A\varphi\|_{V'}\right], \quad \forall \varphi \in \mathcal{D}(\Omega).$$

En exprimant alors les normes dans V et V' de façon standard, et en utilisant la définition de X^*, (2) peut encore s'écrire

(3) $$\langle X^*AX\varphi, \bar{\varphi}\rangle \leq C \langle A(1-\Delta)\varphi, \bar{\varphi}\rangle, \quad \forall \varphi \in \mathcal{D}(\Omega)$$

La proposition sera alors conséquence du lemme suivant.

<u>Lemme</u> : <u>Soient P(D) et Q(x,D) deux opérateurs différentiels d'ordre 2m, positifs (au sens de [7]), une condition nécessaire de validité d'une inégalité du type:</u>

(4) $$\langle Q(x,D)\varphi, \bar{\varphi}\rangle \leq C \langle P(D)\varphi, \bar{\varphi}\rangle, \quad \forall \varphi \in \mathcal{D}(\Omega)$$

est que $P_{2m}(\xi) = 0$, $\xi \in R^n$, implique $Q_{2m}(x,\xi) = 0$, pour tout $x \in \bar{\Omega}$.

Démonstration: Supposons que $\xi = (1,0,\dots,0)$ vérifie $P_{2m}(\xi) = 0$ et $Q_{2m}(x_0,\xi) = 0$, avec $x_0 = (0,\dots,0)$. Alors $P(\xi)$ ne contient pas de terme de degré 2m en ξ_1, et $Q(x,\xi) = a_{2m}(x)\xi_1^{2m} + R(x,\xi)$, où R ne contient pas de terme de degré 2m en ξ_1. Soit U un voisinage de x_0 dans Ω dans lequel $|a_{2m}(x)| \geq e > 0$. On considère deux fonctions $\Psi \in \mathcal{D}(R)$ et $\theta \in \mathcal{D}(R^{n-1})$ telles que supp $\Psi(x_1)\theta(x_2,\dots,x_n) \subset U$, et la suite de fonctions de $\mathcal{D}_U(R^n)$ définie par

$$\varphi_p(x) = p^{1-m}\,\Psi(px_1)\,\theta(x_2,\dots,x_n)$$

On voit facilement que la suite de nombres réels $\langle P(D)\varphi_p, \bar{\varphi}_p\rangle$ est bornée, alors que la suite $\langle Q(x,D)\varphi_p, \bar{\varphi}_p\rangle$ est un infiniment grand avec p.

Remarque 1: La proposition 1 ne donne aucune condition sur X lorsque A est elliptique. On peut effectivement montrer que tout champ de vecteurs sur $\bar{\Omega}$ est régulier dans ce cas.

Venons en maintenant au cas où A(D) est un opérateur hyperbolique positif à caractéristi ques d'ordre 2 au plus, on fera aussi l'hypothèse que A est homogène. On peut alors écrire $A(D) = [B(D)]^2$, B étant un opérateur strictement hyperbolique homogène.

Avant d'énoncer le résultat dans ce cas, il faut introduire une notation. Soit Ω une variété différentiable de dimension n, X un champs de vecteurs sur Ω, α une forme différentielle de degré 1, on notera $\langle X,\alpha\rangle$ le produit intérieur $i_X\alpha$. Son expression en coordonnées locales est, pour $X = \sum_{i=1}^n a_i(x)\frac{\partial}{\partial x_i}$ et $\alpha = \sum_{i=1}^n \alpha^i(x)\,dx_i$, $\langle X,\alpha\rangle(x) = \sum_{i=1}^n a_i(x)\,\alpha^i(x)$; c'est une fonction différentiable sur Ω. Dans le cas où Ω est un ouvert de R^n, on identifie le vecteur $\xi \in R^n$ avec la section $\xi : x \longmapsto (x,\xi)$ du fibré (trivial) $T^*(\Omega)$, d'où la notation $\langle X,\xi\rangle$.

Proposition 2 : Soit A(D) hyperbolique homogène positif à caractéristiques d'ordre 2 au plus, si X est régulier pour A dans Ω et si $\xi \in R^n$, $\xi \neq 0$, vérifie $A(\xi) = 0$, alors $\langle X,\xi\rangle$ est constant sur les bicaractéristiques de B associées à ξ .

Démonstration : Soit $L_m(x,\xi)$ la partie principale du commutateur BX-XB, on a d'une part:

$$(5) \qquad L_m(x,\xi) = \overrightarrow{\mathrm{grad}}_\xi\, B(\xi) . \overrightarrow{\mathrm{grad}_x \langle X,\xi\rangle} \ ;$$

d'autre part la proposition 1 implique, si X est régulier pour A, qu'en chaque point $x \in \bar{\Omega}$ $L_m(x,\xi)$ est divisible par $B(\xi)$ (car B est strictement hyperbolique), autrement dit il existe une fonction $\lambda \in \mathcal{E}(\bar{\Omega})$ telle que :

(6) $\qquad L_m(x,\xi) = \lambda(x)\, B(\xi), \qquad \forall\, x \in \bar{\Omega}.$

Il ne reste plus qu'à rassembler (5) et (6) pour mettre en évidence le résultat.

<u>3- Applications et Remarques</u>: Les champs de vecteurs réguliers pour A dans Ω forment un espace vectoriel qui, dans une certaine mesure caractérise la régularité du problème de Dirichlet. On notera, pour tout $x \in \bar{\Omega}$, $R_x(A)$ le sous espace vectoriel de $T_x(\bar{\Omega})$ engendré par ces champs de vecteurs, $r_x(A)$ sa dimension, $\rho_x(A)$ la dimension de l'espace vectoriel engendré par les champs de vecteurs vérifiant la condition de la proposition 2 ($r_x(A) \leq \rho_x(A) \leq n$ si $x \in \Omega$, $r_x(A) \leq \rho_x(A) \leq n-1$ si $x \in \partial\Omega$). On va exploiter la proposition 2 sur un exemple.

<u>Exemple:</u> Ω est un ouvert du plan R^2 et $A(D) = D_1^2\, D_2^2$. Alors la proposition 2 s'énonce: Les champs de vecteurs réguliers pour A sont de la forme $X = a(x_1)\frac{\partial}{\partial x_1} + b(x_2)\frac{\partial}{\partial x_2}$.

En un point x_o caractéristique de $\partial\Omega$ deux cas peuvent se produire:

a) Ω est convexe au voisinage de x_o, alors en écrivant que X, de la forme ci-dessus, est tangent au bord on constate qu'il est nécessairement nul au point x_o. Par suite en tout point de l'unique bicaractéristique passant par x_o et rencontrant Ω, X est dirigé suivant cette bicaractéristique.

b) Ω n'est pas convexe au voisinage de x_o, alors les deux bicaractéristiques passant par x_o rencontrent Ω, et la condition de tangence de X montre que, sur celle qui est tangente au bord, le résultat précédent est encore vrai.

Soit alors b une bicaractéristique passant par un point caractéristique de $\partial\Omega$ telle que X soit dirigé suivant b. Supposons que b rencontre $\partial\Omega$ en un autre point y_o, non caractéristique; la tangence de X implique alors sa nullité en y_o. Par suite X est dirigé suivant l'autre bicaractéristique b' passant par y_o, en tout point deb'. Autrement dit, il y a réflexion de la condition sur X sur le bord.

Ces considérations permettent le calcul de $\rho_x(A)$ (on verra dans [2], qu'en fait, dans le cas présent, $r_x(A) = \rho_x(A)$ pour tout x). Donnons quelques exemples typiques:

(i) $\Omega = \{(x_1,x_2);\ t_o < x_1+x_2 < t_1\}$ (bande spatiale, Cf [1], ch.III), alors $\rho_x(A) = 2$ pour tout $x \in \Omega$, $\rho_x(A) = 1$ pour $x \in \partial\Omega$.

(ii) $\Omega = \left\{(x_1,x_2);\ x_1 \geq x_2^2\right\}$, alors $\rho_x(A) = 2$ pour $x \in \Omega$, $x_2 \neq 0$; $\rho_x(A) = 1$ pour $x \in \Omega$ avec $x_2 = 0$, où bien $x \in \partial\Omega$, $x_2 \neq 0$; $\rho_x(A) = 0$ pour $x = (0, 0)$.

(iii) Dans [1] on a donné un exemple d'ouvert, non-convexe au voisinage du seul point caractéristique de sa frontière, avec une solution singulière sur la bicaractéristique correspondante, montrant qu'aucun gain de régularité n'était possible transversalement à cette bicaractéristique.

(iv) On peut construire un exemple (domaine limité par une ellipse) où l'ensemble des points d'intersection des bicaractéristiques portant X est dense dans Ω, et où, par suite, X est nécessairement nul identiquement. (Cf. [2] et [4]).

Remarque 2: Quelques considérations heuristiques permettaient de prévoir les résultats ci-dessus. D'abord les résultats de propagation des singularités (Cf. [5]) pour un opérateur à caractéristiques réelles simples, et le fait (Cf. [1]) que, pour le problème de Dirichlet étudié, le nombre des conditions de nullité au bord indépendantes diminue aux points caractéristiques du bord, font que les bicaractéristi ues issues des points caractéristiques du bord sont un domaine naturel pour les singularités des solutions. Enfin, en tout point non-caractéristique, le nombre des conditions au bord est suffisant pour que les singularités se réfléchissent sur le bord.

BIBLIOGRAPHIE

1 Authier M.: Problème de Dirichlet pour des opérateurs hyperboliques de type positif. Ann. Sc. Sup. Pisa (1971) (à paraître)

2 Authier M.: Régularité des solutions de problèmes de Dirichlet hyperboliques par la méthode des quotients différentiels (à paraître).

3 Bolley P. Camus J.: Certains résultats de régularité des problèmes elliptiques variationnels par la méthode des quotients différentiels. Univ. Rennes (1969)

4 Hadamard J.: Le problème de Dirichlet pour les équations hyperboliques. J. Chinese math. Soc. 2 (1937) p.620.

5 Hörmander L.: On the existence and the regularity of solutions of linear partial differential equations. Enseign. Math. 17 (1971) p 99.

6 Lions J.L. Magenes E.: Problèmes aux limites non-homogènes T 1. Dunod (1968).

7 Schwartz L.: Sous espaces hilbertiens d'espaces vectoriels topologiques et noyaux associés. J. Anal. Math. Jerusalem 13 (1964) p. 115.

Symmetric Relations on a Hilbert Space

Ch. Bennewitz

1. INTRODUCTION. In some recent papers ([2], [3], [4], [5], [6]) from Uppsala problems related to the eigenvalue problem $Su = \lambda Tu$, where S and T are formally symmetric, ordinary differential operators, have been treated. The equation $Su = T\hat{u}$ defines a differential relation between u and $\hat{u}$, and this seems to be the natural setting for a spectral theory concerning S and T.

This paper outlines an abstract theory of symmetric linear relations that was inspired by the above concrete situation, but also can be applied, e.g. to nondensely defined symmetric operators. The first treatment of a spectral theory for relations seems to be by Arens [1], who, however, does not treat the resolvent operator of a maximal symmetric relation.

2. LINEAR RELATIONS. Let H be a Hilbert space and S a linear subset of $H \times H = H^2$. We will call S a linear relation on H. If S is a subspace of H^2 we say that S is a closed linear relation on H.

Define the boundary operator $\mathcal{U}$ by

$$\mathcal{U}(u, v) = (iv, iu) \quad \text{for} \quad (u, v) \in H^2 .$$

Clearly $\mathcal{U}$ is defined everywhere, is isometric, selfadjoint and involutary. Orthogonal complement in a Hilbert space is denoted by $\ominus$ and $\oplus$ means direct sum of orthogonal subspaces. We define the adjoint relation S^* of a linear relation S, by

$$S^* = H^2 \ominus \mathcal{U} S .$$

Clearly S^* is a closed linear relation on H. An immediate consequence is

<u>Theorem 1</u>: The closure $\bar{S}$ of S equals $\bar{S} = S^{**}$, $S \subset T \Rightarrow T^* \subset S^*$.

A subspace S will be called <u>symmetric</u> if $S \subset S^*$ and <u>selfadjoint</u> if $S^* = S$.

3. <u>EXTENSIONS OF SYMMETRIC RELATIONS</u>. If S is a symmetric relation $\bar{S}$ is a closed symmetric extension of S. Hence, from now on <u>we assume that already S is closed</u>. Further, from Theorem 1 it follows that if S_0 is a symmetric extension of S, then $S \subset S_0 \subset S_0^* \subset S^*$. Hence, any symmetric extension of S is a restriction of S^*. Elements of H^2 are denoted by capitals U, V, W etc. Corresponding to the deficiency spaces of a symmetric operator we define

$$D_+ = \{U \in S^* : \mathcal{U}U = U\},$$

$$D_- = \{U \in S^* : \mathcal{U}U = -U\}.$$

For a closed symmetric relation we have the basic

<u>Theorem 2</u>: $S^* = S \oplus D_+ \oplus D_-$.

<u>Proof</u>: An easy check shows that S, D_+ and D_- are orthogonal. Denote by $\langle \cdot , \cdot \rangle$ the scalar product in H^2. Let $U \in S^* \ominus \{S \oplus D_+ \oplus D_-\}$. Then for any $V \in S$ we have $\langle V, U \rangle = 0$ so that $\mathcal{U}U = \mathcal{U}^{-1}U \in S^*$ and consequently $(\mathcal{U} + I)U \in S^*$, where I is the identity operator. But $\mathcal{U}(\mathcal{U} + I)U = (\mathcal{U}^2 + \mathcal{U})U = (\mathcal{U} + I)U$ so that $(\mathcal{U} + I)U \in D_+$. But if $W \in D_+$ we get $\langle (\mathcal{U} + I)U, W \rangle = \langle U, (\mathcal{U} + I)W \rangle = 2 \langle U, W \rangle = 0$ so that $(\mathcal{U} + I)U = 0$. Hence $U \in D_-$ and therefore $U = 0$ which concludes the proof.

We now can characterize the symmetric extensions of a closed symmetric relation.

<u>Theorem 3</u>: If S_0 is a symmetric extension of the closed symmetric relation S, then $S_0 = S \oplus D$, where D is a subset of $D_+ \oplus D_-$ which is closed if S_0 is.

Furthermore, if we write elements of D as $U_+ + U_-$ with $U_+ \in D_+$ and $U_- \in D_-$, then the set $\{(U_+, U_-) \in D_+ \times D_-;\ U_+ + U_- \in D\}$ is the graph of an isometry of part of D_+ onto part of D_-. Conversely, every such set D gives rise to a symmetric extension $S_0 = S \oplus D$ of S.

Proof: The proof is obvious after noting that if $U_+ + U_- \in D$, $V_+ + V_- \in D$ with U_+ and $V_+ \in D_+$ and U_- and $V_- \in D_-$, then $\langle U_+, V_+\rangle = \langle U_-, V_-\rangle$ if and only if $\langle U_+ + U_-, \mathcal{U}(V_+ + V_-)\rangle = 0$.

Corollary: If S is a closed symmetric relation, then S is maximal precisely when either $D_+ = \{0\}$ or $D_- = \{0\}$. Every symmetric relation has a maximal symmetric extension.

4. SPECTRAL THEORY. In this section we assume that S is a maximal symmetric relation in H, that $D_- = \{0\}$ and that λ is a nonreal number with $\operatorname{Im} \lambda > 0$.

If we let $\langle \cdot, \cdot\rangle$ denote also the scalar product in H, and if $U = (u, \mu u) \in S$ it follows that $0 = \langle U, \mathcal{U} U\rangle = i^{-1}(\mu - \bar{\mu}) \langle u, u\rangle$. Since μ is nonreal we get $u = 0$. Furthermore, $S^* = S \oplus D_+$ so that $U \in S^*$ has a unique representation $U = U_0 + U_+$ with $U_0 \in S$ and $U_+ \in D_+$. Then $\langle U, \mathcal{U} U\rangle = \langle U_+, U_+\rangle \geq 0$. If, with $\operatorname{Im} \lambda > 0$, $U = (u, \bar{\lambda} u) \in S^*$ we get that $0 \leq \langle U, \mathcal{U} U\rangle = -i^{-1}(\lambda - \bar{\lambda}) \langle u, u\rangle \leq 0$ so that $u = 0$.

Consequently, we can define operators, called resolvent operators, in the following way. First put, for a nonreal μ,

$$D_\mu = \{u \in H : (v, \mu v + u) \in S \text{ for some } v \in H\}.$$

From the previous reasoning it follows that if $(v, \mu v + u) \in S$, then v is uniquely determined by u. Hence we can define R_μ by

$$R_\mu u = v \text{ if } (v, \mu v + u) \in S \text{ for any } u \in D_\mu.$$

Similarly, put

$$D'_{\bar{\lambda}} = \{u \in H : (v, \bar{\lambda} v + u) \in S^* \text{ for some } v \in H\}.$$

Again v is uniquely determined by u and we define the operator $R'_{\bar\lambda}$ by putting

$$R'_{\bar\lambda} u = v \text{ if } (v, \bar\lambda v + u) \in S^* \quad \text{for any } u \in D'_{\bar\lambda} .$$

We also need to define

$$\tilde S = \{u \in H : (u, \dot u) \in S \text{ for some } \dot u \in H\}$$

and

$$\tilde S^* = \{u \in H : (u, \dot u) \in S^* \text{ for some } \dot u \in H\}.$$

Obviously $\tilde S \subset \tilde S^*$.

<u>Theorem 4</u>: $R_\mu D_\mu = \tilde S$ for any nonreal μ. $R'_{\bar\lambda} D'_{\bar\lambda} = \tilde S^*$ when $\operatorname{Im}\lambda > 0$.

The proof is immediate.

<u>Theorem 5</u>: R_μ is a closed operator for all nonreal μ. $R'_{\bar\lambda}$ is a closed operator. $D_\lambda = D'_{\bar\lambda} = H$ and $R^*_\lambda = R'_{\bar\lambda}$, $(R'_{\bar\lambda})^* = R_\lambda$. Finally, $\|R_\mu\| \leq |\operatorname{Im}\mu|^{-1}$, $\|R'_{\bar\lambda}\| \leq (\operatorname{Im}\lambda)^{-1}$.

<u>Proof</u>: Since S and S^* are closed it follows that R_μ and $R'_{\bar\lambda}$ are closed operators. Now, if $U = (v, \bar\lambda v + u) \in S^*$ we have as before that $\langle U, \mathcal{U} U\rangle \geq 0$. However,

$$\langle U, \mathcal{U} U\rangle = 2 \operatorname{Im} \langle u, v\rangle - i^{-1}(\lambda - \bar\lambda) \langle v, v\rangle$$

so that $\langle U, \mathcal{U} U\rangle \geq 0$ together with the Cauchy-Schwarz inequality gives $|v| \leq (\operatorname{Im}\lambda)^{-1} |u|$. Similarly, if $U = (v, \mu v + u) \in S$ we have $\langle U, \mathcal{U} U\rangle = 0$ and it follows $|v| \leq |\operatorname{Im}\mu|^{-1} |u|$. Hence $\|R'_{\bar\lambda}\| \leq (\operatorname{Im}\lambda)^{-1}$ and $\|R_\mu\| \leq |\operatorname{Im}\mu|^{-1}$. Since R_μ and $R'_{\bar\lambda}$ are closed, bounded operators, they have closed domains. To show that $D_\lambda = D'_{\bar\lambda} = H$ it is then sufficient to show that D_λ, $D'_{\bar\lambda}$ are <u>dense</u> in H. Assume that $(w, w^*) \in H^2$ is such that

$$\langle R_\lambda u, w\rangle = \langle u, w^*\rangle \quad \text{for all } u \in D_\lambda .$$

Hence, by adding $i\lambda(R_\lambda u, w^*)$ to both sides,

$$0 = \langle R_\lambda u, -i(\bar\lambda w^* + w)\rangle + \langle \lambda R_\lambda u + u, iw^*\rangle =$$

$$= \langle (R_\lambda u, \lambda R_\lambda u + u), \mathcal{U}(w^*, \bar\lambda w^* + w)\rangle .$$

Since it is easily seen that $(R_\lambda u, \lambda R_\lambda u + u)$ takes all values in S, it follows that $(w^*, \bar\lambda w^* + w) \in S^*$ and hence that w^* is uniquely determined by w. It follows that R_λ is densely defined and that $R^*_\lambda = R'_{\bar\lambda}$. Similarly one proves that $R'_{\bar\lambda}$ is densely defined and has the adjoint R_λ which concludes the proof.

<u>Theorem 6</u>: $R_\mu - R_\nu = (\mu - \nu)R_\mu R_\nu$ for all nonreal μ and ν.

<u>Proof</u>: Assume that $(v_1, \mu v_1 + u) \in S$ and $(v_2, \nu v_2 + u) \in S$ so that $u \in D_\mu \cap D_\nu$ and $R_\mu u = v_1$, $R_\nu u = v_2$. Then

$(v_1 - v_2, \mu(v_1 - v_2) + (\mu - \nu)v_2) = (v_1, \mu v_1 + u) - (v_2, \nu v_2 + u) \in S$

so that $v_2 \in D_\mu$ and $R_\mu - R_\nu \subset (\mu - \nu)R_\mu R_\nu$. Now, assume that $(v_2, \nu v_2 + u) \in S$ and $(w, \mu w + v_2) \in S$ so that $u \in D_\nu$ and $R_\nu u \in D_\mu$. Then

$((\mu - \nu)w + v_2, \mu((\mu - \nu)w + v_2) + u) = (v_2, \nu v_2 + u) + (\mu - \nu).(w, \mu w + v_2) \in S$

so that $u \in D_\mu$ and the theorem is proved.

<u>Remark</u>: If $\operatorname{Im} \mu > 0$, $\operatorname{Im} \nu > 0$ one also has $R'_\mu - R'_\nu = (\mu - \nu)R'_\mu R'_\nu$.
It follows in the usual way that R_λ and $R'_{\bar\lambda}$ are analytic functions of λ and $\bar\lambda$ respectively. Theorems 5 and 6 show that if S is selfadjoint, so that $R'_{\bar\lambda} = R_{\bar\lambda}$, then R_λ is a normal operator whose spectral resolution gives information about S. For R_μ to be the resolvent operator of a maximal symmetric operator only the invertibility of R_μ is missing. However, this can be

achieved by restricting R_μ appropriately, roughly by orthogonalizing with respect to an "eigenspace" belonging to the "eigenvalue" ∞. Put

$$H(\infty) = \{u \in H : (0, u) \in S\}$$

<u>Theorem 7</u>: $H(\infty) = H \ominus \tilde{S}$.

<u>Proof</u>: Let $(v, \dot{v}) \in S$. Then $\langle (v, \dot{v}), \mathcal{U}(0, u)\rangle = i \langle v, u\rangle$ so that $(0, u) \in S^*$ precisely when $u \in H \ominus \tilde{S}$. However, if $(0, u) \in S^*$ there is a $v \in \tilde{S}$ such that $V = (v, \lambda v + u) \in S$. Hence
$0 = \langle V, \mathcal{U}V\rangle = i^{-1}(\lambda - \bar{\lambda}) \langle v, v\rangle + i(\langle v, u\rangle - \langle u, v\rangle) = i^{-1}(\lambda - \bar{\lambda}) \langle v, v\rangle$
since $\langle u, v\rangle = 0$. Hence $v = 0$ and it follows that $u \in H(\infty)$.

Next put $H_0 = \tilde{\tilde{S}} = H \ominus H(\infty)$.

Note that from the proof of Theorem 7 it follows that $\tilde{\tilde{S}}^* = \tilde{\tilde{S}} = H_0$.

Restriction to a set Z is denoted by $|Z$.

<u>Theorem 8</u>: $R_\mu|(D_\mu \cap H_0)$ is invertible with range $\tilde{S}$ for all nonreal μ,

$R'_{\bar{\lambda}}|H_0$ is invertible with range $\tilde{S}^*$ for $\operatorname{Im} \lambda > 0$ (when $D_- = \{0\}$).

Now put $\overset{\circ}{S} = S \cap H_0^2$, $\overset{\circ}{R}_\mu = R_\mu |(D_\mu \cap H_0)$, $\overset{\circ}{R}{}'_{\bar{\lambda}} = R'_{\bar{\lambda}} | H_0$.

<u>Theorem 9</u>: $\overset{\circ}{S}$ is the graph of a densely defined, maximal symmetric operator on H_0. The resolvent operators of $\overset{\circ}{S}$ are $\overset{\circ}{R}_\mu$ and $\overset{\circ}{R}{}'_{\bar{\lambda}}$

$(\overset{\circ}{S})^* = (S^*)^\circ$. $D_+ = \overset{\circ}{D}_+$, $D_- = \overset{\circ}{D}_-$.

The proofs of theorems 8 and 9 consist of simple verifications.

<u>REFERENCES</u>

[1] Arens, R: Operational calculus of linear relations Pac. J. Math. 11 (1961) pp 9-23.

[2] Bennewitz, C: Generalization of some statements by Kodaira (to appear in the Proc. Royal Soc. Edinburgh).

[3] Bennewitz, C and Pleijel Å: Selfadjoint extensions of ordinary differential operators. (to appear in the Proc. of the Coll. on Math. Analysis, Jyväskylä, Finland 1970).

[4] Emanuelsson, K: Spectral theory of one formally symmetric ordinary differential operator with respect to another. Mimeographed preprint, Department of Mathematics, Uppsala University, Sweden.

[5] Pleijel, Å: Spectral theory of pairs of ordinary formally self-adjoint differential equations. (to appear in the Journal of the Indian Math. Society).

[6] Pleijel, Å: Green's functions for pairs of formally self-adjoint ordinary differential operators. (to appear in the Proc. of this conference).

Fixed-Point Theorems for Fréchet Spaces

J. Bona

In considering nonlinear partial differential equations defined on functions whose domain is an unbounded subset of Euclidean space, one is led to search for solutions u of operator equations of the form

$$u = Au. \tag{1}$$

Typically, A will be an integral operator. It often happens that while A is nonlinear, it nevertheless presents the zero function as a solution, and what is required is a second, nontrivial solution.

Many of the powerful methods of attacking equations like (1) appear not to be applicable in these circumstances. The contraction mapping principle is difficult to apply in the presence of the trivial solution, unless one has knowledge of an approximation to the nontrivial solution. Further, in the usual Banach spaces of functions used in analyzing (1), A is not a compact operator, owing to the unboundedness of the underlying Euclidean domain. In particular the results of Krasnosel'skii [K] on mappings of Banach spaces ordered by a cone are not applicable.

It very often happens, however, that the operators A <u>are</u> compact in wider function spaces, which are generally not normed spaces, but more general metric linear spaces. It is therefore of interest to extend the 'cone theorems' to a broader class of spaces, and this note contributes a result in this direction.

The proofs of the cone theorems in a Banach space setting are accomplished by Krasnosel'skii [K, ch. 4] by intricate arguments which appear difficult to extend. However, a recent and simplified proof of

Krasnosel'skii's result has been obtained by Benjamin [B, appendix 1], and the method proposed there can be generalized by using the degree theory in locally convex topological linear spaces developed by Nagumo [N].

We let E denote a Fréchet space (complete, separable metric linear space, over $\mathbb{R}$) and K a cone in E ($K \subset E$ is not the zero element alone, is convex, closed, $x \in K \Rightarrow \alpha x \in K$ for $\alpha > 0$, and $x, -x \in K \Rightarrow x = 0$). Let d denote the metric in E. Without loss of generality, we can suppose d has the form

$$d(x,y) = \sum_{j=1}^{\infty} 2^{-j} p_j(x-y)/(1+p_j(x-y)), \tag{2}$$

where $p_1, p_2, \ldots$ is an increasing sequence of pseudo-norms [T, ch. 8]. Further, we may suppose that for some $w \in K$, $p_1(w) > 0$. Let $A: K \to K$ be a continuous mapping which is 'compact' in the sense that $A(S_r(0) \cap K)$ is a precompact set, for $r<1$, where $S_r(0) = \{x \in E: d(x,0) < r\}$.

It follows from Dugundgi's theorem [H, theorem 14.1, p. 57] that A can be extended to a continuous mapping $\tilde{A}$ of E into K such that $\tilde{A}(S_r(0))$ is precompact, for $r<1$ (apply the extension theorem sequentially to the closed sets $S_{r_n}(0)$ where $r_n \uparrow 1$ and use the fact that the convex hull of a precompact set is precompact). Define $\Phi = I - \tilde{A}$. Then solutions of (1) for A correspond to zeros of Φ.

Suppose G is an open subset of E which $\tilde{A}$ maps to a precompact set and for which $\Phi \neq 0$ on ∂G $(= \overline{G} \setminus G)$. Then the rotation $\mathrm{rot}[\Phi, G]$ of Φ on G is an integer defined, in a locally convex linear space setting, by Nagumo [N]. In Nagumo's notation, $\mathrm{rot}[\Phi, G] = A[0, G, \Phi]$. This integer obeys all the usual rules of the usual rotation in n-dimensional Euclidean space [N, §4]. Here is our result.

THEOREM. Let K be a cone in the Fréchet space E as above and let $A: K \to K$ be a continuous and compact mapping. Suppose that for $0<r<R<1$,

(i) $u - Au \notin K$ for $u \in K$ and $d(u,0) = r$,

(ii) $Au - u \notin K$ for $u \in K$ and $d(u,0) = R$.

then A has a solution ϕ to (1) such that $r<d(\phi,0)<R$.

REMARK. The same conclusion may be drawn if instead of (i) and (ii), we suppose

(iii) $Au - u \notin K$ for $u \in K$ and $d(u,0) = r$,

(iv) $u - Au \notin K$ for $u \in K$ and $d(u,0) = R$.

The proof is a trivial modification of the proof given below assuming (i) and (ii).

Proof. We let $\tilde{A}$ and Φ be as defined above. First, if $d(u,0) = r$, $u - \tilde{A}u \notin K$. If $u \in K$, this is provided by (i) since then $\tilde{A}u = Au$. If $u \notin K$, and $u - \tilde{A}u = v \in K$, then $u = \tilde{A}u + v \in K$, a contradiction.

Next, there is a $v \in K$ such that $v \notin \overline{\Phi(S_r(0))}$. In fact, a large positive multiple of the element $w \in K$ such that $p_1(w)>0$ works. For if $\overline{\Phi(S_r(0))}$ contains the entire half ray $\{\alpha w: \alpha \geq 0\}$, then since $d(\alpha w,0) \to 1$ as $\alpha \to +\infty$ (from (2) and the condition $p_1(w) > 0$), there would be a sequence $u_n - \tilde{A}u_n$ such that $d(u_n - \tilde{A}u_n, 0) \to 1$. From (2), this can happen only if $p_1(u_n - \tilde{A}u_n) \to +\infty$. We can assume that $\tilde{A}u_n \to y$ in the metric d (by compactness of $\tilde{A}$) and hence that $\tilde{A}u_n \to y$ for the pseudo-norm p_1. It then follows that $p_1(u_n) \to +\infty$, which implies $d(u_n,0) \to 1$ from (2), a contradiction because $u_n \in \overline{S_r(0)}$.

Consider the homotopy $M: \overline{S_r(0)} \times [0,1] \to E$ given by $M(u,t) = \tilde{A}u - tv$. For $u \in \partial S_r(0)$, $u - M(u,t) \neq 0$, $0 \leq t \leq 1$. For $u - M(u,t) = 0 \Rightarrow \Phi(u) = tv \in K$, which we have shown cannot hold. For fixed t, M is a compact mapping of $\overline{S_r(0)}$, since $\tilde{A}$ is, and for fixed u, M is uniformly continuous in t. The conditions of [N, theorem 7, p. 504] are met, and we may conclude that the rotation of $I - M(u,0)$ and $I - M(u,1)$ are the same on $S_r(0)$. Thus,

$$\mathrm{rot}[\Phi, S_r(0)] = \mathrm{rot}[\Phi - v, S_r(0)]. \tag{3}$$

The right side of (3) is zero, for if not, then by [N, theorem 5, p. 503], $\Phi - v$ would have a zero in $S_r(0)$, a contradiction to our choice of v.

Finally, consider the homotopy F: $\overline{S_R(0)} \times [0,1] \to E$ defined by $F(u,t) = t\tilde{A}u$. If $u - F(u,t) = 0$ for $u \in \partial S_R(0)$, then $u = t\tilde{A}u \in K$ (and hence $t \neq 0$). Thus $\tilde{A}u = Au$ and so $t(Au-u) = (1-t)u \in K \Rightarrow Au - u \in K$ which contradicts (ii). Again, for fixed t, F is compact on $\overline{S_R(0)}$, and for fixed u, F is uniformly continuous in t. Thus we derive that

$$\text{rot}[\Phi, S_R(0)] = \text{rot}[I, S_R(0)]$$

and the latter rotation is 1 [N, p. 503].

Now, let $T = \{x \in E: r < d(x,0) < R\}$. T is an open set, and $T \cap S_r(0) = \emptyset$, $\overline{T \cup S_r(0)} = \overline{S_R(0)}$, $T \cup \overline{S_r(0)} \subset S_R(0)$ and $\Phi \neq 0$ on $\partial S_r(0)$ or ∂T by (i) and (ii). It follows from [N, theorem 6, p. 503] that

$$\text{rot}[\Phi, S_R(0)] = \text{rot}[\Phi, T] + \text{rot}[\Phi, S_r(0)],$$

from which we conclude $\text{rot}[\Phi, T] = 1$. Hence Φ has a zero in T, [N, theorem 5], and this is the required result.

REFERENCES

[B] T.B. Benjamin. "A unified theory of conjugate flows", Phil. Trans. Roy. Soc., vol. 269, A. 1201, July 1971, p. 49.

[H] S.-T. Hu. Theory of retracts, Wayne State U. Press, 1965, Detroit.

[K] M.A. Krasnosel'skii. Positive solutions of operator equations, Noordhoff, 1964, Groningen.

[N] M. Nagumo. "Degree of mapping in convex linear topological spaces", Amer. J. Math. vol. 73, 1951, p. 497.

[T] F. Treves. Topological vector spaces, distributions and kernels, Academic Press, 1967, London.

Boundary Value Problems Involving the Thomas-Fermi Equation

J. S. Bradley

The existence and uniqueness of solutions of boundary-value problems is proved for the Thomas-Fermi equation and certain generalizations. The main results are contained in the following Theorem.

Theorem 1. *Suppose* g *is continuous,* $g(x) > 0$ *for* $x > 0$, *and* g *is integrable over* $(0, a)$ *for* $a > 0$, *but is not integrable on* $(0, \infty)$. *If* $k \geq 1$, $0 \leq A$, $0 \leq B$, $0 < b$, *then there is exactly one solution of the boundary value problem*

(1) $$y'' = g(x)y^k,$$

(2) $$y(0) = A, \quad y(b) = B.$$

Equation (1) becomes the Thomas-Fermi equation if $k = 3/2$, $g(x) = x^{-1/2}$. If the condition (2) were replaced by

(3) $$y(a) = A, \; y(b) = B, \quad a > 0,$$

then Theorem 1 would be similar to a theorem of Lasota and Opial [9] and would actually reduce to a special case of a theorem of Jackson [8]. It should also be remarked that while the equation considered here is not as general as that considered by Bailey, Shampine and Waltman [1], [2], their results do not allow for a singularity at zero.

The existence, uniqueness and continuous dependence on initial values of solutions of an initial value problem

$$y'' = g(x)y^k, \quad y(0) = \alpha \geq 0, \quad y'(0) = \beta$$

is needed in the proof of Theorem 1. Existence and uniqueness is from a slight modification of Theorem 2.3.2 in Hille's text [5], and continuous dependence on initial values can be established by modifying the proof of Theorem 19 in the text by Petrovski [10]. The modification of Petrovski's theorem is accomplished with the aid of a slight alteration of Gronwall's inequality and this is easily obtained from the proof of that inequality in the text by Barrett and Bradley [3], or from Reid [11].

The fact that the boundary-value-problem (1), (2) has at most one solution follows from the following lemma which can be deduced from some work of Hille [6]

Lemma. Assume that y_1, y_2 are solutions of equation (1) and that for a point $x_1 \geq 0$

$$y_1(x_1) \geq y_2(x_1), \quad y_1'(x_1) \geq y_2'(x_1).$$

If y_1 exists on $[x_1, x_2)$, then y_2 exists on $[x_1, x_2)$ and $y_1(x) > y_2(x)$, $y_1'(x) > y_2'(x)$ on $[x_1, x_2)$.

To prove existence we denote by $y(\cdot, m)$ the solution of (1) satisfying the conditions $y(0, m) = A$, $y'(0, m) = m$. Hille [6], [7] has shown that there is a unique number μ such that $y(x, \mu)$ exists on $[0, \infty)$ and $\lim_{x\to\infty} y(x, \mu) = 0$. In particular, $y(b, \mu)$ exists.

It follows that if m is sufficiently near μ, then $y(x, m)$ exists on $[0, b]$ and it follows from the lemma that if $\mu < m$ also, then $y(x, \mu) < y(x, m)$ on $(a, b]$. Let S be the set of points s such that for some $m \geq \mu$, $y(x, m)$ exists on $[0, b]$ and $s = y(b, m) \geq y(b, \mu)$. We want to prove that S does not have an upper bound.

Suppose to the contrary that S is bounded above; then S has a least upper bound, say s_0, and we denote by M the set of numbers $m \geq \mu$ with the property that $y(x, m)$ exists on $[0, b]$. Now all solutions of (1) are concave upward and therefore lie above their tangent lines. It follows that if m_1 is the slope of the line joining the points $(0, A)$ and (b, s_0) then $y(x, m_1)$ does not exist on $[0, b]$, and therefore m_1 does not belong to M and in fact is an upper bound for M. Let ν be the least upper bound of M. Then we can prove that $y(x, \nu)$ exists on $[0, b]$ and $y(b, \nu) = s_0$.

Indeed, suppose that $y(x, \nu)$ does not exist on $[0, b]$. Since $\nu \geq \mu$, $y(x, \nu) \geq y(x, \mu) > 0$ and there is a number $t \in (0, b]$ such that $y(x, \nu)$ exists on $[0, t)$, $y(x, \nu) \to \infty$ and $y'(x, \nu) \to \infty$ as $x \to t$. (See Hartman [4].) Therefore we can choose a point $x_1 \in (0, t)$ with the property that $y(x_1, \nu) > 2s_0$ and $y'(x_1, \nu) > 2$. Then we can choose a number $m < \nu$ such that $y(x_1, m) > s_0$ and $y'(x_1, m) > 1$. Since $m < \nu$, $m \in M$ and therefore $y(x, m)$ exists on $[0, b]$. But because $y(\cdot, m)$ is a solution of (1), $y'(x, m)$ is increasing as x increases, and it follows that $y'(x, m) > 1$ on $[x_1, b]$ and that $y(x, m) > s_0$ on $[x_1, b]$, contradicting the assumption that s_0 is an upper bound for S. Therefore $y(x, \nu)$ exists on $[0, b]$, and then it follows from the definition of ν and of s_0 that $y(b, \nu) = s_0$.

We can now reach a contradiction of the assumption that S has an upper bound. There is a number $m > \nu$ such that $y(x, m)$ exists on $[0, b]$. But since $m > \nu$, Lemma 2 implies that $y(b, m) > y(b, \nu) = s_0$, contrary to the statement that s_0 is an upper bound for S.

But $y(b, m)$ is a continuous function of m and therefore, since S is not bounded above, there is an m such that $y(b, m) = B$ for any $B \geq y(b, \mu)$.

For the case $B < y(b, \mu)$ we consider the family of solutions $u(\cdot, m)$ of (1) defined by the conditions that $u(b, m) = B$, $u'(b, m) = m$. It follows that if $m \leq y'(b, \mu)$, then $u(x, m)$ will exist on $[0, b]$ and $u(0, m) < A$. An argument similar to that above will show that $u(0, m)$ is not bounded above as $m \to -\infty$ and therefore there is a value of m for which $u(0, m) = A$.

R E F E R E N C E S

1. P. Bailey, L. Shampine and P. Waltman, Existence and uniqueness of solutions of the second-order boundary value problem, Bull. Amer. Math. Soc. 72 (1966) 96-98; MR 32 # 1392.

2. P. Bailey, L.Shampime and P. Waltman, The first and second boundary value problems for nonlinear second-order differential equations, J. Differential Equations, 2 (1966), 399-411; MR 34 # 7871.

3. John H. Barrett and John S. Bradley, Ordinary Differential Equations, Intext Educational Publishers, Scranton, Pa. 18515, 1972.

4. Philip Hartman, Ordinary Differential Equations, John Wiley and Sons, Inc., New York, 1964.

5. Einar Hille, Lectures on Ordinary Differential Equations, Addison-Wesley, Reading, Mass., 1969.

6. Einar Hille, Some aspects of the Thomas-Fermi equation, J. Analyse Math., 23 (1970), 147-170.

7. Einar Hille, Aspects of Emden's equation, Journal of the Faculty of Science, University of Tokyo, Sec. I, Vol. 17 (1970), 11-30.

8. Lloyd K. Jackson, Subfunctions and second-order ordinary differential inequalities, Advances in Math. 2 (1968), 307-363; MR 37 # 5462.

9. A. Lasota and Z. Opial, Colloq. Math. 18 (1967), 1-5; MR 36 # 2871.

10. I. G. Petrovski, Ordinary Differential Equations, Prentice-Hall, Inc., 1966.

11. W. T. Reid, Ordinary Differential Equations, John Wiley and Sons, Inc., 1971.

The Initial Value Problem for Operators which can be Factorized into a Product of Klein-Gordon Operators

D.W. Bresters

1. Introduction

Let L denote the following hyperbolic differential operator:

$$L = \prod_{j=1}^{p} \left(\Delta - \frac{1}{c_j^2}\frac{\partial^2}{\partial x_0^2} - m_j^2 \right)^{r_j} \qquad (1)$$

where Δ denotes $\frac{\partial^2}{\partial x_1^2} + \cdots + \frac{\partial^2}{\partial x_n^2}$ while c_j and m_j are real numbers and r_j a positive integer for $j=1,2,\ldots p$. Without loss of generality we may assume that $c_1 \geqslant c_2 \geqslant c_3 \geqslant \ldots \geqslant c_p$. The operator is of order 2ℓ where ℓ denotes $\sum_{j=1}^{p} r_j$. In this lecture we consider the following initial value problem:

$$L\, u(x_0,x) = 0 \qquad x_0>0 \qquad (2)$$

$$\lim \left(\frac{\partial}{\partial x_0}\right)^i u(x_0,x) = \varphi_i(x) \qquad i=0,1,\ldots 2\ell-1 \qquad (3)$$

where x denotes the set of space coordinates $(x_1,x_2,\ldots x_n)$ and x_0 is the time coordinate. We shall look for solutions $u(x_0,x)$ in S' (the space of tempered distributions). The initial data are functions or distributions. Special cases of this problem have been studied by several authors.

2. Preliminaries

Theorem 1. Let P be an arbitrary non-degenerate quadratic form in the variables $x_0,x_1,\ldots x_n$. Define for $\ell=1,2,\ldots$

$$(m^2+P\pm i0)^{-\ell} = \lim_{\varepsilon\downarrow 0}\,(m^2+P\pm i\varepsilon P')^{-\ell}$$

where P' is a positive definite quadratic form in $x_0,x_1,\ldots x_n$. The following relation then holds if $m\neq 0$

$$(m^2+P\pm i0)^{-\ell} = (m^2+P)^{-\ell} \mp \frac{(-1)^{\ell-1} i\pi}{(\ell-1)!}\,\delta^{(\ell-1)}(m^2+P) \qquad (4)$$

The distribution $(m^2+P)^{-\ell}$ is defined as the result of the application of a sequence of suitable differential operators on $(m^2+P)^{-1}$. The latter is defined as a Cauchy Principal Value. $(m^2+P)^{-\ell}$ thus defined satisfies the equation $(m^2+P)^{\ell}\, u(x_0,x)=1$.

For m=0 relation (4) only holds for $2\ell \neq n+1, n+3, \ldots$.

<u>Theorem 2.</u> Let L be a hyperbolic differential operator with constant coefficients and of order m with respect to x_0. The Cauchy Problem:

$$\begin{aligned} &L\, u(x_0,x)=0 && x_0>0 \\ &\lim \left(\tfrac{\partial}{\partial x_0}\right)^i u(x_0,x)=\varphi_i(x) && i=0,1,\ldots m-1 \end{aligned} \qquad (5)$$

can be solved by obtaining first a "fundamental solution of the Cauchy Problem" $G(x_0,x)$ which satisfies:

$$L\, G(x_0,x)=\delta(x_0,x) \qquad (6)$$

$$G(x_0,x)=0 \quad \text{for } x_0<0 \qquad (7)$$

The solution of problem (5) is then given by a linear combination (depending on the specific form of L) of convolutions between the initial data $\varphi_i(x)$ and $G(x_0,x)$ and its time derivatives.

<u>Remark.</u> As a consequence of Th.2 we must then look among the fundamental solutions of L for those which vanish for $x_0<0$. To satisfy this latter condition we use the following theorem of the Paley-Wiener type.

<u>Theorem 3.</u> Let $g(\sigma,\sigma_0+i\tau_0)$ be analytic in the upper half of the complex (σ_0,τ_0) plane for any set of real numbers $\sigma=(\sigma_1,\sigma_2,\ldots\sigma_n)$. In any region $\tau_0>\delta>0$ let it be possible to majorate g as follows:

$$|g(\sigma,\sigma_0+i\tau_0)|<C_\delta \prod_{i=1}^{n}(1+\sigma_i^2)^{p_i}\,|\sigma_0+i\tau_0|^{p_0}$$

where $p_0,p_1,\ldots p_n$ are positive integers independent of δ and C_δ is a constant which may depend on δ. Finally assume that $\lim\limits_{\tau_0\downarrow 0} g(\sigma,\sigma_0+i\tau_0)$ exists in the distributional sense in S'.

Then $\lim\limits_{\tau_0\downarrow 0} g(\sigma,\sigma_0+i\tau_0)$ is the n+1-dimensional Fourier Transform of a distribution in S' which vanishes for $x_0<0$.

With the help of the preceding theorems one may solve the Cauchy Problem for the iterated Klein-Gordon Operator (see reference)

The result is stated in Theorem 4 .

Theorem 4. The unique fundamental solution of the Cauchy Problem for the iterated Klein-Gordon Operator

$$L = \left(\Delta - \frac{\partial^2}{\partial x_0^2} - m^2\right)^{\ell} \qquad (8)$$

is given by

$$\Delta_{n,\ell}(x_0,x) = F^{-1}\left[\left\{k^2 + m^2 - (k_0 + i0)^2\right\}^{-\ell}\right] \qquad (9)$$

which is equivalent to

$$\begin{aligned} \Delta_{n,\ell}(x_0,x) &= 2\,F^{-1}\left[(m^2 + k^2 - k_0^2)^{-\ell}\right] \quad & x_0>0 \\ &= 0 & x_0<0 \end{aligned} \qquad (10)$$

The Fourier Transform in (10) can be calculated by means of Th.1. The distribution $\left\{k^2 + m^2 - (k_0 + i0)^2\right\}^{-\ell}$ is defined as the limit for $\varepsilon \downarrow 0$ of the holomorfic function $\left\{k^2 + m^2 - (k_0 + i\varepsilon)^2\right\}^{-\ell}$ and satisfies the conditions of Th.3. A solution in the classical sense of the problem (2),(3) for this operator is found to exist when the initial data $\varphi_i(x)$ are $3\ell + \left[\frac{n}{2}\right] - i - 1$ times continuously differentiable. Here $\left[\frac{n}{2}\right]$ denotes entier$(\frac{n}{2})$.

3. An operator consisting of two different factors

As the general case mentioned in section 1 can not be treated in the space available for this lecture, we restrict ourselves to:

$$L = \left(\Delta - \frac{1}{c_1^2}\frac{\partial^2}{\partial x_0^2} - m_1^2\right).\left(\Delta - \frac{1}{c_2^2}\frac{\partial^2}{\partial x_0^2} - m_2^2\right)$$

As in the previous section one may show that a fundamental solution of the Cauchy Problem for this operator is given by

$$G(x_0,x) = F^{-1}\left[\left\{(k^2 + m_1^2 - \gamma_1(k_0 + i0)^2\right\}^{-1} . \left\{(k^2 + m_2^2 - \gamma_2(k_0 + i0)^2\right\}^{-1}\right]$$

where $\gamma_i = c_i^{-2}$. To ensure the existence of the distribution on the right hand side we assume that the hyperboloids $k^2 + m_i^2 - \gamma_i k_0^2 = 0$ $(i=1,2)$ do not intersect. This is the case for $m_1^2 c_1^2 \geq m_2^2 c_2^2$.

It follows that we may then write:

$$G(x_o,x) = {}_1\Delta_{n,1} * {}_2\Delta_{n,1} \qquad (11)$$

where ${}_1\Delta_{n,1}$ and ${}_2\Delta_{n,1}$ denote the fundamental solutions of the Cauchy Problems for each of the factors (see Th.4)This result,though simple, is of no great use in applications because of the convolution product which is not so easily calculated.Several methods to obtain a more practical result are available:

1.Resolution in partial fractions of $\tilde{G}(k_o,k) = F\left[G(x_o,x)\right]$

2.Use of the Feynmann parameter formula on $\tilde{G}(k_o,k)$

3.Complex variable methods.

The first method yields convolutions of ${}_1\Delta_{n,1}$ and ${}_2\Delta_{n,1}$ with a distribution depending on either x_o or $r= (x_1^2+ x_2^2+\ldots\ldots x_n^2)^{\frac{1}{2}}$ only, which may be calculated by method 3.The distributions occurring in the convolutions are then represented by means of holomorphic functions which permits an easy caculation of the convolution.(see the lecture submitted for this conference by de Roever)This method is still in development and I will therefore restrict this lecture to the second method which I have found most fruitful.

For two real numbers a and b we have:

$$\frac{1}{a.b} = \int_0^1 \frac{d\lambda}{\{a\lambda + b(1-\lambda)\}^2}$$

For the two factors in $\tilde{G}(k_o,k)$ we tentatively write

$$\lim_{\varepsilon\downarrow 0} \frac{1}{\gamma_2-\gamma_1}\int_{\gamma_1}^{\gamma_2}\left\{k^2-\chi(k_o+ i\varepsilon)^2+\mu^2\right\}^{-2} d\chi \qquad (12)$$

where $\mu^2=m_1^2\frac{\gamma_2-\chi}{\gamma_2-\gamma_1} - m_2^2\frac{\gamma_1-\chi}{\gamma_2-\gamma_1}$ .This indeed equals

$$\lim_{\varepsilon\downarrow 0} \left\{k^2-\gamma_1(k_o+ i\varepsilon)^2+ m_1^2\right\}^{-1}\cdot\left\{k^2-\gamma_2(k_o+ i\varepsilon)^2+ m_2^2\right\}^{-1} = \tilde{G}(k_o,k)$$

In (12) we may interchange integration and taking the limit;we obtain

$$G(x_o,x) = \frac{1}{\gamma_2-\gamma_1}F^{-1}\left[\int_{\gamma_1}^{\gamma_2}\left\{k^2-\chi(k_o+ i0)^2+\mu^2\right\}^{-2} d\chi\right]$$

It is easily shown that integration with respect to a parameter and Fourier Transforming may also be interchanged and we obtain finally

$$G(x_0,x)= \frac{1}{\gamma_2-\gamma_1}\int_{\gamma_1}^{\gamma_2} {}_{\chi}\Delta_{n,2}\, d\chi \tag{13}$$

where we denote by ${}_{\chi}\Delta_{n,2}$ the fundamental solution of the Cauchy Problem for the operator $(\Delta-\chi\frac{\partial^2}{\partial x_0^2}-\mu^2)^2$ which we already met in Th.4.

As an example we consider the propagation of elastic waves in isotropic media, which is described by

$$(\Delta-\gamma_1\frac{\partial^2}{\partial x_0^2}).(\Delta-\gamma_2\frac{\partial^2}{\partial x_0^2})\, u(x_0,x)=0$$

The initial value problem in this case has fundamental solution (n=3)

$$D(x_0,x)= \frac{1}{\gamma_2-\gamma_1}\int_{\gamma_1}^{\gamma_2} {}_{\chi}D_{3,2}\, d\chi = \frac{1}{\gamma_2-\gamma_1}\int_{\gamma_1}^{\gamma_2}\frac{\theta(\frac{x_0^2}{\chi}-x^2)}{\sqrt{\chi}}\, d\chi$$

which can easily be reduced to

$$D(x_0,x)= \frac{c_1^2.c_2^2}{4\pi(c_1^2-c_2^2)}\left[\left(\frac{1}{c_2}-\frac{1}{c_1}\right)\theta(c_2^2x_0^2-x^2)+\left(\left|\frac{x_0}{x}\right|-\frac{1}{c_1}\right)\left\{\theta(c_1^2x_0^2-x^2)-\theta(c_2^2x_0^2-x^2)\right\}\right]$$

This result coincides with the solution obtained in Courant-Hilbert by the method of spherical means.

Formula (13) can be generalized for the operator (1) from section 1, and then involves integration with respect to p-1 parameters. It provides an excellent way to study several properties of the solution of the initial value problem (2),(3) for operator (1). It turns out that this problem has a unique solution which is a solution in the classical sense if the initial data $\varphi_i(x)$ are $2\ell+\left[\frac{n}{2}\right]+q-i-1$ times continuously differentiable. (here $q = \max r_j$). In the case where $m_j=0$ (j=1,2...p) one may prove a theorem on the occurrence of gaps in the domain of dependence of an arbitrary point in (x_0,x)-space.

References

Bresters, D.W. Initial Value Problems for iterated wave operators. Thesis. Twente University of Technology. Enschede. Holland. Contains a list of publications on this subject.

An Oscillation Theorem for Linear Systems with more than one Degree of Freedom

C. Conley

1. Consider the equation $y''(x) - \lambda^2 Q(x)y(x) = 0$ where $y \in R^n$, Q is a real $n \times n$ symmetric matrix and λ is a non-negative real parameter.

1.1 Theorem. Suppose there exists an open interval (x_0, x_1) and a point $\bar{x} \in (x_0, x_1)$ such that: (a) $x \notin (x_0, x_1)$ implies $Q(x) > 0$ and (b) $Q(\bar{x}) < 0$. Then there exist infinitely many values of λ such that the equation admits a solution bounded on the whole real line.

2. The associated $2n$ dimensional system can be written:

2.1
$$y' = \lambda z \qquad \text{or} \qquad w' = \lambda JSw$$
$$z' = \lambda Q(x) y$$

where $y, z \in R^n$; $w \in R^{2n}$, I is the $n \times n$ identity matrix and:

$$w = \begin{pmatrix} y \\ z \end{pmatrix} ; \quad J = \begin{pmatrix} 0 & I \\ -I & 0 \end{pmatrix} ; \quad S(x) = \begin{pmatrix} -Q(x) & 0 \\ 0 & I \end{pmatrix} .$$

Solutions of 2.1 are denoted $w(x, \lambda)$. Define $L_0(\lambda)$ $[L_1(\lambda)]$ to be the set of values $w(x_0, \lambda)$ $[w(x_1, \lambda)]$ corresponding to solutions with y bounded as $x \to -\infty$ $[x \to +\infty]$. Define: $C_0 \equiv \{w \mid y \cdot z > 0$ or $w = 0\}$ and $C_1 \equiv \{w \mid y \cdot z < 0$ or $w = 0\}$. A Lagrangian subspace of R^{2n} means an n-dimensional subspace L such that $w_1, w_2 \in L$ implies $(w_1, Jw_2) = 0$.

2.2 Lemma. For all $\lambda > 0$, $L_0(\lambda)$ and $L_1(\lambda)$ are Lagrangian subspaces of R^{2n} contained respectively in C_0 and C_1; they depend continuously on λ.

Proof. Their definition implies $L_0(\lambda)$ and $L_1(\lambda)$ are linear subspaces. Since $(y \cdot z)' = \lambda(z \cdot z + y \cdot Qy)$ is positive outside (x_0, x_1), either $w(x, \lambda) \in C_1$ for all $x \geq x_1$ or $w(x, \lambda)$ is in C_0 for all large x. Since $(y \cdot y)' = \lambda(2y \cdot z)$, the former is the case if and only if y is bounded as $x \to +\infty$.

From 2.1, for solutions w_1 and w_2, $(w_1\ J\,w_2) = y_1 z_2 - y_2 z_1$ is independent of x; if they stay in C_1, the expression must come arbitrarily close to zero ($z \cdot z$ is integrable near $+\infty$ in this case). Therefore, it must be zero. Thus $L_1(\lambda)$ is contained in a Lagrangian plane.

Boundary points of $C_1 \setminus 0$ are strict exit points in the sense of T. Wazewski [5], and $C_1 \setminus 0 \approx$ (the real line) $\times$ (an n-1 sphere) $\times$ (an n disk).

Wazewski's theorem thus implies $L_1(\lambda)$ is at least n-dimensional. Since C_1 contains no n+1 dimensional subspaces, $L_1(\lambda)$ is a Lagrangian subspace. If the λ solution through some point of $C_1 \times x_1$ leaves C_1, the same is true for nearby λ. Thus $L_1(\lambda)$ is continuous in λ. The case of $L_0(\lambda)$ is similar.

3. Let W be a $2n \times n$ matrix and associate to W the $n \times n$ matrices Y and Z just as y and z are associated to w. If the columns of W span a Lagrangian plane, then $W^T JW = 0$; if they are orthonormal, then $W^T W = I$. The two equations together imply $U = Y + iZ \in \mathcal{U}(n)$ = the unitary group. U_1 and U_2 represent the same Lagrangian subspace in this way if and only if $U_1 = U_2 R$ where $R \in \mathcal{O}(n)$ = the real orthogonal group. Thus, $\mathcal{U}(n)$ is a fibre bundle over the space, $\mathcal{L}$, of Lagrangian planes with fibre $\mathcal{O}(n)$ (cf. [1]).

Each U can be written R exp(iS) where R is real orthogonal and S is real symmetric ([2]). Let $\psi_1, \ldots, \psi_n$ denote the eigenvalues of S mod π. These depend only on the coset to which U belongs (thus on L). Then $(\cos 2\psi_k, \sin 2\psi_k)$ are the naturally corresponding pairs of eigenvalues of the commuting real symmetric matrices $\mathrm{Re}(U^T U)$ and $\mathrm{Im}(U^T U)$.

3.1 Lemma. Let U_α represent L_α, $\alpha = 0, 1$. Then $L_0 \cap L_1 \neq \{0\}$ if and only if $\mathrm{Im}(U_1^{-1} U_0)$ is singular, or in other words some corresponding $\psi_k \equiv 0 \bmod \pi$.

Proof. If $\xi \neq 0$ is in the null space, then $\eta = U_1^{-1} U_0 \xi$ is real; then $U_1 \eta = U_0 \xi$ and corresponds to a vector in $L_0 \cap L_1$.

3.2 Lemma. Let $U_0(\lambda)$ be a lift to $\mathcal{U}(n)$ of the path $L_0(\lambda)$ in $\mathcal{L}$. Let $\gamma_0(\lambda)$ be a lift to R^1 of the path $\lambda \mapsto \det U_0(\lambda)$ considered as a map into the circle $\{\theta \bmod \pi\}$ (rather than $\{\theta \bmod 2\pi\}$).

Then $\gamma_0(\lambda)$ is bounded. The same holds for $\gamma_1(\lambda)$ defined in terms of a lift $U_1(\lambda)$ of $L_1(\lambda)$ and for any lift $\gamma(\lambda)$ of $\det U_1^{-1}(\lambda)\, U_0(\lambda)$.

Proof. Since $L_0(\lambda) \subset C_0$, there exists L, represented, say, by U, such that $L \cap L_0(\lambda) = \{0\}$ for all λ. Since U is constant in λ, any lifts $\gamma_0(\lambda)$ of $\det U_0(\lambda)$ and $\gamma(\lambda)$ of $\det U^{-1} U_0(\lambda)$ are bounded or unbounded together.

Corresponding to $(U^{-1} U_0(\lambda))^T (U^{-1} U_0(\lambda))$, choose n-continuous paths $(\cos 2\psi_k(\lambda), \sin 2\psi_k(\lambda))$ in the circle $\{\theta \bmod \pi\}$. Lift these to R^1 so that $\gamma(\lambda) = \Sigma\, \psi_k(\lambda)$. Since no ψ_k can be congruent to 0 mod π (by 3.1), $\gamma(\lambda)$ must be bounded. The case of γ_1 is similar, and γ is then a difference of bounded lifts.

4. The lift $U_0(\lambda)$ determines a path $W(\lambda)$ of $2n \times n$ matrices. Let $W(x,\lambda)$ be the family of paths determined by $W(x_0,\lambda) = W(\lambda)$; $d/dx \quad W(x,\lambda) = J\,S(x)\,W(x,\lambda)$. A family $L(x,\lambda)$ is then determined (the path of $L_0(\lambda)$ determined by the equation) and a lift $U(x,\lambda)$ can be chosen. As before, real valued functions $\psi_k(x,\lambda)$ and $\gamma(x,\lambda) = \Sigma\,\psi_k(x,\lambda)$ can be chosen corresponding to $U(x,\lambda)$. Also, we can assume $\gamma(x_0,\lambda) = \gamma_0(\lambda)$.

Observe that if W_1 and W_2 represent L, and given that Y_1 and Y_2 are non-singular, $\Sigma \equiv Z_1 Y_1^{-1}$ is equal to $Z_2 Y_2^{-1}$ and has eigenvalues $\tan \psi_k$, $k = 1,\dots,n$, the ψ_k corresponding as above to L.

From 2.1 the function $\Sigma(x,\lambda)$ so associated to $W(x,\lambda)$ (when possible) satisfies $\Sigma' = \lambda(Q - \Sigma^2)$. A theorem on continuous curves of eigenvalues corresponding to a differentiable curve of symmetric matrices (a mild extension of one such in [4]) implies that the paths $\psi_k(x,\lambda)$ in R (which are associated to $U(x,\lambda)$) must be continuously differentiable from the right. From the above equation, the derivatives must be given by

$$\psi_k' = \lambda\,(\mu(x)\cos^2 \psi_k - \sin^2 \psi_k).$$

Here $\mu(x)$ is an eigenvalue of $P_k\,Q(x)\,P_k$ where P_k is the spectral projector of Σ associated to $\tan \psi_k$. For $Q(x) < 0$, $\mu(x) < 0$.

If $\psi_k \equiv \pi/2 \bmod \pi$, then $\psi_k' = -\lambda$; thus, ψ_k cannot increase through values congruent to $\pi/2 \bmod \pi$. Also, if $Q(\bar{x}) < 0$, then in some neighborhood of $\bar{x}$, $\psi_k'(x,\lambda) < -c\lambda$ for some $c > 0$. In particular, all the ψ's decrease across this neighborhood by an amount "proportional" to λ.

It follows that $\psi_k(x_1,\lambda) - \psi_k(x_0,\lambda)$, and so $\gamma(x_1,\lambda) - \gamma(x_0,\lambda)$, tend to $-\infty$ as λ goes to $+\infty$. Since $\gamma(x_0,\lambda) = \gamma_0(\lambda)$, it is bounded and so $\gamma(x_1,\lambda) \to -\infty$ as $\lambda \to \infty$.

Now define $\tilde{U}(x,\lambda) = U_1^{-1}(\lambda)\,U(x,\lambda)$ and let $\tilde{\psi}_k(x,\lambda)$, $\tilde{\gamma}(x,\lambda) = \Sigma\,\tilde{\psi}_k$ be the associated functions. Since $\tilde{\gamma}$ differs from $\gamma - \gamma_1$ by a constant and since γ_1 is bounded, $\tilde{\gamma}(x_1,\lambda) \to -\infty$ as $\lambda \to \infty$. But this is possible only if for some k, $\tilde{\psi}_k(x_1,\lambda) \to -\infty$ as $\lambda \to -\infty$.

In particular, for infinitely many values of λ, $\tilde{\psi}_k(x_1,\lambda) \equiv 0 \bmod \pi$. This means $U_1(\lambda)$ and $U(x_1,\lambda)$ represent (nontrivially) intersecting Lagrangian planes and so bounded solutions of the equation exist for these λ.

One can weaken the conditions on S in several ways. Less trivial extensions can be proved in cases where not all eigenvalues of Q go negative (however, one must be able to "separate" those which don't from those which do for all x). A more interesting problem is that where $y \in \mathbb{C}^n$; $s = S_1 + iS_2$, S, S_1 and S_2 symmetric. More generally, one might ask what forms other than the symplectic one admit such a treatment.

BIBLIOGRAPHY

[1] V. I. Arnold, "Funct. Ana. and Its App. (Translation of Funktsional'nyi Analiz i Ego Prilozheniya) Vol. 1, pp. 1-13.

[2] F. R. Gantmacher, Matrix Theory, Vol. II; Chelsea Publishing Co. New York, N. Y. (1964).

[3] W. J. Harris, Jr. and Y. Sibuya, "Asymptotic distribution of eigenvalues," Proceedings of the N. R. L. -M. R. C. Conference on Ordinary Differential Equations, Washington, D. C. (June 14-23, 1971) Academic Press.

[4] T. Kato, Perturbation Theory for Linear Operators, Springer-Verlag, New York (1966).

[5] T. Wazewski, "Sur un principe topologique de l'examen de l'allure asymptotique des integrals des equations differentielles," Proc. Internat. Congr. Mathematicians, Vol. III, Amsterdam (1954).

Some Existence Results for Nonlinear Steklov Problems

J. M. Cushing

We deal here briefly with the following nonlinear problem (N): show there exist (real) values of μ for which $Au = 0$ on D, $u_n = \mu f(x, u, Du, \mu)$ on ∂D has a nontrivial (i.e., $\not\equiv 0$) solution. Here $Au = \sum_{i,j=1}^{m} D_i(a_{ij}(x)D_j u)$, $D_i = \partial/\partial x_i$, $x = (x_1, \ldots, x_m)$; Du is an arbitrary first partial of u; and u_n is the outward conormal derivative of u on the boundary ∂D of a bounded, smooth (say $C^{2+\alpha}$) domain in Euclidean m-space. The coefficients $a_{ij}(x) = a_{ji}(x)$ are in $C^{1+\alpha}(\bar{D})$, $\bar{D} = D + \partial D$; and A is elliptic: $\sum a_{ij}(x)\xi_i\xi_j \geq c\sum \xi_i^2$, $c = \text{const.} > 0$, $x \in D$, $\xi = (\xi_i) \in R^m$. We consider classical solutions $u(x) \in C^{2+\alpha}(\bar{D})$ and assume

H1: $f(x, y, z, \mu) \equiv a(x)y + g(x, y, z, \mu)$, $a(x) \in C^{1+\alpha}(\partial D)$, $\int_{\partial D} a\,dx \neq 0$,

$g(x, y, z, \mu) \in C^1(\partial D \times R \times R \times R)$, where $g = o((|y|^2 + |z|^2)^{1/2})$,

uniformly in x for bounded μ intervals, near $y = z = 0$.

Thus, $u \equiv 0$ is a solution for all $\mu \in R$.

Nonlinear eigenvalue problems have been studied in many contexts and the general, local bifurcation theory for completely continuous operators can be found in the well known book of Krasnosel'skii [1]. More recently, the important results in [2, 3] have extended these local results globally as well as studied in detail many aspects of the existence question. Our task here will only be to formulate problem (N) so that these general results apply and to discuss some applications. We expect, of course, bifurcation to occur at $(0,\mu)$ for $\mu \in r(L)$, the spectrum of the linearized problem (L): $Au = 0$, $u_n = \mu a(x)u$. Note that $0 \in r(L)$ and that (N) has the solution branch (u, o), $u \equiv$ const. This is really the source of our difficulties in formulating (N) as an operator equation.

Let B be a sub-Banach space of $C^{1+\alpha}(\bar{D})$ under the usual norm $\|u\|_{1+\alpha}$. Any solution $(u,\mu) \in B \times R$ necessarily satisfies the condition $\int_{\partial D} f(x, u, Du, \mu)\, dx = 0$. To reformulate (N) as an operator equation we do the following: given $(u,\mu) \in B \times R$, denote by $k = k(u,\mu)$ a constant satisfying the equation

$$\text{(H)} \qquad \int_{\partial D} f(x, u+k, Du, \mu)\, dx = 0$$

and by $v = F(u,\mu) \in C^{2+\alpha}(\bar{D})$ the unique solution to the Neumann problem $Lv = 0$, $v_n = \mu f(x, u+k, Du, \mu)$ satisfying $\int_{\partial D} v\, dx = 0$ (see [4, 5]). Thus, $F(u,\mu)$ maps $B \times R \to C^{1+\alpha}(\bar{D})$ provided $k(u,\mu)$ defined by (H) is well-defined. For the moment, we assume

H2: $k(u,\mu): S(\varepsilon) \times R \to R$ with $k(0,\mu) = 0$ is well-defined and continuous on some ball $S(\varepsilon)$ at the origin of radius $0 < \varepsilon \leq +\infty$ in $C^{1+\alpha}(\bar{D})$;

H3: F maps $B \times R \to B$.

If $(u,\mu) \in B \times R$ satisfies (O): $u = \mu F(u,\mu)$ then it is clear that $u^* = u + k$ satisfies (N). Conversely, it can easily be shown that if u^* solves (N) for $\mu \neq 0$ then $u = u^* - k$, $k = s^{-1} \int_{\partial D} u^*\, dx$, $s = \int_{\partial D} dx$ solves the operator equation (O). Thus, (N) and (O) are equivalent for $\mu \neq 0$. Known a priori estimates (see [4, 5]) imply that F is completely continuous and that any solution to (O) in fact lies in $C^{2+\alpha}(\bar{D})$. Moreover, if $w = Lu \in C^{2+\alpha}(\bar{D})$ denotes the unique solution of $Aw = 0$, $w_n = a(x)(u+d)$ satisfying $\int_{\partial D} w\,dx = 0$ where d is the constant such that $\int_{\partial D} a(u+d)\,dx = 0$, then it can be shown that $F(u,\mu) = \mu Lu + G(u,\mu)$, $G(u,\mu) = o(\|u\|_{1+\alpha})$ near 0 on bounded μ intervals and that problem (L) is equivalent to $u = \mu Lu$ for $\mu \neq 0$. The general results of [1-3] are now seen to apply to (O) and we can conclude (see [2]) that <u>if H1 - H3 hold then bifurcating from any eigenvalue $(0,\mu^*)$, $\mu^* \in r(L) - \{0\}$ of odd multiplicity is a continuum of nontrivial solutions $(u,\mu) \in B \times R$ of (N) connecting $(0,\mu^*)$ to either $\partial(B(\varepsilon) \times R)$ or to $(0,\mu^{**})$ for some $\mu^{**} \in r(0) - \{0\}$, $\mu^{**} \neq \mu^*$.</u>

Concerning H2 and H3, we note: (i) H3 is satisfied for $B = C^{1+\alpha}(\bar{D})$, and an application of the implicit function theorem [6] shows that H2 is always satisfied for some $\varepsilon > 0$ sufficiently small; (ii) it can be shown that $\varepsilon = +\infty$ in H2 with $B = C^{1+\alpha}$ in H3 if the condition $f_y(x,y,z,\mu) > 0$ for all $(x,y,z,\mu) \in \partial D \times R \times R \times R$ holds; (iii) if certain symmetry assumptions are made on D, a_{ij}, and f, then H2 is satisfied with $\varepsilon = +\infty$ (by taking $k = 0$ in (H) for all u) if B is taken to be the subspace of odd functions in $C^{1+\alpha}$ with respect to the symmetry on D (see [5]); (iv) the example $f \equiv u - u^2$ easily shows that (H) may not, in general, have a solution k for all $u \in C^{1+\alpha}$.

An application deals with Levi-Civita's theory of permanent water waves on deep water [7] where the existence of such waves is reduced to the equivalent problem $L = \Delta$ (Laplace's operator), $m = 2$, D = unit disk centered at the origin and $f \equiv \exp(-3Tu)\sin u$ where Tu is the harmonic conjugate of u vanishing at the origin. Nontrivial solutions in $C^{2+\alpha}$ are desired which vanish on the horizontal axis. (Because of the presence of Tu, this problem is not strictly included in the remarks above. Its presence imposses no difficulty, however, when the above techniques are applied. Tu is continuous under $\|u\|_{1+\alpha}$.) Here $\mu = g\lambda/2\pi c^2$ where g is the acceleration due to gravity, λ is the wavelength and c the speed of the water wave, and u is the angle that the flow vector makes with the horizontal. Choosing $B = B_i = \{u \in C^{1+\alpha} :$ u vanishes for all polar angles $\theta = \pm j\pi/i$, $j = 0, 1, \ldots, i$ for all $0 \leq r \leq 1\}$, i = positive integer, it can be shown that H3 and H2 are satisfied (with $\varepsilon = +\infty$). Thus, from each $(0, i)$ (the eigenvalues of the linearized problem) there branches an unbounded continuum of solution $(u, \mu) \in B_i \times \mathbf{R}$; and consequently, the local branches shown to exist by Levi-Civita exist globally. Moreover, it can be shown that the only solutions branching from $(0, i)$, $i \geq 2$, lying in B_1 actually lie in $B_i \subset B_1$. This latter statement justifies a conjecture of Levi-Civita to the effect that only solutions branching from $(0, 1)$ are of interest physically. Recent numerical work (by the author at IBM T. J. Watson Research Center) on this problem indicates that although $\|u\|_{1+\alpha}$ is unbounded along the solution branch from $(0, 1)$ the norm $\max_{\partial D} |u|$ is bounded (as are the values of μ); in fact, as $\|u\|_{1+\alpha} \to +\infty$, $\max |u|$ seems to tend to a value near .5 radians (the numerical work is not complete). This, together with the numerical calculation of the boundary values of the solution (which indicates the shape of the wave), suggests that Levi-Civita's theory predicts the empirically observed "peaking" of such waves at an angle of $\pi/6$ [8].

Waves on water of finite depth can also be treated. Here D is the annulus $R^{-1} \leq r \leq R$, $R > 1$.

REFERENCES

1. M. A. Krasnosel'skii, Topological Methods in the Theory of Nonlinear Integral Equations, MacMillan Co., N.Y., 1964.

2. Paul H. Rabinovitz, Some global results for nonlinear eigenvalue problems, J. Func. Anal. 7 (1971), 487-513.

3. Michael G. Crandall and Paul H. Rabinowitz, Bifurcation from simple eigenvalues, J. Func. Anal 8 (1971), 321-340.

4. O. A. Ladyzhenskaya and N. Uraltseva, Linear and Quasilinear Elliptic Equations, Academic Press, New York/London, 1968.

5. J. M. Cushing, Some existence theorems for nonlinear eigenvalue problems associated with elliptic equations, Arch. Rat. Mech. Anal. 42 (1971), 63-76.

6. T. H. Hildebrandt and L. M. Graves, Implicit functions and their differentials in general analysis, Trans. AMS 29 (1927), 127-153.

7. T. Levi-Civita, Determination rigoureuse des on des permanentes d'ampleur fini, Math. Ann. 93 (1925), 264-314.

8. J. J. Stoker, Water Waves, Interscience Publishers Inc., New York, 1957.

The Deficiency Indices and Spectrum Associated with Self-Adjoint Differential Expressions Having Complex Coefficients

M.S.P. Eastham and K. Unsworth

1. Introduction

Let l be a self-adjoint linear differential expression of order n defined in an x-interval I. Then there are real-valued functions $p_r(x)$ and $q_r(x)$ such that l can be written as

$$l = \sum_{r=0}^{k} \frac{d^r}{dx^r} p_r(x) \frac{d^r}{dx^r} + i \sum_{r=0}^{K} \left(\frac{d^{r+1}}{dx^{r+1}} q_r(x) \frac{d^r}{dx^r} + \frac{d^r}{dx^r} q_r(x) \frac{d^{r+1}}{dx^{r+1}} \right),$$

where $k = \frac{1}{2}n$, $K = k - 1$ if n is even and $k = \frac{1}{2}(n - 1)$, $K = k$ if n is odd (see, e.g., [3,§2.7]). We assume that the leading coefficient of l is nowhere zero in I. In the Hilbert space $L^2(I)$ we associate with l the symmetric differential operator L_0 which is defined by $L_0 f = lf$ and whose domain D_0 consists of functions f with compact support in the interior of I and n continuous derivatives in I. The closure L_1 of L_0 is called the minimal closed symmetric operator associated with l (cf. [2,p.1226], [9,p.28]). The deficiency indices N_+, N_- of L_1 are the number of linearly independent $L^2(I)$ solutions of the differential equation

$$ly(x) = \lambda y(x) \qquad (1.1)$$

for $\operatorname{im} \lambda > 0$ and $\operatorname{im} \lambda < 0$ respectively. The values of N_+ and N_- do not depend on the choice of λ in the appropriate half-plane ([2,p.1232], [6]).

If $q_r(x) = 0$ for all r, in which case n is even, l only has real-valued coefficients. Then any solution $y(x)$ of (1.1) gives rise to a solution $z(x) = \bar{y}(x)$ of $lz(x) = \bar{\lambda} z(x)$, from which it follows that $N_+ = N_-$. Hence L_1 has self-adjoint extensions, not necessarily unique ([2,pp.1295-6]).

If, on the other hand, not all the $q_r(x)$ are zero, it is not necessarily the case that $N_+ = N_-$. Indeed, even in the simple example where $l = id^3/dx^3$ and I is $[0,\infty)$, $N_+ = 2$ and $N_- = 1$ [5]. The possibility of unequal deficiency indices for even-order l was first proved in [10]. We always have $0 \leqslant N_+, N_- \leqslant n$ but, if I has a regular end-point, then (i) $\frac{1}{2}n \leqslant N_+, N_- \leqslant n$ if n is even and (ii) $k + 1 \leqslant N_+ \leqslant n$, $k \leqslant N_- \leqslant n$ (or vice versa) if n is odd and $n = 2k + 1$ ([5], [7]).

Here, we consider l when not all the $q_r(x)$ are zero and I is $[0,\infty)$ or $(-\infty,\infty)$. As above, cases where $N_+ = N_-$ are of interest because then L_1 has self-adjoint extensions. Cases where $N_+ \neq N_-$ are also of interest because few such examples are known. We give below a new class of expressions l with $N_+ \neq N_-$.

We remark in passing that, even when $N_+ \neq N_-$, there is associated with l a symmetric operator having equal deficiency indices in a larger Hilbert space $H \supset L^2(I)$. It is known that $H \ominus L^2(I)$ is infinite-dimensional [11]. We do not pursue this aspect here.

2. The deficiency indices

Theorem 1 Let I be $(-\infty,\infty)$. If $q_r(x) = q_r(-x)$ and either (i) $p_r(x) = p_r(-x)$ for all r or (ii) $p_r(x) = - p_r(-x)$ for all r, then L_1 has equal deficiency indices.

Let $y(x)$ satisfy (1.1). In Case (i), the change of variable from x to $-x$ gives

$$lz(x) = \bar{\lambda} z(x), \tag{2.1}$$

where $z(x) = \bar{y}(-x)$. Since $z(x)$ is $L^2(I)$ if and only if $y(x)$ is, we obtain $N_+ = N_-$. In Case (ii), instead of (2.1), we have

$$lz(x) = - \lambda z(x),$$

where $z(x) = y(-x)$. The equality of N_+ and N_- follows again.

If I is $[0,\infty)$, the above transformation from x to $-x$ is, of course, not possible and we know of no general criterion for $N_+ = N_-$. However, under suitable conditions on the coefficients of l, the asymptotic form of the solutions of (1.1) as $x \to \infty$ can be found and then N_+ and N_- can be determined. This method has been developed in [1, 8, 12, 15, 16] when l has real coefficients and in [13] when l has non-real coefficients (see also [17]). Here we mention that the method of [1] is adapted in [14] to prove :

Theorem 2 Let I be $[0,\infty)$ and $n = 3$. Let $p_1(x) = ax^{\alpha}$, $p_0(x) = 0$, $q_1(x) = 1$, $q_0(x) = bx^{\beta}$, where a, b, α, β are constants.

(i) If $\beta > 1$ and $\alpha < \frac{1}{2}\beta$, then $N_+ = N_- = 2$ or $N_+ = N_- = 3$ according as $b < 0$ or $b > 0$.

(ii) If $2/5 < \beta \leq 1$ and $\alpha < 3\beta/8 - \frac{1}{4}$, then $N_+ = 2$ and $N_- = 1$.

3. The spectrum

In this section we take I to be $(-\infty,\infty)$ and we suppose that $N_+ = N_-$. Then L_1 has a self-adjoint extension L and we denote the spectrum of L by S. A real number μ is in S if there is a sequence $\{f_m\}$ in the domain of L such that $\|f_m\| = 1$ and, as $m \to \infty$,

$$\|Lf_m - \mu f_m\| \to 0 \tag{3.1}$$

(cf. [9,§3]). The method given below works for general n but, for the purpose of illustrating it, we take the case where

$$n = 3,\ p_0(x) = 0,\ q_1(x) = 1. \tag{3.2}$$

Theorem 3 Let (3.2) hold and, as either $x \to \infty$ or $x \to -\infty$, let

$$|p_1(x)| \geq a|x|^{\alpha},\ |q_0(x)| \geq b|x|^{\beta}\quad (a > 0,\ b > 0), \tag{3.3}$$

$$p_1^{(r)}(x) = O(|x|^{\alpha-r}),\ q_0^{(r)}(x) = O(|x|^{\beta-r})\quad (0 \leq r \leq 3). \tag{3.4}$$

(i) If $0 < \beta < 1$, $\alpha < 2\beta$, and $b > 0$, then $S = (-\infty,\infty)$.

(ii) If $0 < \alpha < 2$, $2\beta < \alpha$, and $a > 0$, then S contains $[0,\infty)$ or $(-\infty,0]$ according as $p_1(x) < 0$ or $p_1(x) > 0$ in (3.3).

We give the proof for $x \to \infty$, the proof when $x \to -\infty$ being similar. Let $\{a_m\}$ be a sequence of real numbers such that $a_m \to \infty$ as $m \to \infty$ and let

$$h_m(x) = \begin{cases} (1 - \{(x - 2a_m)/a_m\}^2)^4 & (|x - 2a_m| \leq a_m) \\ 0 & (\text{otherwise}). \end{cases}$$

Then we define $\quad f_m(x) = b_m\{\exp iQ(x)\}h_m(x),$

where b_m is the normalization constant making $\|f_m\| = 1$ and $Q(x)$

is a real-valued function to be defined below. Functions f_m of this form have been used previously in [4] in connexion with the spectrum of second- and fourth-order operators with real coefficients. As in [4], we have

$$b_m = O(a_m^{-\frac{1}{2}}), \quad h_m^{(r)}(x) = O(a_m^{-r}) \tag{3.5}$$

as $m \to \infty$. Now f_m is in the domain of L_0 and hence

$$\begin{aligned} Lf_m(x) - \mu f_m(x) &= L_0 f_m(x) - \mu f_m(x) \\ &= \{Q'^3(x) - p_1(x)Q'^2(x) - 2q_0(x)Q'(x) - \mu\} f_m(x) + \\ &\quad + (\text{terms involving derivatives of } Q', p_1, q_0, h_m). \end{aligned} \tag{3.6}$$

The first bracket on the right-hand side determines the choice to be made for $Q'(x)$. In Case (i), we define

$$Q'(x) = -\mu/\{2q_0(x)\}$$

for any real μ. Then, by (3.3), (3.4) and (3.5), the first bracket on the right of (3.6) reduces to

$$f_m(x)\{O(x^{-3\beta}) + O(x^{\alpha-2\beta})\} = o(a_m^{-\frac{1}{2}})$$

as $m \to \infty$. The second bracket is also easily verified to be $o(a_m^{-\frac{1}{2}})$. It follows from this that, since the integration implied in (3.1) is only carried out over the range $|x - 2a_m| \leqslant a_m$, (3.1) does hold. Hence μ is in S and, since μ is arbitrary, S is $(-\infty, \infty)$.

In Case (ii), we wish to define

$$Q'(x) = \{-\mu/p_1(x)\}^{\frac{1}{2}}$$

but, in order to get a real-valued $Q(x)$, we need to take $\mu \geqslant 0$ if $p_1(x) < 0$ and $\mu \leqslant 0$ if $p_1(x) > 0$. In these circumstances (3.1) holds again as above, giving (ii) of the theorem.

The theorem covers cases where the coefficients of l are large and, in this connexion, we point out that the restriction $\beta < 1$ in (i) of the theorem is best possible. This is seen by considering the example where $p_1(x) = 0$ and $q_0(x) = |x|^\beta$ $(\beta > 1)$. As indicated by Part (i) of Theorem 2, all solutions of (1.1) are $L^2(-\infty, \infty)$ in this example and hence the corresponding spectrum is entirely discrete [2, p.1402].

REFERENCES

1. A. Devinatz, Quart. J. Math., to appear.

2. N. Dunford and J. T. Schwartz, Linear operators, Part 2 (Interscience, 1963).

3. M. S. P. Eastham, Theory of ordinary differential equations (Van Nostrand Reinhold, 1970).

4. M. S. P. Eastham and A. A. El-Deberky, J. London Math. Soc., 2 (1970), 257-66.

5. W. N. Everitt, Quart. J. Math., 10 (1959), 145-55.

6. W. N. Everitt, Quart. J. Math., 13 (1962), 217-20.

7. W. N. Everitt, Proc. London Math. Soc., to appear.

8. M. V. Fedorjuk, Trans. Moscow Math. Soc. (Amer. Math. Soc. translation) 15 (1966), 333-86.

9. I. M. Glazman, Direct methods of qualitative spectral analysis of singular differential operators (I. P. S. T., 1965).

10. J. B. McLeod, Quart. J. Math., 17 (1966), 285-90.

11. I. M. Michael, Canadian Math. Bull., to appear.

12. M. A. Naimark, Linear differential operators (Ungar, 1968).

13. R. A. Shirikyan, Diff. Equations, 3 (1967), 1942-56 (Russian).

14. K. Unsworth, London Univ. Ph.D. thesis, in preparation.

15,16. P. W. Walker, J. Diff. Equns., 9 (1971), 108-32, 133-40.

17. A. D. Wood, J. London Math. Soc., 3 (1971), 96-100.

The Generalized Ahlfors-Heins Theorem in Certain D-Dimensional Cones

M. Essén and J. Lewis

1. Notation

Let Ω be an open set in $\underline{R}^d$, $d \geq 2$, and let $\partial\Omega$ be its boundary. Let $|PQ|$ denote the euclidean distance between the points P and Q and let $|P| = |PO|$, where O is the origin. Also let e(P) be the radial projection of P onto the unit sphere. If the function u is defined in Ω and $Q \in \partial\Omega$, we define

$$u(Q) = \limsup u(P), \quad P \to Q, \quad P \in \Omega.$$

We also introduce

$$M(r) = M(r,u) = \sup u(P), \quad |P| = r, \quad P \in \Omega.$$

We denote the points in $\underline{R}^3$ by (x,y,z). Let θ be the angle between the positive x-axis and OP and let $|P| = r$. If α is given, $0 < \alpha < \pi$, let $K(\alpha)$ be the cone $\{P \mid 0 \leq \theta < \alpha\}$ and let $S(\alpha)$ be the polar cap $K(\alpha) \cap \{P \mid r = 1\}$. Spherical coordinates in R^3 are given by

$$\begin{cases} x = r\cos\theta \\ y = r\sin\theta\cos\varphi \\ z = r\sin\theta\sin\varphi \end{cases}$$

where $0 \leq \theta < \pi$ and $0 \leq \varphi < 2\pi$. Relative to this system, the Laplace operator Δ may be written

$$(1.1) \quad \Delta = r^{-2}\frac{\partial}{\partial r}\left(r^2\frac{\partial}{\partial r}\right) + r^{-2}\delta$$

where

$$\delta = \frac{1}{\sin\theta}\frac{\partial}{\partial\theta}\left(\sin\theta\frac{\partial}{\partial\theta}\right) + \frac{1}{\sin^2\theta}\frac{\partial^2}{\partial\varphi^2}$$

Consider in $S(\alpha)$ the boundary value problem

$$(1.2) \quad \delta g + \mu g = 0, \quad g = 0 \text{ on } \partial S(\alpha),$$

where we assume that g is continuous in $\overline{S(\alpha)}$ and δg is continuous in $S(\alpha)$. Let $\mu_1 > 0$ be the first eigenvalue and Ψ_1 the unique corresponding eigenfunction satisfying

$$(1.3) \quad \Psi_1(1,0,0) = 1 .$$

Next let ρ denote the positive root of the equation

$$(1.4) \quad \rho(\rho + 1) = \mu_1 .$$

We also consider for fixed λ, $0 < \lambda < 1$, and for the same class of functions as in (1.2), the boundary value problem

$$\delta g + \rho\lambda(\rho\lambda + 1)g = 0, \quad g = 1 \text{ on } \partial S(\alpha) .$$

This problem has a unique solution, which we denote by Ψ_λ. It follows from the minimum principle (cf. [3, p. 326]) and from the fact that μ_1 is the smallest eigenvalue of (1.2) that

$$(1.5) \quad 1 < \Psi_\lambda(P) < \Psi_\lambda(1,0,0) = C(\lambda)^{-1}, \quad P \in S(\alpha), \quad P \neq (1,0,0) .$$

With $C(\lambda)$ as in (1.5), we define

$$H_\lambda(P) = C(\lambda)\, \Psi_\lambda(e(P))\, r^{\rho\lambda}, \quad P \in K(\alpha) .$$

It follows from (1.1) that H_λ is harmonic in $K(\alpha)$. Moreover, from (1.5) we see that

$$(1.6) \quad H_\lambda(1,0,0) = 1$$

$$(1.7) \quad H_\lambda(Q) = C(\lambda) H_\lambda(|Q|,0,0) = C(\lambda) \max_{|P|=|Q|} H_\lambda(P), \quad Q \in \partial K(\alpha)$$

2. The main result

Let u be subharmonic in $K(\alpha)$. Suppose for given λ, $0 < \lambda < 1$, that u satisfies

$$(2.1) \quad u(Q) \leq C(\lambda) M(|Q|), \quad Q \in \partial K(\alpha) - \{0\},$$

$$(2.2) \quad u(Q) < \infty, \quad Q \in \partial K(\alpha) .$$

Theorem 1: Let λ be given, $0 < \lambda < 1$. Let $u \not\equiv -\infty$ be subharmonic in $K(\alpha)$ and satisfy (2.1) and (2.2). If

(2.3) $u(P_o) > 0$ for some $P_o \in K(\alpha)$

and if $\liminf_{r\to\infty} r^{-\lambda\rho} M(r) < \infty$, then for some B, $0 < B < \infty$, we have

A. Except when $e(P) \in S(\alpha)$ belongs to a set of capacity zero,

(2.4) $\lim_{r\to\infty} r^{-\lambda\rho} u(re(P)) = B\, C(\lambda)\, \Psi_\lambda\ (e(P))$.

B. The limit in (2.4) holds uniformly in $K(\alpha_1)$, $0 < \alpha_1 \le \alpha$, when $P \to \infty$ in $K(\alpha_1)$ outside a set of spheres F_o, whose radii r_i and distances R_i of their centers from the origin satisfy

(2.5) $\sum_{i=1}^{\infty} (r_i/R_i)^2 < \infty$.

This result is true in $\underline{R}^d$, $d \ge 2$, if we define ρ as the positive root of the equation

(1.4′) $\rho(\rho + d - 2) = \mu_1$

and if (2.5) is replaced by

(2.5′) $\sum_{i=1}^{\infty} (r_i/R_i)^{d-1} < \infty$.

For simplicity, we here consider $d = 3$. Only an outline of the proof is given. A more complete version will appear elsewhere.

Letting $\lambda \to 1$, we obtain an important limiting case. Indeed, if we put

$$C(1) = \lim_{\lambda\to 1} C(\lambda) = 0,$$

and if Ψ_1 is as in (1.3), our result is true also when for $\lambda = 1$. In fact, the case $\lambda = 1$, $\alpha_1 < \alpha$, was considered in $\underline{R}^2$ by Ahlfors and Heins [1]. The extension to the case $\lambda = 1$, $\alpha_1 = \alpha$ was given by Hayman [12]. In higher dimensions the case $\lambda = 1$, $\alpha_1 < \alpha$, was first treated by J. Lelong-Ferrand ([9], [10], [11]). The extension to the case $\lambda = 1$, $\alpha_1 = \alpha$, is due to Azarin [2].

Actually in $\underline{R}^2$, our result holds for $0 < \lambda < 2$ if $K(\alpha)$ is a half-plane and $C(\lambda) = \cos(\pi\lambda/2)$. The case $0 < \lambda < 2$, $\alpha_1 < \alpha$, can be found in Essén [7] and Lewis [15].

A preliminary result in $\underline{R}^3$ is given in Essén [8].

We note that Dahlberg [4] has considered subharmonic functions u, which satisfy the hypotheses of Theorem 1 in more general cones. For the cones $K(\alpha)$, his conclusion is that $r^{-\lambda\rho} M(r)$ tends to a positive limit as $r \to \infty$. This conclusion also follows from our result as is easily seen.

In general, we do not have an explicit formula for the constant $C(\lambda)$. However, if $d = 2$ and $K(\alpha)$ is a half-plane, then as mentioned above $C(\lambda) = \cos(\pi\lambda/2)$. If $d = 3$ and $\alpha = \pi/2$, it is known that

$$C(\lambda)^{-1} = \int_0^\infty t^{1+\lambda}(1+t^2)^{-3/2}\,dt \qquad \text{(cf. Essén [8])}$$

3. A harmonic majorant and a convolution inequality

Lemma 1: Let h be subharmonic and bounded above in $K(\alpha) \cap \{|P| < R\}$, $0 < R < \infty$. If

$$h(Q) \le C(\lambda)\, M(|Q|, h)^+, \quad 0 < |Q| < R,$$

then $M(r,h)^+$ is a non-decreasing function, which is also a convex function of r^{-1} on $(0,R)$. In particular, if $\frac{d}{dr} M(r,h)^+$ is the left hand derivative of $M(r,h)^+$, then $r^{+2} \frac{d}{dr} M(r,h)^+$ is a non-decreasing function on $(0,R)$.

Proof: For a corresponding lemma in $\underline{R}^2$, see Lewis [14, Lemma 1]. For $d > 2$, the proof is essentially the same. In the last part of the lemma we use Heins [13, p. 79, ex. 1].

We now claim that

$$(3.1)\quad 0 < \sup_{r>0} r^{-\lambda\rho} M(r) \leq C(\lambda)^{-1} \lim_{r\to\infty} \inf r^{-\lambda\rho} M(r) < \infty$$

where $M(r) = M(r, u)$.

To prove (3.1), consider

$$h(P) = u(P) - R^{-\lambda\rho} C(\lambda)^{-1} M(R)^{+} H_{\lambda}(P),\quad P \in K(\alpha) \cap \{|P| < R\}.$$

It follows from (2.1), (2.2) and (1.7) that h satisfies the hypotheses of Lemma 1. Hence $M(r,h)^{+}$ is non-decreasing. Since $h(Q) \leq 0$, $Q \in K(\alpha) \cap \{|P| = R\}$, h is non-positive and hence

$$r^{-\lambda\rho} M(r) \leq C(\lambda)^{-1} R^{-\lambda\rho} M(R)^{+},\quad 0 < r < R .$$

Since (2.3) holds, (3.1) is proved.

Next we shall study a particular harmonic majorant of u. For this purpose, let $g(\cdot\, ,Q)$ denote Green's function for $K(\alpha)$ with pole at Q. Also let $\frac{\partial g}{\partial n}(P,Q)$ denote the inner normal derivative of $g(P,\cdot)$ evaluated at the point $Q \in \partial K(\alpha)$. We note that for $a > 0$,

$$(3.2)\quad \frac{\partial g}{\partial n}(aP, aQ) = a^{-2} \frac{\partial g}{\partial n}(P,Q),\quad P \in K(\alpha),\quad Q \in \partial K(\alpha) .$$

We also introduce

$$B(t,\theta) = \sin\alpha \int_0^{2\pi} \frac{\partial g}{\partial n}(e(P),Q)\, d\varphi$$

where the spherical coordinates of Q and P are (t,α,φ) and (r,θ,φ'), respectively. Using (3.2), we see that

$$\sin\alpha \int_0^{2\pi} \frac{\partial g}{\partial n}(P,Q)\, d\varphi = r^{-2} B(t/r,\ \theta) .$$

Now consider

$$(3.3)\quad v(P) = \frac{C(\lambda)}{4\pi} \int_0^{\infty} M(|Q|)^{+} \frac{\partial g}{\partial n}(P,Q)\, d\sigma(Q)$$

$$= \frac{C(\lambda)}{4\pi} \int_0^{\infty} M(rs)^{+} B(s,\theta)\, s\, ds ,$$

where σ denotes Lebesgue measure on $\partial K(\alpha)$, $P \in K(\alpha)$ (and the spherical coordinates of P are (r,θ,φ')) .

Using (3.1), we prove that

$$v(P) \le H_\lambda(P) \sup_{r>0} r^{-\lambda\rho} M(r)^+, \quad P \in K(\alpha).$$

It is clear that v is a harmonic majorant of u. Applying first the monotone convergence theorem and secondly the fact that $r^{+2} \frac{d}{dr}(M(r)^+)$ is nondecreasing (cf. Lemma 1), we see that $\frac{d}{dr}(r^2 \frac{\partial v}{\partial r}) \ge 0$. Since $\Delta v = 0$, it follows that $\delta v \le 0$. Once more using the argument leading up to (1.5), we see that

$$(3.4) \quad v(r,0,0) = \max v(P), \quad |P| = r, \quad P \in \overline{K(\alpha)}.$$

Since v is a harmonic majorant of u, it is clear from (3.3) and (3.4) that

$$(3.5) \quad r^{-\lambda\rho} M(r)^+ \le r^{-\lambda\rho} v(r,0,0) = \frac{C(\lambda)}{4\pi} \int_0^\infty (rs)^{-\lambda\rho} M(rs)\, B(s,0) s^{1+\lambda\rho}\, ds.$$

After the change of variables $r = e^x$, $s = e^{-y}$, we obtain

$$(3.6) \quad \Phi(x) \le \Phi * L(x), \quad x \in \underline{R},$$

where

$$(3.7) \quad \Phi(x) = e^{-\lambda\rho x} M(e^x)^+,$$

$$L(x) = \frac{C(\lambda)}{4\pi} e^{-(2+\lambda\rho)x} B(e^{-x}, 0),$$

$$\int_{-\infty}^{\infty} L(x)\, dx = H_\lambda(1,0) = 1 \qquad \text{(cf. (1.6))}.$$

The above convolution inequality has been studied by Essén ([5], [6], [7]). In particular, if L satisfies

$$(3.8) \quad \int_{-\infty}^{\infty} y\, L(y)\, dy \neq 0,$$

$$(3.9) \quad \int_x^{\infty} L(y)\, dy = \underline{O}(L(x)), \quad x \to \infty,$$

$$\int_{-\infty}^{x} L(y)\, dy = \underline{O}(L(x)), \quad x \to -\infty,$$

it follows from Essén [7, Lemmas 5.2 and 5.3] that

(3.10) $\lim_{x\to\infty} \Phi(x) = \lim_{r\to\infty} r^{-\lambda\rho} M(r)^+$ exists,

(3.11) $\int_0^\infty (\Phi - \Phi * L)(y)dy = \int_1^\infty (M(t)^+ - v(t,0,0))\, t^{-1-\lambda\rho} dt > -\infty$.

To prove (3.8), we note that there exists a bounded solution of (3.6) for which there is strict inequality (take e.g $\Phi(x) = e^{-\rho\lambda x}(H_\lambda(e^x, 0, 0) - C)^+$, where C is a positive constant) and hence (3.8) holds (cf. Essén [5, p. 13]).

As to (3.9), it is an obvious consequence of (b) and (d) of the following lemma (cf. Azarin [2, Lemma 1]). If f and g are non-negative functions in $K(\alpha)$, let us say that $f \approx g$ if there exist positive constants c_1 and c_2 only depending on α such that $c_1 f \le g \le c_2 f$.

<u>Lemma 2:</u> Let Ψ_1 and ρ be as in (1.3) and (1.4). Then if $0 < 2|P| \le |Q|$, we have

(a) $\Psi_1(e(P))\,\Psi_1(e(Q))\,(|P|/|Q|)^\rho |Q|^{-1} \approx g(P,Q) \quad P, Q \in K(\alpha)$

(b) $\Psi_1(e(P))\,\Psi_1(e(Q))\,(|P|/|Q|)^\rho |Q|^{-2} \approx \frac{\partial g}{\partial n}(P,Q), \quad P \in K(\alpha), \; Q \in \partial K(\alpha)$.

If $0 < 2|Q| \le |P|$, we have

(c) $\Psi_1(e(P))\,\Psi_1(e(Q))\,(|Q|/|P|)^\rho |P|^{-1} \approx g(P,Q), \quad P, Q \in K(\alpha)$

(d) $\Psi_1(e(P))\,\Psi_1(e(Q))\,(|Q|/|P|)^\rho |P|^{-2} |Q|^{-1} \approx \frac{\partial g}{\partial n}(P,Q), \quad P \in K(\alpha), \; Q \in \partial K(\alpha)$.

4. The final proof

Using (3.10) and (3.1), we see that

(4.1) $\lim r^{-\lambda\rho} M(r) = B, \quad 0 < B < \infty$.

From the definition of v, it follows that for $e(P) \in S(\alpha)$,

$$(4.2)\quad r^{-\lambda\rho} v(re(P)) \to B\, C(\lambda)\, \Psi_\lambda(e(P)),$$

uniformly as $r \to \infty$. Hence it suffices to prove (A) and (B) of Theorem 1 for $-p = u - v$. The function p is superharmonic and non-negative in $K(\alpha)$. Moreover, from Lemma 1, (2.1) and (2.3), we see that for r large enough, there exists $P_r \in K(\alpha)$, $|P_r| = r$, such that $u(P_r) = M(r,u) = M(r)^+$. Hence

$$(4.3)\quad p(P_r) = v(P_r) - u(P_r) \le v(r,0,0) - M(r)^+ .$$

From (1.6) and (4.1) - (4.3), we find that

$$(4.4)\quad \lim_{r\to\infty} r^{-\lambda\rho} p(P_r) = 0,$$

and that for any δ, $0 < \delta < \alpha$, there exists T such that

$$(4.5)\quad P_r \in K(\delta), \qquad r > T .$$

Using (3.11), (4,3) and (4.5), we deduce that if δ is given, $0 < \delta < \alpha$,

$$(4.6)\quad \int_T^\infty w(r)\, r^{-1-\lambda\rho}\, dr < \infty,$$

where $w(r) = \inf p(P)$, $|P| = r$, $P \in K(\delta)$.

Using (4.4), we obtain as in Azarin [2, Theorem 1] a Riesz representation formula for the non-negative, superharmonic function p

$$(4.7)\quad p(P) = \int_{\partial K(\alpha)} \frac{\partial g}{\partial n}(P,Q)\, d\gamma(Q) + \int_{K(\alpha)} g(P,Q)\, d\xi(Q), \quad P \in K(\alpha).$$

there γ and ξ are positive Borel measures on $\partial K(\alpha)$ and $K(\alpha)$, respectively.

A consequence of (4.6) and of Lemma 2 is that

$$(4.8)\quad \int_{K_1(\alpha)} \Psi_1(e(Q))\, |Q|^{\rho}\, d\xi(Q) + \int_{K_2(\alpha)} \Psi_1(e(Q))\, |Q|^{-1-\lambda\rho}\, d\xi(Q) < \infty,$$

$$(4.9)\quad \int_{\partial K_1(\alpha)\cap\partial K(\alpha)} \frac{\partial \Psi_1}{\partial n}(e(Q))\, |Q|^{\rho-1}\, d\gamma(Q) + \int_{\partial K_2(\alpha)\cap\partial K(\alpha)} \frac{\partial \Psi_1}{\partial n}(e(Q))\, |Q|^{-2-\lambda\rho}\, d\gamma(Q) < \infty,$$

where $K_1(\alpha) = K(\alpha) \cap \{|Q| < T/2\}$ and $K_2(\alpha) = K(\alpha) \cap \{|Q| \geq T/2\}$.

We define

$$d\eta(Q) = \begin{cases} \Psi_1(e(Q))\,|Q|^{-1-\lambda\rho}\,d\xi(Q), & Q \in K_2(\alpha) \\ \dfrac{\partial \Psi_1}{\partial n}(e(Q))\,|Q|^{-2-\lambda\rho}\,d\gamma(Q), & Q \in \partial K_2(\alpha) \cap \partial K(\alpha) \\ \Psi_1(e(Q))\,|Q|^{\rho}\,d\xi(Q) & Q \in K_1(\alpha) \\ \dfrac{\partial \Psi_1}{\partial n}(e(Q))\,|Q|^{\rho-1}\,d\gamma(Q), & Q \in \partial K_1(\alpha) \cap \partial K(\alpha) \end{cases}$$

and $N(P,Q)$ for $P \in K(\alpha)$ by

$$N(P,Q)\,d\eta(Q) = \begin{cases} \dfrac{\partial g}{\partial n}(P,Q)\,d\gamma(Q), & Q \in \partial K(\alpha) \\ g(P,Q)\,d\xi(Q), & Q \in K(\alpha) \end{cases}$$

By (4.8) and (4.9)

$$(4.10) \quad \int_{\overline{K(\alpha)}} d\eta(Q) < \infty$$

and $$p(P) = \int_{\overline{K(\alpha)}} N(P,Q)\,d\eta(Q)\,.$$

Corresponding formulas for the case $\lambda = 1$ can be found in Hayman [12, p. 117] and Azarin [2, p. 131].

Once (4.10) has been proved, the method of Hayman and Azarin can be applied also when $0 < \lambda < 1$. Let us discuss the proof of (B) in Theorem 1. We need the following definition. (cf. [12, p. 120] and [2, p. 133]).

Definition: Let ε be a fixed positive number. Then the point $P \in K(\alpha)$ is said to be λ-normal (ε) if for $0 < h < |P|/2$, we have

$$\int_{|Q-P|<h} d\eta(Q) < \varepsilon(h/|P|)^2$$

Lemma 3: (cf. Azarin [2, Lemma 6]). If P is λ-normal (ε) and $|P| > T$ (cf. (4.5)), then

$$I_1(P) = \int_{D(P)} N(P,Q)\, d\eta(Q) < \left[\text{Const.}\ \varepsilon + \underline{o}(1)\right] |P|^{\lambda\rho}, \quad P \to \infty$$

where $D(P) = \left\{ Q \in \overline{K(\alpha)} \mid 1/2 \leq |P|/|Q| \leq 2 \right\}$. The positive constant depends only on α.

Using Lemma 2 and arguing as in Hayman [12, § 3], it can be shown that $(p - I_1)(P) = \underline{o}(|P|^{\lambda\rho})$ uniformly in $K(\alpha)$ as $P \to \infty$. The last step in the proof of (B) in Theorem 1 is given by the following Lemma of Azarin [2, Lemma 7].

Lemma 4: The set $\Delta(\varepsilon)$ of points not λ-normal (ε) may be covered by a system $F(\varepsilon)$ of spheres $\{G_i\}$ whose radii $\{r_i\}$ and distances from $\{R_i\}$ from their centers to the origin satisfy (2.5).

References

1. L. Ahlfors and M. Heins, Questions of regularity connected with the Phragmén-Lindelöf principle, Ann. of Math. 50 (1949), 341 - 346.

2. V. Azarin, Generalization of a theorem of Hayman on subharmonic functions in an m-dimensional cone, Amer. Math. Soc. Transl. (2) 80 (1969), 119 - 138.

3. R. Courant and D. Hilbert, Methods of mathematical physics, vol. II, Wiley-Interscience 1962.

4. B. Dahlberg, Growth properties of subharmonic functions, thesis, University of Göteborg, 1971.

5. M. Essén, Note on "A theorem on the minimum modulus of entire functions" by Kjellberg, Math. Scand. 12 (1963), 12 - 14.

6. M. Essén, A generalization of the Ahlfors-Heins Theorem, Bull. Amer. Math. Soc. 75 (1969), 127 - 131.

7. M. Essén, A generalization of the Ahlfors-Heins theorem, Trans. Amer. Math. Soc. 142 (1969), 331 - 344.

8. M. Essén, The generalized Ahlfors-Heins theorem in R^3, to appear in a volume with talks given at the conference on mathematical analysis at Jyväskylä, Finland, August 1970 (Springer).

9. J. Lelong-Ferrand, Étude au voisinage de la frontière des fonctions subharmoniques positives dans un demi-espace, Ann. Sci. École Norm. Sup. (3) 66 (1949), 125 - 159.

10. J. Lelong-Ferrand, Étude des fonctions subharmoniques positives dans un cylindre ou dans un cône, C.R. Acad. Sci. Paris 229 (1949), 340 - 341.

11. J. Lelong-Ferrand, Extension du théorème de Phragmén-Lindelöf-Heins aux fonctions sousharmoniques dans un cône ou dans un cylindre, C.R. Acad. Sci. Paris 229 (1949), 411 - 413.

12. W. Hayman, Questions of regularity connected with the Phragmén-Lindelöf principle, J. Math. pure appl. (9) 35 (1956), 115 - 126.

13. M. Heins, Selected topics in the classical theory of functions of a complex variable, New York, Holt, Rinehart and Winston, 1962.

14. J. Lewis, Subharmonic functions in certain regions, to appear in Trans. Amer. Math. Soc.

15. J. Lewis, On Essén's generalization of the Ahlfors-Heins theorem, to appear in Trans. Amer. Math. Soc.

On a Lower Bound of a Second-Order, Linear Differential Operator with an Integrable-p coefficient

W.N. Everitt

1. We consider here a differential operator generated by the differential expression $M[\cdot]$ with

$$M[y] = -y'' + qy \text{ on } [0, \infty) \qquad ({}' = \frac{d}{dx}) \tag{1.1}$$

where the coefficient q is real-valued on $[0, \infty)$ and for an index p in the range $1 \leq p < \infty$ satisfies the condition

$$q \in L^p(0, \infty). \tag{1.2}$$

Here [and) denote a closed and open end point of an interval of the real line R respectively; $L^p(0, \infty)$ denotes the set of complex-valued, Lebesgue measurable functions $\{f\}$ defined on the half-line $[0, \infty)$ which satisfy

$$\|f\|_p = \left\{\int_0^\infty |f|^p\right\}^{\frac{1}{p}} < \infty. \tag{1.3}$$

We recall that f may or may not be bounded on $[0, \infty)$ if (1.3) is satisfied.

The differential expression $M[\cdot]$ generates a family $\{T_\alpha : \alpha \in [0, \pi)\}$ of unbounded, self-adjoint differential operators in the Hilbert function space $L^2(0, \infty)$; given $\alpha \in [0, \pi)$ define the domain $D(T_\alpha)$ by

$f \in D(T_\alpha)$ if (i) $f \in L^2(0, \infty)$ (ii) f' is absolutely continuous on $[0, X]$ for all $X > 0$ (iii) $M[f] \in L^2(0, \infty)$ (iv) $f(0) \cos \alpha + f'(0) \sin \alpha = 0$

and then $T_\alpha : D(T_\alpha) \to L^2(0, \infty)$ by

$$T_\alpha f = M[f] \quad (f \in D(T_\alpha)).$$

The self-adjointness of T_α follows from the analysis in the book by Naimark, see [6] sections 17.5 and 18.3, and the fact that the condition (1.2) on the coefficient q implies that $M[\cdot]$ is in the limit-point condition at ∞; the latter result was first proved by Hartman and Wintner, see [5], and recently, using different methods, by Patula and Wong in [7] and Everitt, Giertz and Weidmann in [2].

The self-adjoint operator T_α is said to be bounded below in $L^2(0, \infty)$ if there is a real number k such that

$$(T_\alpha f, f) \geq k(f, f) \quad (f \in D(T_\alpha)) \tag{1.4}$$

where $(\cdot, \cdot)$ is the inner-product in $L^2(0, \infty)$; see the book by Glazman [4], section 1.4 of Chapter I, for the implications of the property on the spectrum of T_α.

It is known, from a far-reaching result in [4] section 27, that all the differential operators $T_\alpha (\alpha \in [0, T))$ are bounded below in $L^2(0, \infty)$ when the coefficient q satisfies the condition (1.2). We are concerned here with an explicit formula for the lower bound of the operator T_0 in terms of the $L^p(0, \infty)$ norm $\|q\|_p$ of the coefficient q.

We have the result: let the differential expression $M[\cdot]$ be given by (1.1) and let the coefficient q satisfy the condition (1.2); let the differential operator T_0 be defined as above; then

$$(T_0 f, f) \geq -k(p) \{\|q\|_p\}^{\frac{2p}{2p-1}} (f, f) \quad (f \in D(T_0)) \tag{1.5}$$

where $\quad k(p) = (2p - 1)\, 2^{-\frac{2p-2}{2p-1}} p^{-\frac{2p}{2p-1}} \quad (p \in [1, \infty)). \tag{1.6}$

It is worth noting that the lower bound in (1.5) has the following values:

$p = 1$	$-\|q\|_1^2$
$p = 2$	$-\frac{3}{4}\{\|q\|_2\}^{\frac{4}{3}}$
$p = \infty$	$-\|q\|_\infty$.

We do not discuss the case $p = \infty$ here, although the analysis extends, but it is of interest to note that the result in this case is consistent with obtaining the lower bound of T_0 by known methods when $q \in L^{\infty}(0, \infty)$, i.e. when q is bounded on $[0, \infty)$.

2. The proof of the above result depends on the availability of the Dirichlet formula, so called, for T_0; details may be found in the forthcoming paper [2]. We have: let $f \in D(T_0)$ then both f' and $|q|^{\frac{1}{2}}f$ are in $L^2(0, \infty)$, $\lim_{\infty} f = \lim_{\infty} f' = 0$ and

$$\int_0^{\infty} f'g' + \int_0^{\infty} qfg = [fg']_0^{\infty} + \int_0^{\infty} fM[g]$$

$$= \int_0^{\infty} f.T_0 g \quad (f,\ g \in D(T_0)). \tag{2.1}$$

Define $M_0(\cdot)$ on $D(T_0)$ by

$$M_0(f) = \sup\{|f(x)| : x \in [0, \infty)\} \quad (f \in D(T_0)) \tag{2.2}$$

so that from the results in [2] we have $0 \leqslant M_0(f) < \infty$ for all $f \in D(T_0)$. Integration by parts gives

$$|f(x)|^2 = \int_0^x \{f\bar{f}' + f'\bar{f}\} \leqslant 2\int_0^x |ff'| \quad (x \geqslant 0)$$

for all $f \in D(T_0)$, since then $f(0) = 0$. This yields

$$M_0(f)^2 \leqslant 2\|f\|_2 \|f'\|_2 \quad (f \in D(T_0)). \tag{2.3}$$

In the Dirichlet formula (2.1) replace f by $\bar{f}$ and take $g = f$ to give

$$\int_0^{\infty} |f'|^2 + \int_0^{\infty} q|f|^2 = (T_0 f,\ f) \quad (f \in D(T_0)),$$

which yields

$$(T_0 f,\ f) \geqslant \int_0^{\infty} |f'|^2 - \int_0^{\infty} |q||f|^2 \quad (f \in D(T_0)). \tag{2.4}$$

With $p \in [1, \infty)$ given, let s be the conjugate index, i.e. $p^{-1} + s^{-1} = 1$. We give details only for the case $1 < p < \infty$; the case $p = 1$ requires only minor modification; thus $1 < s < \infty$ also.

From the Hölder inequality we obtain

$$\int_0^\infty |q||f|^2 \leq \|q\|_p \| |f|^2 \|_s \qquad (f \in D(T_0));$$

note that $2s > 2$ and $f \in L^{2s}(0, \infty)$ since $f \in L^2(0, \infty)$ and f is bounded on $[0, \infty)$ from (2.3). This gives

$$\int_0^\infty |q||f|^2 \leq \|q\|_p M_0(f)^{\frac{2s-2}{s}} \left\{ \int_0^\infty |f|^2 \right\}^{\frac{1}{s}} \quad (f \in D(T_0)) \tag{2.5}$$

From (2.3, 4 and 5) we now have

$$(T_0 f, f) \geq \int_0^\infty |f'|^2 - \|q\|_p \left\{ M_0(f)^2 \right\}^{\frac{s-1}{s}} \left\{ \int_0^\infty |f|^2 \right\}^{\frac{1}{s}}$$

$$\geq \int_0^\infty |f'|^2 - 2^{\frac{s-1}{s}} \|q\|_p \left\{ \int_0^\infty |f'|^2 \right\}^{\frac{s-1}{2s}} \left\{ \int_0^\infty |f|^2 \right\}^{\frac{s+1}{2s}} \tag{2.6}$$

valid for all $f \in D(T_0)$.

<u>We now need: let $1 < P < \infty$, $1 < Q < \infty$ with $P^{-1} + Q^{-1} = 1$; let A, B, C be positive real numbers; let α, β be non-negative real numbers; $\sigma(P, Q) = P^{\frac{1}{P}} Q^{\frac{1}{Q}}$; then</u>

$$A\alpha^P - \sigma(P, Q) B \alpha\beta + C\beta^Q \geq 0 \quad (\alpha \geq 0, \beta \geq 0) \tag{2.7}$$

<u>if and only if</u>

$$A^{\frac{1}{P}} C^{\frac{1}{Q}} \geq B. \tag{2.8}$$

This generalises a well-known result when $P = Q = 2$; for a proof see [3] and references therein.

Apply this result as follows: let

$$P = \frac{2s}{s-1}, \quad Q = \frac{2s}{s+1}, \text{ i.e. } P^{-1} + Q^{-1} = 1$$

$$\alpha = \left\{ \int_0^\infty |f'|^2 \right\}^{\frac{1}{P}}, \quad \beta = \left\{ \int_0^\infty |f|^2 \right\}^{\frac{1}{Q}}$$

$$A = 1 \qquad B = 2^{\frac{s-1}{s}} \cdot \|q\|_p \{\sigma(P, Q)\}^{-1}$$

and then choose C so that $A^{\frac{1}{P}} C^{\frac{1}{Q}} = B$, i.e. $C = B^Q$. We then obtain from (2.7), since (2.8) is satisfied,

$$\int_0^\infty |f'|^2 - 2^{\frac{s-1}{s}} \|q\|_p \left\{ \int_0^\infty |f'|^2 \right\}^{\frac{s-1}{2s}} \left\{ \int_0^\infty |f|^2 \right\}^{\frac{s+1}{2s}} +$$

$$\left[2^{\frac{s-1}{s}} \|q\|_p \{\sigma(P, Q)\}^{-1} \right]^Q \int_0^\infty |f|^2 \geq 0 \tag{2.9}$$

valid for all $f \in D(T_0)$.

Now subtract and add the last term on the left-hand side of (2.9) to the right-hand side of (2.6) to give, using (2.9),

$$(T_0 f, f) \geq -\left[2^{\frac{s-1}{s}} \|q\|_p \{\sigma(P, Q)\}^{-1} \right]^Q (f, f) \qquad (f \in D(T_0))$$

and a calculation then shows that (1.5 and 6) are satisfied.

3. A self-adjoint operator which is bounded below has its spectrum, a closed subset of the real line, bounded below by the same bound; this if $\sigma(T_0)$ is the spectrum of T_0 then

$$\lambda \geq -k(p) \{ \|q\|_p \}^{\frac{2p}{2p-1}} \qquad (\lambda \in \sigma(T_0)). \tag{3.1}$$

The operators T_α ($\alpha \in (0, \pi)$) are also bounded below in $L^2(0, \infty)$ but may have at most one isolated eigenvalue below the bound given by (3.1); details of this result may be found in [3].

It does not seem to be known if the lower bound for T_0 given by (1.5 and 6) is best possible.

It is known that the essential spectrum of T_0 is unbounded above; see [4], theorem 25 of section 31. It does not seem to be known if the condition (1.2) on the coefficient q implies that the eigenvalues of T_0 are bounded above in the same way as they are bounded below. Additional conditions on q do give an upper bound for the eigenvalues of T_0; the outstanding result in this direction is due to Borg, see [1].

REFERENCES

1. Göran Borg, 'On the point spectra of $y'' + (\lambda - q)y = 0$' Amer. Journ. Math. 73 (1951) 122-6.
2. W. N. Everitt, M. Giertz and J. Weidmann, 'Some remarks on a limit-point and separation criterion for second-order, linear differential expressions', (to appear in Math. Annalen).
3. W. N. Everitt, 'On the spectrum of a second-order, linear differential equation with an integrable-p coefficient' (to appear in Applicable Analysis).
4. I. M. Glazman, Direct methods for the qualitative spectral analysis of singular differential operators (Israel Program for Scientific Translations, Jersusalem, 1965).
5. P. Hartman and A Wintner, 'A criterion for the non-degeneracy of the wave equation', Amer. Journ. Math. 71 (1949) 206-13.
6. M. A. Naimark, Linear differential operators: Part II (Ungar, New York, 1968).
7. W. T. Patula and J. S. W. Wong, 'On a limit-point criterion of second-order, ordinary differential equations', (to appear in Math. Annalen).

Some Dirichlet and Separation Results for Certain Differential Expressions on Finite Intervals

W.N. Everitt and M. Giertz

For a given interval I on the real line we associate with each real-valued function q defined on I a subspace $D_1 = D_1(q)$ of $L^2(I)$ by

$$f \in D_1 \quad \underline{\text{if}} \quad \begin{array}{ll} \text{(i)} & f \in L^2(I), \\ \text{(ii)} & f' \in AC_{loc}(I), \\ \text{(iii)} & M[f] = -f'' + qf \in L^2(I). \end{array} \qquad (1)$$

We may regard D_1 as the maximal domain of those linear operators in $L^2(I)$ which can be generated by the differential expression $M[\cdot]$ introduced in (iii) of (1).

The object of this talk is to discuss sufficient conditions on q in order that either the <u>Dirichlet result</u>

$$f' \in L^2(I) \quad \underline{\text{and}} \quad |q|^{\frac{1}{2}} f \in L^2(I) \qquad (f \in D_1), \qquad (2)$$

or the <u>separation result</u>

$$f'' \in L^2(I) \quad \underline{\text{and}} \quad qf \in L^2(I) \qquad (f \in D_1), \qquad (3)$$

holds true in the cases when the interval I is finite and open. If we assume that $q \in L_{loc}(I)$ or $q \in L^2_{loc}(I)$ when considering (2) and (3) respectively, it is only the behavuour of q in neighbourhoods of the end points of the interval I that decides whether or not these statements are valid.

We begin by recalling the corresponding classical Dirichlet result if I is finite and closed or infinite and closed at any finite end point; it states that (2) holds true for each q satisfying

(i) $q: I \to (-\infty, \infty)$

(ii) $q \in L_{loc}(I)$ (4)

(iii) q <u>is bounded below near any open end points of</u> I.

This is a consequence of the following identity, valid for any number k and for all interior points x and y of I, here f^* denotes the complex conjugate of f,

$$\int_x^y \{f^*M[f] + k|f|^2\} = -(f^*f')(y) + (f^*f')(x) + \int_x^y \{|f'|^2 + (q+k)|f|^2\} \qquad (f \in D_1). \tag{5}$$

Assuming k to be real and chosen so that $q(x) + k > 0$ near the end points a and b of I, we let y tend to the right end point b. If b is finite, so that I is closed at b, then $(f^*f')(y)$ must tend to a finite limit since (ii) of (1) and (4) require, respectively, f' to be absolutely continuous and q to be integrable up to b. On the other hand, if $b = +\infty$ then (5) forces $(f^*f')(y)$ to tend to a finite limit or to $+\infty$. But now any non-zero limit contradicts the assumption $f \in L^2(I)$. Using similar arguments as $x \to a$ we conclude from (5) that (2) is valid.

The above argument breaks down if one of the end points, say b, is finite and I is open at b. In this situation $\lim(f^*f')(y)$ $(y \to b)$ does not contradict $f \in L^2(I)$, and there are, in fact, examples of coefficients q which satisfy (4) but for which the Dirichlet result does not hold.

The separation problem (3) for infinite intervals, again closed at any finite end point, has been studied by Atkinson [1] and by Everitt-Giertz in [2] and [3]. Among other things this problem is relevant to the question when the operator theoretic sum of two self-adjoint operators in $L^2(I)$ is also self-adjoint. We refer to [2] for further details.

In order to ensure the validity of (2) or (3) for a finite open interval $I = (a, b)$ it is, essentially, necessary to impose some control condition on $|q'q^{-3/2}|$ near the end points a and b. For the Dirichlet result (2) it is sufficient to require that q, in some subset $I_1 \subset I$, of the form $I_1 = (a, a_1) \cup (b_1, b)$ with $a < a_1 \leq b_1 < b$, satisfies

$$\begin{aligned} &\text{(i)} \quad q \in AC_{loc}(I_1), \\ &\text{(ii)} \quad |q'(x)| \leq (4/\sqrt{3})\,|q(x)|^{3/2} \quad (\underline{\text{for almost all } x \in I_1}), \end{aligned} \tag{6}$$

in addition to (4). This result is best possible in the sense that the constant $4/\sqrt{3}$ in (6) is critical; if we replace it by some number $c > 4/\sqrt{3}$ then there are coefficients q satisfying (4) and the "new" (6) for which the Dirichlet result (2) is false. A more complete discussion and proofs of the above statements are given in [5].

As for the separation result (3), there exist sufficient conditions on q analogous to (4) and (6). In fact, (3) holds true if q satisfies the basic condition

$$\begin{aligned} &\text{(i)} \quad q: I \to (-\infty, \infty) \\ &\text{(ii)} \quad q \in L^2_{loc}(I), \end{aligned} \tag{7}$$

and if, in addition, there exists a subset $I_1 \subset I$ as in (6) such that

$$\begin{aligned} &\text{(i)} \quad q \ \underline{\text{is bounded below on}}\ I_1, \\ &\text{(ii)} \quad q \in AC_{loc}(I_1), \\ &\text{(iii)} \quad \text{for some } c \in (0, 2) \\ &\qquad |q'(x)| \leq c\,|q(x)|^{3/2} \quad (\text{for almost all } x \in I_1). \end{aligned} \tag{8}$$

There are examples which show that the corresponding statement is false if the constant c in (iii) of (8) is taken to be greater than $4/\sqrt{3} \sim 2.3$.

The above separation results are proved in [4], where the proof is based on inequalities of the form

$$a_0\|qf\|^2 + a_1\|q^{\frac{1}{2}}f'\|^2 + a_2\|f''\|^2 \leq \|M[f]\|^2 \quad (f \in D_1), \tag{9}$$

where a_0, a_1 and a_2 denote real, positive numbers. Explicit inequalities of the form (9) are given, which are valid for each coefficient q which is positive on I, unbounded above near the end points and satisfies (7) and (8), with I in place of I_1, for any specified $c \in (0, 2)$. To exemplify, we mention the following single special cases:

$$(1 - c^2/4)\|qf\| \leq \|M[f]\| \quad (f \in D_1),$$

$$(2 - c)\|q^{\frac{1}{2}}f'\| \leq \|M[f]\| \quad (f \in D_1),$$

$$\{\min(1, 4/c^2 - 1)\|f''\| \leq \|M[f]\| \quad (f \in D_1),$$

and, if $c \in (0, \surd 2]$ and $\delta \in [c/2, 1/c]$,

$$(1 - \delta c)\|qf\|^2 + (2 - c/\delta)\|q^{\frac{1}{2}}f'\|^2 + \|f''\|^2 \leq \|M[f]\|^2 \quad (f \in D_1).$$

R E F E R E N C E S

1. F. V. Atkinson, "On some results of Everitt and Giertz", (to appear in Proc. Royal Soc. Edin.)
2. W. N. Everitt and M. Giertz, "Some properties of the domains of certain differential operators", Proc. London Math. Soc. (3) 23 (1971) 301-327
3. ——, "Some remarks on a separation and limit-point criterion of second-order, ordinary differential expressions", (to appear in Math. Annalen.)
4. ——, "Inequalities and separation for certain differential operators", (to be published)
5. ——, "A Dirichlet type result for differential operators on finite intervals", (to be published).

On Certain Parabolic Differential Equations and an Equivalent Variational Problem

W. Förster

1. Introduction

Most partial differential equations in applied mathematics are derived from simplified problems. To describe complex problems, it is sometimes necessary to use a number of partial differential equations. In such a case, it is important to clarify the interrelation between these equations. It will be shown in this paper that the Navier-Stokes equation, the heat conduction equation, and the diffusion equations for a mixture of N fluids are equivalent to a certain variational problem.

2. A variational problem

We state the following variational problem

$$\delta S = 0 \quad , \tag{1}$$

with the action function

$$S = \iiiint_a^b L \, d^4x \quad , \tag{2}$$

and the additional conditions

$$ds \geq 0 \quad , \tag{3}$$

$$\partial_\mu [\overset{H}{M} u^\mu + \overset{H\mu}{\nu}] = 0 \qquad H = 1, 2, \ldots , N \ . \tag{4}$$

In (2) L is the Lagrangian density. (3) expresses the non-decrease of the entropy s . (4) states conservation of the mass $\overset{H}{M}$ for the component H , ∂_μ is a differential operator, u_μ is the 4-velocity, and the additional term $\overset{H}{\nu}_\mu$ takes care of dissipative processes. For a Greek index appearing twice, summation from 1 to 4 is understood.

3. Thermodynamics

In order to give a precise meaning to (3), we have to introduce thermodynamic quantities. For details we refer to [1]. We assume a functional relation of the form

(5) $e = e(s, v, \overset{1}{M}, \ldots, \overset{N}{M})$, where e is the internal energy and v is a volume. We give the following definitions:

(6) $\left\{\frac{\partial e}{\partial s}\right\}_{v, \overset{H}{M}} = T$, with T the temperature,

(7) $\left\{\frac{\partial e}{\partial v}\right\}_{s, \overset{H}{M}} = -p$, with p the pressure,

(8) $\left\{\frac{\partial e}{\partial \overset{H}{M}}\right\}_{s, v, \overset{K}{M} \neq \overset{H}{M}} = \overset{H}{\mu}$, with $\overset{H}{\mu}$ the chemical potential of the component H,

(9) $e + pv = w$, with w the heat function,

(10) $w - Ts = \Phi$, with Φ the thermodynamic potential.

From additivity requirements for Φ, we obtain

(11) $\Phi = \sum_{H} \overset{H}{\mu}\overset{H}{M}$

We use:

a) the first law of thermodynamics, which states the conservation of energy in a system at rest,

b) the second law of thermodynamics, which states that the entropy of an isolated system can never decrease.

We give the following thermodynamic identities:

(12) $de = T\,ds - p\,dv + \sum_{H} \overset{H}{\mu} d\overset{H}{M}$,

(13) $d(\overset{H}{\mu}/T) = (1/\overset{H}{M})[(-w/T^2)dT + (v/T)dp - \sum_{K \neq H} \overset{K}{M}\, d(\overset{K}{\mu}/T)]$.

4. Conservation law form of the problem

If we require the Lagrangian density in (2) to be invariant under an infinitesimal transformation, e.g. under

$$\bar{x}_\nu = x_\nu + \epsilon\, \xi^\nu (x_1, x_2, x_3, x_4) \quad , \qquad (14)$$

with ϵ a small parameter, we can obtain a set of conservation laws. This is known as Noether's theorem [2]. Details are discussed in [3]. The conservation laws have the form

$$\partial_\nu T_\mu^{\ \nu} = 0 \qquad \mu, \nu = 1, 2, 3, 4 \ . \qquad (15)$$

For viscous fluids, the energy-momentum tensor is given by

$$T_{\mu\nu} = w\, u_\mu u_\nu - p\, g_{\mu\nu} + \tau_{\mu\nu} \quad , \qquad (16)$$

with the thermodynamic quantities taken per unit volume, and with the metric tensor

$$g_{\mu\nu} = \begin{bmatrix} -1 & 0 & 0 & 0 \\ 0 & -1 & 0 & 0 \\ 0 & 0 & -1 & 0 \\ 0 & 0 & 0 & 1 \end{bmatrix} \ . \qquad (17)$$

The pressure p is connected to the internal energy e by the equation of state. Details are given in [1] and [4].

A fluid element is at rest when the momentum of the element is zero and when its energy is expressible in terms of the other thermodynamic quantities by the usual thermodynamic relations. This leads to the conditions

$$\tau_{\mu\nu}\, u^\nu = 0 \qquad (18)$$

and

$$\overset{H}{\nu}_\mu\, u^\mu = 0 \qquad H = 1, 2, \ldots , N \ . \qquad (19)$$

The forms of the tensor $\tau_{\mu\nu}$ and of the vector $\overset{H}{\nu}_\mu$ can be established from the second law of thermodynamics. Using the thermodynamic formulae given in section 3, and equation (18) it can be shown that

$$\partial_\mu \left[u^\mu s - \sum_H (\overset{H}{\mu}/T)\, \overset{H\mu}{\nu} \right] = - \sum_H \overset{H\mu}{\nu}\, \partial_\mu (\overset{H}{\mu}/T) - (1/T)\, \tau_\nu{}^\mu\, \partial_\mu u^\nu \quad . \quad (20)$$

The left-hand side is the 4-divergence of the entropy flux, the right-hand side is the increase in entropy due to dissipative processes. The requirement of a positive right-hand side in (20) and the conditions (18) and (19) determine uniquely the forms of $\tau_{\mu\nu}$ and $\overset{H}{\nu}_\mu$.

$$\tau_{\mu\nu} = - \eta\, (\partial_\mu u_\nu + \partial_\nu u_\mu - u_\mu u^\kappa \partial_\kappa u_\nu - u_\nu u^\kappa \partial_\kappa u_\mu)$$
$$- \{\xi - (2/3)\eta\}(\partial_\kappa u_\kappa)(g_{\mu\nu} - u_\mu u_\nu) \quad (21)$$

with η and ξ the viscosity coefficients.

$$\overset{H}{\nu}_\mu = - \overset{H}{\text{const}}\, \{(\overset{H}{M} T)/w\}^2\, [\partial_\mu (\overset{H}{\mu}/T) - u_\mu u^\kappa \partial_\kappa (\overset{H}{\mu}/T)] \quad (22)$$

The variational problem of section 2 is now reduced to (15) subject to (4). Condition (3) is now already included in $T_{\mu\nu}$.

Remark: The following invariant scalar gives the Lagrangian density

$$L = T_{\mu\nu}\, u_\lambda\, u_\kappa\, g^{\mu\lambda}\, g^{\nu\kappa} = e \quad . \quad (23)$$

Because of (18) $\tau_{\mu\nu}$ leads to a zero-contribution to L, but nevertheless assures that (3) is satisfied.

5. Special cases

Between the velocity

$$\underset{\sim}{v} = (v_1, v_2, v_3) \quad (24)$$

and the 4-velocity u_ν, we have the relation

$$u_i = v_i/\{(1 - \underset{\sim}{v}^2)^{\frac{1}{2}}\} \quad , \quad i = 1, 2, 3 \quad , \quad (25)$$

with

$$\underset{\sim}{v}^2 = v_1^2 + v_2^2 + v_3^2 \quad . \quad (26)$$

For small velocities, that is for

$$\underset{\sim}{v}^2 \ll 1 \quad , \quad (27)$$

it can be shown that:

a) the first two terms in (16) lead to Euler's equations of fluid dynamics,

b) terms containing $u_i\, u_j$ with $i,j = 1, 2, 3$ can be neglected,

c) $\tau_{\mu\nu}$ given by (21) leads together with the first two terms in (16) to the Navier-Stokes equation,

d) for $N = 1$, that is a homogeneous fluid, we can obtain the heat conduction equation from (4), (22), (16) and using (13) and (15),

e) for $N = 2$ and constant temperature and constant pressure, we can obtain in a similar manner the diffusion equation.

6. Conclusion

Starting with a variational problem, then using a few thermodynamic identities, and imposing a mild restriction on the velocities, we have outlined how the Navier-Stokes equation, the heat conduction equation, and the diffusion equation are interrelated. It was shown that these parabolic partial differential equations are equivalent to a certain variational problem.

References

[1] L.D. Landau and E.M. Lifshitz, Statistical Physics, Pergamon Press, London, 1959.

[2] E. Noether, Invariante Variationsprobleme, Nachr. Ges. Göttingen (math.-phys. Kl.), 1918, pp. 235 - 257.

[3] R. Courant and D. Hilbert, Methods of Mathematical Physics, Vol. I, Interscience, New York, 1953.

[4] L.D. Landau and E.M. Lifshitz, Fluid Mechanics, Pergamon Press, Oxford, 1963.

Turning Points in Second-Order Elliptic Singular Perturbation Problems

P.P.N. de Groen

§1. Introduction

In almost all studies of the asymptotic behaviour for $\varepsilon \downarrow 0$ of the solution of the Dirichletproblem

$$\varepsilon L_2 \Phi + L_1 \Phi = 0$$

on a bounded domain G in $\mathbb{R}^2$, where L_2 is a 2nd order elliptic and L_1 a first order operator, the characteristics of L_1 are smooth curves, c.f. [1,2,3 and 5]. In other words, the system of ordinary partial differential equations associated with L_1 has no singular points inside G or on its boundary. If singular points inside G are admitted, they in general give rise to free boundary layers and the problems become much more difficult.

In this contribution some points will be clarified for the case, when L_1 has exactly one singular point inside G, which is a node or a saddle-point. For the case of the singular line, we refer to de Jager [4]. Because of the amount of calculations involved, we will treat only some simple prototype problems, giving an idea of what is happening in the general case. We always assume that a solution of the problem exists and is unique. The proofs of the validity of the approximations are based on estimates of the remainder, using barrierfunctions and the maximumprinciple for elliptic differential equations, cf [1].

§2. The Node

Let G be the unit disk in $\mathbb{R}^2$, $L_2 = \Delta$ and $L_1 = x\frac{\partial}{\partial x} + y\frac{\partial}{\partial y}$.

We try to approximate the solution Φ_ε of

(1) $\quad L_\varepsilon \Phi_\varepsilon = \varepsilon L_1 \Phi_\varepsilon + L_2 \Phi_\varepsilon = 0, \; \Phi_\varepsilon(1,\phi) = f(\phi)$

where f is of class C^2 and (r,ϕ) are polar coordinates. The characteristics of L_1 are straight lines pointing to the origin, along which the solution u of the reduced equation $L_1 u = 0$ is constant. We see that u cannot satisfy the boundary conditions and also be continuous at the origin. It appears that we have to take $u(r,\phi) = f(\phi)$ and that a boundary layer arises around the origin. In fact we will prove the following:

(2) $$\Phi_\varepsilon(r,\phi) = f(\phi) + O(\varepsilon\, r^{-2}).$$

With the barrierfunction $\lambda\, \varepsilon\, r^{-2}$ we estimate $\Phi_\varepsilon - f$ on the annulus $2\sqrt{\varepsilon} \le r \le 1$. From the maximum principle it follows that Φ_ε is bounded by $\max|f(\phi)|$; furthermore we have

$$L_\varepsilon r^{-2} = -2\, r^{-2}(1-2\,\varepsilon\, r^{-2}) \le -r^{-2} \quad \text{for } r > 2\sqrt{\varepsilon}.$$

Hence we can choose λ such that

$$|\,\Phi_\varepsilon - f\,| < \lambda\,\varepsilon\, r^{-2} \quad \text{for } r = 1 \;\&\; r = 2\sqrt{\varepsilon}$$

and $$L_\varepsilon\, \lambda\, \varepsilon\, r^{-2} < L_\varepsilon(\Phi_\varepsilon - f) = \varepsilon\, f''r^{-2} < -\, L_\varepsilon\, \lambda\, \varepsilon\, r^{-2},$$

from which (2) follows.

For this problem the exact solution is the infinite sum

$$S(\rho,\phi) = \sum_{k=-\infty}^{\infty} a_k\, \rho^{|k|}\; {}_1F_1(\tfrac{1}{2}|k|, |k|+1, -\rho^2)\; e^{ik\phi} \int_{-\pi}^{\pi} f(\theta) e^{-ik\theta}\, d\theta$$

where $\rho = r/\sqrt{\varepsilon}$ and $a_k^{-1} = 2\pi\varepsilon^{-\frac{1}{2}|k|}\; {}_1F_1(\tfrac{1}{2}|k|, |k|+1, -\varepsilon^{-1})$.

S is the general representation of the boundary layer solution, for it is easily seen that if G is circled (i.e. $(r,\phi) \in G \Longleftrightarrow (\lambda\, r,\phi) \in G$ for every $\lambda \in [0,1]$) and if the boundary value is a C^2-function of the angular variable ϕ, then Φ_ε can be represented by $S(r\sqrt{\varepsilon},\phi) + O(\varepsilon)$ for $r \le 1$ and by $f(\phi) + O(\varepsilon)$ for $r \ge r_o > o$. Due to the focussing effect of L_1 in this case there is no loss at all of boundary conditions.

However, for the equation

(3) $$\varepsilon\, \Delta\, \Phi - r\frac{\partial \Phi}{\partial r} = 0 \quad \text{with } \Phi(1,\phi) = f(\phi)$$

things are completely different. If for instance $f(\phi) = \cos 2\,\phi$, then

$$\Phi_\varepsilon(r,\phi) = a\, r^{-2} \cos 2\,\phi\, \{\exp(\tfrac{r^2}{2\varepsilon}) - \tfrac{r^2}{2\varepsilon} - 1\} \text{ with } a^{-1} = \exp(\tfrac{1}{2\varepsilon}) - \tfrac{1}{2\varepsilon} - 1$$

is the exact solution. It is easily seen that this is asymptotically equal to zero almost everywhere, except in a neighbourhood of the boundary. The sign of the coefficient of $r\frac{\partial}{\partial r}$ is decisive whether or not Φ_ε converges to the discontinuous solution of the reduced equation satisfying the boundary conditions. In this it is analogous to the behaviour of the solution of the ordinary differential equation

$$\varepsilon u'' + \lambda x u' = f(x) \quad \text{with } u(1) = \alpha,\ u(-1) = \beta;$$

c.f. Wasow [6] example 5.

§3. The Saddlepoint

Let G be the square $|x| \leq 1$, $|y| \leq 1$, $L_1 = x\frac{\partial}{\partial x} - y\frac{\partial}{\partial y}$ and $L_2 = \Delta$.

We try to give an approximation to the solution Φ_ε of

(4a) $\quad L_\varepsilon \Phi_\varepsilon = \varepsilon L_2\Phi + L_1\Phi = 0$

(4b) $\quad \Phi_\varepsilon(x,\pm1) = f_\pm(1)$ and $\Phi_\varepsilon(\pm1,y) = g_+(y) \pm g_-(y)$

where f_+, f_-, g_+ and g_- are of class C^3 and satisfy the relations

(4c) $\quad f_+(\pm1) = g_+(1) \pm g_-(1)$ and $f_-(\pm1) = g_+(-1) \pm g_-(-1)$.

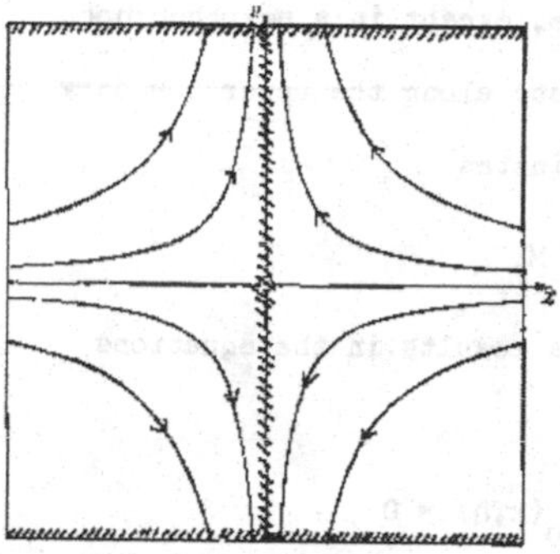

The characteristics are the family of hyperbola's xy = constant, along which the solution u of the reduced equation Lu = 0 is constant. It is obvious that u cannot satisfy the boundary conditions both along the lines $x = \pm1$ and along $y = \pm1$.
It appears that boundary layers of width $O(\varepsilon)$ arise along the sides $y = \pm1$ and that a free boundary layer of with $O(\sqrt{\varepsilon})$ lies along the line $x = 0$. In order to prove this, we use the barrierfunction ψ defined by

(5) $$\psi(x) = \frac{1}{\varepsilon}\int_x^1 \int_0^t \exp\frac{s^2-t^2}{2\varepsilon}\, ds\, dt + 1$$

which is positive on $-1 < x < 1$ and for which the following holds:

(6) $\quad \psi(x) = 1 - \log|x| + O(x^{-1}\sqrt{\varepsilon}\log\varepsilon)$

$\quad |\psi(x)| \leq 2 + \sqrt{\varepsilon}\,|\log\varepsilon|$

$\quad L_\varepsilon\psi = -1.$

The solution of the reduced equation, satisfying the boundary condition along $x = \pm 1$, is $g_+(xy) + \frac{x}{|x|} g_-(xy)$; it is discontinuous at $x = 0$, the value jumping from $-g_-(0)$ to $+ g_-(0)$. This discontinuity can be removed by multiplying g_- by $\mathrm{erf}(x/\sqrt{\varepsilon})$, being a solution of the reduced equation in the local coördinates $(\xi, y) = (x/\sqrt{\varepsilon}, y)$. For the resulting function

$$u_o(x,y) = g_+(xy) - g_-(xy)\,\mathrm{erf}(x/\sqrt{\varepsilon})$$

we easily conclude

$$L_\varepsilon u_o = \sqrt{8\,\varepsilon/\pi}\, y\, g'(xy) \exp(-\frac{x^2}{2\varepsilon}) + O(\varepsilon).$$

Due to the fact that

$$L_\varepsilon\, y \exp(-\frac{x^2}{2\varepsilon}) = -\,2y \exp(-\frac{x^2}{2\varepsilon})$$

we obtain, by defining

$$u_1(x,y) = \sqrt{2/\pi}\; g'(0) y \exp(-\frac{x^2}{2\varepsilon})$$

the improved result

$$L_\varepsilon\,(u_o + \sqrt{\varepsilon}\, u_1) = \sqrt{8\,\varepsilon/\pi}\,(g'(xy) - g'(0))\, y \exp(-\frac{x^2}{2\varepsilon}) + O(\varepsilon) = O(\varepsilon).$$

Conjecturing that $u_o + \sqrt{\varepsilon}\, u_1$ approximates Φ everywhere, except in a neighbourhood of the upper and lower boundary, we stretch the ordinate along the upper boundary by taking $\varepsilon\,\eta = 1-y$, such that we have in local coördinates

$$L_\varepsilon = \frac{1}{\varepsilon}\,(\frac{\partial^2}{\partial\eta^2} + \frac{\partial}{\partial\eta}) + x\frac{\partial}{\partial x} - \eta\frac{\partial}{\partial\eta} + \varepsilon\frac{\partial^2}{\partial x^2} = \frac{1}{\varepsilon} M_o + M_1$$

For the boundary layer solution $v_o + \varepsilon\, v_1 + \ldots..$ this results in the equations

$$M_o v_o(x,\eta) = 0$$

with $v_o(x,o) = f_+(x) - u_o(x,1) - \sqrt{\varepsilon}\, u_1(x,1)$ and $\lim_{\eta\to\infty} v_o(x,\eta) = 0$

and $\qquad M_o v_1(x,\eta) = -\,M_1 v_o(x,\eta)$ $\qquad$ (modulo $O(\varepsilon)$)

with $v_1(x,o) = o$ and $\lim_{\eta\to\infty} v_1(x,\eta) = 0$.

We find:

$$v_o(x,\eta) = \{f_+(x) - u(x,1) - \sqrt{\varepsilon}\, u_1(x,1)\} \exp - \eta,$$

$$M_1 v_o = \eta\, v_o + e^{-\eta}\,\{x(f_+' - g_+' - g_-'\,\mathrm{erf}\, x/\sqrt{\varepsilon}) - g_-'(0)\sqrt{2\,\varepsilon/\pi}\exp(-\frac{x^2}{2\varepsilon})\} + O(\varepsilon)$$

$$v_1 = (\tfrac{1}{2}\,\eta^2 + \eta) v_o + \eta\, e^{-\eta}\,\{x(f_+' - g_+' - g_-'\,\mathrm{erf}\, x/\sqrt{\varepsilon}) - \sqrt{2\varepsilon/\pi}\, g_-'(0) \exp(-\frac{x^2}{2\varepsilon})\}$$

and straightforward calculation shows that

$$M_1 v_1 = O(1).$$

Similarly, taking $\varepsilon\,\zeta = 1+y$, the boundary layer term along the lower boundary is $w_o(x,\zeta) + \varepsilon\, w_1(x,\zeta)$, where

$$w_o(x,\zeta) = \{f_-(x) - u_o(x,-1) - \sqrt{\varepsilon}\, u_1(x,-1)\} \exp - \zeta$$

and for w_1 an analogous expression.

From (4c) we conclude

$$v_o(\pm 1,\eta) + \varepsilon\, v_1(\pm 1,\eta) = O(\varepsilon) \text{ and } w_o(\pm 1,\zeta) + \varepsilon\, w_1(\pm 1,\zeta) = O(\varepsilon)$$

With aid of the barrierfunction $\varepsilon\,\psi$ and the maximum principle we find clearly the result

$$\Phi_\varepsilon = u_o + \sqrt{\varepsilon}\, u_1 + v_o + w_o + \varepsilon(v_1+w_1) + O(\varepsilon \log \varepsilon)$$

uniformly on G.

Because of the simplicity of equation and domain the solution of the problem is relatively simple; it already becomes much more difficult when we have $L_1 = \lambda\, x \frac{\partial}{\partial x} - y \frac{\partial}{\partial y}$ with $\lambda \neq 1$, for then the characteristics are the curves $x^\lambda y =$ = constant and also $g_+(x^\lambda y)$ may cease to be smooth at $x = 0$. In a more general setting also the interaction of the boundary layers, where they meet, is much more difficult. Still the general picture of Φ_ε, converging to the discontinuous solution of the reduced equation satisfying a part of the boundary conditions, remains valid. It is hoped to present a more detailed investigation in a subsequent paper.

<u>References</u>

1. W. Eckhaus & E.M. de Jager, Asymptotic solutions of singular perturbation problems for linear differential equations of elliptic type. Arch. Rat. Mech Anal. 23, 26-36 (1966).

2. J. Grasman, On the birth of boundary layers. Mathematisch Centrum Amsterdam, Tract 36.

3. P.P.N. de Groen, An elliptic boundary value problem with nondifferentiable parameters. Arch. Rat. Mech. Anal. 42 (1971), 169-183.

4. E.M. de Jager, Singular elliptic Perturbations of vanishing first order differential operators. This Lecture Notes.

5. M.I. Viskik & L.A. Lyusternik, Regular degeneration and boundary layer for linear differential equations with small parameters. Uspehi Math. Nauk 12 (1957) [A.M.S. Transl. Ser. 2, 20 (1962)]

6. W. Wasow, The capriciousness of singular perturbations. Nieuw Archief voor Wiskunde XVIII (1970), 190-210.

Remarks on Interpolation in Spaces with Weights

H.P. Heinig

A measurable function w is called a weight function if it is positive almost everywhere. f belongs to the weighted Lebesgue spaces $L^p(w)$, $1 \leq p \leq \infty$, if $wf \in L^p(X)$, and we write the norm

$$\|f\|_{p,w} = \left\{\int_X |f(x)|^p \, w(x)^p \, dx\right\}^{1/p} \qquad 1 \leq p < \infty ,$$

with the usual modification when $p = \infty$.

E.M. Stein [3] has proved the following interpolation theorem:

Theorem 1. Let v and w be weight functions. If T is a linear operator defined on simple functions satisfying

$$\|Tf\|_{q_i,w_i} \leq M_i \|f\|_{p_i,v_i} \qquad i = 0,1$$

then

$$\|Tf\|_{q,w} \leq M_0^{1-t} M_1^t \|f\|_{p,v} ,$$

where $1/p$ and $1/q$ lie on the lines $(1/p_0, 1/p_1)$ and $(1/q_0, 1/q_1)$, respectively, $v = v_0^{1-t} v_1^t$, $w = w_0^{1-t} w_1^t$ and $0 \leq t \leq 1$.

In this note we extend Theorem 1 to weighted Lorentz spaces $L^v(p,q)$

and for the special case when Theorem 1 applies it is shown that certain linear operators are continuous in weighted L^p spaces. An application is also given for the Laplace transform.

To generalize Theorem 1 to weighted L(p,q) spaces, we recall briefly some notions.

Let f be a complex valued measurable function on a σ- finite measure space (X,μ), where μ is assumed non negative. If

$$D_f(y) = \mu\{x : |f(x)| > y\},$$

then the non increasing rearrangement of f is defined by

$$f^*(x) = \inf\{y > 0 : D_f(y) \leq x\}.$$

We define

$$L(p,q) = \{f : \|f\|^*_{pq} < \infty,\ 0 < p,q \leq \infty\},$$

where

$$\|f\|^*_{pq} = \begin{cases} \left\{(p/q)\int_0^\infty x^{q/p-1} f^*(x)^q\, dx\right\}^{1/q} & 0 < p,q < \infty \\ \sup_{0<x} x^{1/p} f^*(x) & 0 < p \leq \infty,\ q = \infty. \end{cases}$$

Let w be a weight function, then $f \in L^w(p,q)$, $0 < p,q \leq \infty$, if and only if $wf \in L(p,q)$.

With these definitions we can now prove

<u>Theorem 2</u>. Let T be a linear operator defined on simple functions.

If

$$\|Tf\|^*_{\bar{p}_i\bar{q}_i,w_i} \leq M_i\|f\|^*_{p_iq_i,v_i} \qquad i = 0,1,$$

then T satisfies

$$(1) \quad \|Tf\|^*_{\bar{p}\bar{q},w} \leq M_o^{1-t}\, M_1^t\, \|f\|^*_{pq,v} \quad ,$$

where $1/p$, $1/q$, $1/\bar{p}$, $1/\bar{q}$ lie on the lines $(1/p_o,1/p_1)$, $(1/q_o,1/q_1)$, $(1/\bar{p}_o,1/\bar{p}_1)$, $(1/\bar{q}_o,1/\bar{q}_1)$ respectively, and $v = v_o^{1-t}\, v_1^t$, $w = v_o^{1-t}\, v_1^t$ for $0 < t < 1$.

Proof. Let E be a measurable subset of X, where for given $\epsilon > 0$, $v_i \geq \epsilon$, $i = 0,1$. If T' is the linear operator defined for f on E and zero otherwise, then the family $\{U_z\}$ $z = x + iy$, $0 \leq x \leq 1$, defined by

$$U_z(f) = w_o^{1-z}\, w_1^z\, T'(fv_o^{z-1}\, v_1^{-z})$$

is an analytic family of operators of admissible growth (for a definition of this term see for example [2] or [3]).

Note that

$$\|U_{iy}(f)\|^*_{\bar{p}_o\bar{q}_o} = \|w_o^{1-iy}\, w_1^{iy}\, T'(fv_o^{iy-1}\, v_1^{-iy})\|^*_{\bar{p}_o\bar{q}_o}$$

$$\leq \|w_oT'(fv_o^{iy-1}\, v_1^{-iy})\|^*_{\bar{p}_o\bar{q}_o}$$

$$\leq M_o\, \|fv_o^{iy-1}\, v_1^{-iy}\, v_o\|^*_{p_oq_o}$$

$$\le M_o \|f\|^*_{p_o q_o}$$

and

$$\|U_{1+iy}(f)\|^*_{\bar p_1 \bar q_1} = \|w_o^{-iy}\ w_1^{1+iy}\ T'(fv_o^{iy}\ v_1^{-1+iy})\|^*_{\bar p_1 \bar q_1}$$

$$= \|w_1 T'(fv_o^{iy}\ v_1^{-1+iy})\|^*_{\bar p_1 \bar q_1}$$

$$\le M_1 \|fv_o^{iy}\ v_1^{-1+iy}\ v_1\|^*_{p_1 q_1}$$

$$= M_1\ \|f\|^*_{p_1 q_1}.$$

Hence by the interpolation theorem of Y. Sagher [3]

$$\|U_t(f)\|^*_{\bar p \bar q} \le M_o^{1-t}\ M_1^t\ \|f\|^*_{pq}$$

for simple f. But this is clearly equivalent to

$$\|Tf\|^*_{\bar p \bar q, w} \le M_o^{1-t}\ M_1^t\ \|f\|^*_{pq,v}\ ,$$

where vf is simple and vanishes outside the set E. But E is arbitrary and hence the result follows.

Note that for $q < \infty$ simple functions are dense in L(p,q) and hence in $L^v(p,q)$ [1]. Thus the operator T satisfying (1) has a unique extention to all $f \in L^v(p,q)$, $q < \infty$.

We can now give the following result:

<u>Theorem 3</u>. Let $K(x,y)$ be a non negative symmetric kernel such that

$$\int_{-\infty}^{\infty} K(x,y)dy < \infty .$$

If v and w are weight functions satisfying

$$(2) \qquad \int_{-\infty}^{\infty} K(x,y)w(x)dx \leq C\, v(y),$$

then the linear operator T defined by

$$(Tf)(x) = \int_{-\infty}^{\infty} K(x,y)f(y)dy$$

satisfies

$$(3) \qquad \|Tf\|_{q,w^{1/p}v^{-1/p'}} \leq C\, \|f\|_{p,v^{1/p}w^{-1/p'}}$$

for $1 \leq p,q \leq \infty$, $1/p + 1/p' = 1$.

Proof. If $p = q = 1$

$$\|Tf\|_{1,w} = \int_{-\infty}^{\infty} |Tf(x)|\, w(x)dx \leq \int_{-\infty}^{\infty} w(x)\left(\int_{-\infty}^{\infty} K(x,y)\, |f(y)|dy\right)dx$$

$$= \int_{-\infty}^{\infty} |f(y)|\left(\int_{-\infty}^{\infty} K(x,y)w(x)dx\right)dy \leq C\int_{-\infty}^{\infty} v(y)\, |f(y)|\, dy$$

$$= C\|f\|_{1,v}$$

so that (3) holds in this case. For $p = q = \infty$ we observe that

$$|v^{-1}(x)(Tf)(x)| \le |v^{-1}(x)| \int_{-\infty}^{\infty} K(x,y)\,|f(y)|\,dy$$

$$\le C \int_{-\infty}^{\infty} K(x,y)\,|f(y)|\,dy \Big/ \int_{-\infty}^{\infty} K(y,x)w(y)dy$$

$$\le C\,\|w^{-1} f\|_{\infty} \equiv C\|f\|_{\infty,w^{-1}}.$$

The result now follows on applying Theorem 1.

Let Lf denote the Laplace transform of f:

$$(Lf)(x) = \int_{0}^{\infty} e^{-xy} f(y)dy \qquad x > 0,$$

then we have at once

<u>Corollary</u>. If $f \in L^p(w^{-1/p'}\, v^{1/p})$, $1 \le p \le \infty$, and w and v satisfy $(Lw)(x) \le Cv(x)$, then

$$\|Lf\|_{q,w^{1/p}\,v^{-1/p'}} \le C\|f\|_{p,v^{1/p}\,w^{-1/p'}}.$$

Setting $w(x) = x^a$, $a > -1$, one obtains from the corollary

$$\left(\int_0^{\infty} |(Lf)(x)|^p\, x^{p(a+1)-1}\, dx\right)^{1/p} \le C\left(\int_0^{\infty} |f(x)|^p\, x^{-ap-1}\, dx\right)^{1/p}$$

which for $a = -1/p$, $p>1$ and $a = -1/p'$, reduces to well known results.

References

1 R.A. Hunt; On L(p,q) - Spaces; L'Ensignement Math., 12, 1966 249-276.

2 Y. Sagher; On Analytic Families of Operators; Israel J. Math. 7, 1969; 351-356.

3 E.M. Stein; Interpolation of Linear Operators; Trans. Amer. Math. Soc. 83, 1956; 482-492.

4 E.M. Stein; Interpolation of Operators with Change of Measures; Trans. Amer. Math. Soc. 87, 1958; 159-172.

Perturbed Bifurcation and Buckling of Circular Plates*

J. P. Keener and H. B. Keller

Introduction. Many studies of bifurcation theory include some results on the effect of perturbing the basic problem so that it no longer posesses the trivial solution [8,9]. Such modifications can be very significant in practical problems. For example a perfectly flat circular plate subject to a uniform compressive edge thrust has the "trivial solution" corresponding to a radial contraction. However if small initial deviations from the plane are present or some force (such as gravity) acts normal to the plane of the plate this trivial solution is no longer valid. The available theoretical results simply imply that for sufficiently small perturbations both the perturbed and unperturbed problems have the same number of solution branches and that they are "close" to each other. We shall look more carefully at the effects of such perturbations and describe in some detail how solutions of the unperturbed problem go over into solutions of the perturbed problem. In particular, we determine the locus of "branch points" which contains the bifurcation point. For various elasticity problems the effect of imperfections on buckling loads is thus determined by a rigorous constructive theory. We illustrate for the plate problem mentioned above. Many other significant results have been obtained from this study and will be reported elsewhere [1, 2, 3, 4].

*This work was supported, in part, by U.S. Army Contract No. DAHCO-4-68-C-0006, and, in part, by a fellowship from the Fannie and John Hertz Foundation.

Formal Theory. We indicate here some of the results and methods used in our study but space limitations do not permit detailed proofs. The general nonlinear problem may be formulated as

$$\text{(1)} \qquad G(u, \lambda, \tau) = 0$$

with G on $\mathcal{B} \times \mathbb{R} \times \mathbb{R}$ into $\mathcal{B}$, some appropriate Banach space. It is assumed that

$$\text{(2)} \qquad \text{(a)} \quad G(0, \lambda, 0) = 0 \; ; \qquad \text{(b)} \quad G_\tau(0, \lambda, 0) \neq 0 \; .$$

Here τ is the perturbing parameter and for $\tau = 0$ we assume that (1) exhibits bifurcation. That is for some $\lambda = \lambda_0$, say, the linear problem

$$\text{(3)} \qquad G_u(0, \lambda_0, 0)\, \phi_0 = 0 \; , \qquad \|\phi_0\| = 1 \; ;$$

has a solution ϕ_0 in $\mathcal{B}$. We assume that $\mathcal{N}_0$, the null space of $G_u(0, \lambda_0, 0)$, is one dimensional. Of course G_u, G_τ, $G_{u\lambda}$, etc. are appropriate Fréchet derivatives whose existence and other properties will be assumed as required. Further we assume that $\mathcal{B} = \mathcal{N}_0 \oplus \mathcal{N}_0^{*\perp}$ where $\mathcal{N}_0^*$ is the dual space of $\mathcal{N}_0$ in $\mathcal{B}^*$, the dual of $\mathcal{B}$. Thus if, as we also assume, $\mathcal{N}^*$ is one dimensional, say spanned by ϕ_0^*, then $v \in \mathcal{N}_0^{*\perp}$ iff $\langle \phi_0^*, v \rangle = 0$. By means of the iteration scheme:

$$\text{a)} \qquad \overset{0}{u_0}(\epsilon) = \epsilon\,\phi_0 \; , \qquad \overset{0}{\lambda}(\epsilon) = \lambda_0 \; ;$$

$$\text{(4)} \quad \text{b)} \quad G_u(0, \lambda_0, 0) u_0^{\nu+1}(\epsilon) = G_u(0, \lambda_0, 0) u_0^{\nu}(\epsilon) - G\left(u_0^{\nu}(\epsilon), \lambda^{\nu+1}(\epsilon), 0\right), \quad \langle \phi_0^*, u_0^{\nu+1}(\epsilon) - \epsilon\phi_0 \rangle = 0;$$

$$\text{c)} \qquad \langle \phi_0^*, G_u(0, \lambda_0, 0) u_0^{\nu}(\epsilon) - G\left(u_0^{\nu}(\epsilon), \lambda^{\nu+1}(\epsilon), 0\right) \rangle = 0 \; ,$$

it is established in [6] that for sufficiently small $\epsilon_0 > 0$ and each $|\epsilon|$ in $(0, \epsilon_0]$ the problem (1) has a unique nontrivial solution of the form: $\tau = 0$

$$\text{(5)} \quad \text{a)} \qquad u_0(\epsilon) = \epsilon\,\phi_0 + \epsilon^2 U(\epsilon) \; , \qquad \|U(\epsilon)\| \leq M_0 \; , \; \langle \phi_0^*, U(\epsilon) \rangle = 0 \; ;$$

$$\text{b)} \qquad \lambda(\epsilon) = \lambda_0 + \epsilon \lambda_1(\epsilon) \; , \qquad |\lambda_1(\epsilon)| \leq m_0 \; .$$

This result is proved by contraction maps and in the process we find that

$$(5) \qquad c) \qquad \lambda_1(\epsilon) = -\tfrac{1}{2}\, \frac{\langle \phi_0^*, G_{uu}(0,\lambda_0,0)\phi_0^2\rangle}{\langle \phi_0^*, G_{u\lambda}(0,\lambda_0,0)\phi_0\rangle} + O(\epsilon) = \lambda_1^{\nu}(\epsilon) + O(\epsilon^{\nu})$$

To treat the perturbed case, $\tau \neq 0$, we first seek the locus of branch points (or nonisolated solutions) emanating from the bifurcation point. That is we consider the system:

$$(6) \qquad a) \quad G(u,\mu,\tau) = 0\,, \qquad b) \quad G_u(u,\mu,\tau)\phi = 0\,, \qquad \|\phi\| \neq 0\,;$$

and determine solutions of the form:

$$(7) \qquad \begin{aligned} &a) \quad u(\epsilon) = \epsilon\phi_0 + \epsilon^2 V(\epsilon)\,, \ \|V\| \leq M_1\,; &&c) \quad \phi(\epsilon) = \phi_0 + \epsilon\phi_1(\epsilon)\,, \quad \langle \phi_0^*, V(\epsilon)\rangle = 0\,;\\ &b) \quad \mu(\epsilon) = \lambda_0 + \epsilon\mu_1(\epsilon)\,, \ \|\mu_1\| \leq m_1 &&d) \quad \tau(\epsilon) = \epsilon\tau_0 + \epsilon^2\tau_1(\epsilon)\,, \quad \langle \phi_0^*, \phi_1(\epsilon)\rangle = 0\,. \end{aligned}$$

This is again done by contractions, using the iteration scheme [with obvious initial iterate suggested by (7)] ;

$$(8) \qquad a) \quad \begin{aligned} &G_u(0,\lambda_0,0)u^{\nu+1}(\epsilon) = G_u(0,\lambda_0,0)u^{\nu}(\epsilon) - G\big(u^{\nu}(\epsilon), \mu^{\nu+1}(\epsilon), \tau^{\nu+1}(\epsilon)\big), \langle \phi_0^*, u^{\nu+1}(\epsilon) - \epsilon\phi_0\rangle = 0\,;\\ &G_u(0,\lambda_0,0)\phi^{\nu+1}(\epsilon) = G_u(0,\lambda_0,0)\phi^{\nu}(\epsilon) - G_u\big(u^{\nu}(\epsilon), \mu^{\nu+1}(\epsilon), \tau^{\nu+1}(\epsilon)\big)\phi^{\nu}(\epsilon), \ \langle \phi_0^*, \phi^{\nu+1}(\epsilon) - \phi_0\rangle = 0; \end{aligned}$$

Here $\mu^{\nu+1}$ and $\tau^{\nu+1}$ are determined from the two simultaneous equations:

$$(8) \qquad b) \quad \begin{aligned} &\langle \phi_0^*, G_u(0,\lambda_0,0)u^{\nu}(\epsilon) - G\big(u^{\nu}(\epsilon), \mu^{\nu+1}(\epsilon), \tau^{\nu+1}(\epsilon)\big)\rangle = 0\,;\\ &\langle \phi_0^*, G_u(0,\lambda_0,0)\phi^{\nu}(\epsilon) - G_u\big(u^{\nu}(\epsilon), \mu^{\nu+1}(\epsilon), \tau^{\nu+1}(\epsilon)\big)\rangle = 0\,. \end{aligned}$$

Convergence readily follows (see [3]) for small $|\epsilon|$ and indeed we get that: $\tau_0 = 0$,

$$(9) \qquad \begin{aligned} &a) \quad \mu_1(\epsilon) = -\frac{\langle \phi_0^*, G_{uu}(0,\lambda_0,0)\phi_0^2\rangle}{\langle \phi_0^*, G_{u\lambda}(0,\lambda_0,0)\phi_0\rangle} + O(\epsilon) = \mu_1^{\nu}(\epsilon) + O(\epsilon^{\nu})\,;\\ &b) \quad \tau_1(\epsilon) = \tfrac{1}{2}\, \frac{\langle \phi_0^*, G_{uu}(0,\lambda_0,0)\phi_0^2\rangle}{\langle \phi_0^*, G_{\tau}(0,\lambda_0,0)\phi_0\rangle} + O(\epsilon) = \tau_1^{\nu}(\epsilon) + O(\epsilon^{\nu})\,. \end{aligned}$$

We point out that this locus of branch points will exist only for τ values of one sign, near $\tau = 0$, provided $\tau_1(0) \neq 0$. Also a comparison of (9a) and (5c) reveals that $[\mu(\epsilon)-\lambda_0] = 2[\lambda(\epsilon)-\lambda_0] + O(\epsilon^2)$. Thus the locus of branching solutions departs from the bifurcation point twice as fast as does the locus of bifurcating solutions. If $\mu_1(0) \neq 0$ the branching solutions exist for all λ in some interval with λ_0 as an interior point.

Through each of the branching solutions (7), with ϵ fixed in $0 < |\epsilon| \leq \epsilon_0$, we now determine a unique family of solutions on which τ is a nonzero constant. Each such family has the form:

$$\text{(10)}\quad \begin{aligned} &\text{a)}\quad u(\epsilon,\delta) = u(\epsilon) + \delta\phi(\epsilon) + \delta^2 v(\epsilon,\delta),\ \|v\| \leq M_2,\ \langle \phi^*(\epsilon), v(\epsilon,\delta)\rangle = 0;\\ &\text{b)}\quad \lambda(\epsilon,\delta) = \mu(\epsilon) + \delta\nu(\epsilon,\delta),\quad |\nu(\epsilon,\delta)| \leq m_2;\\ &\text{c)}\quad \tau = \tau(\epsilon); \end{aligned}$$

for all δ in $0 \leq |\delta| \leq \delta_0(\epsilon)$. This is again done by contraction. Specifically with $G_u(\epsilon) \equiv G_u\big(u(\epsilon), \mu(\epsilon), \tau(\epsilon)\big)$ we use the iteration scheme:

$$\text{(11)}\quad \text{a)}\quad G_u(\epsilon)\,u^{k+1}(\epsilon,\delta) = G_u(\epsilon)\,u^k(\epsilon,\delta) - G\big(u^k(\epsilon,\delta), \lambda^{k+1}(\epsilon,\delta), \tau(\epsilon)\big),\ \langle \phi^*(\epsilon), u^{k+1}(\epsilon,\delta) - u(\epsilon) - \delta\phi(\epsilon)\rangle = 0$$

where $\lambda^{k+1}(\epsilon,\delta)$ is the unique root, near $\mu(\epsilon)$, of:

$$\text{(11)}\quad \text{b)}\quad \langle \phi^*(\epsilon), G\big(u^k(\epsilon,\delta), \lambda^{k+1}, \tau(\epsilon)\big) - G_u(\epsilon)\big[u^k(\epsilon,\delta) - u(\epsilon)\big]\rangle = 0.$$

[Here $\phi^*(\epsilon)$ is a basis for the dual of the nullspace, $\mathcal{N}_\epsilon$, of $G_u(\epsilon)$.] We find from this procedure that $\nu^k(\epsilon,0) = 0$ for all k provided $\langle \phi^*(\epsilon), G_\lambda(\epsilon)\rangle \neq 0$. Further it turns out that

$$\text{(12)}\quad \text{a)}\quad \nu(\epsilon,\delta) = -\delta\,\frac{\langle \phi^*(\epsilon), G_{uu}(\epsilon)\phi^2(\epsilon)\rangle}{\langle \phi^*(\epsilon), G_\lambda(\epsilon)\rangle} + O(\delta^2) = \nu^k(\epsilon,\delta) + O(\delta^{k+1})$$

so from (10 b) we find:

$$\text{(12)}\quad \text{b)}\quad \lambda(\epsilon,\delta) = \mu(\epsilon) - \delta^2\nu_\delta(\epsilon,0) + O(\delta^3) = \lambda^k(\epsilon,\delta) + O(\delta^{k+2}).$$

If $\nu_\delta(\epsilon,0) \neq 0$ it follows that, for sufficiently small $|\delta|$, the family of solutions in (10) exist only for λ in some interval with $\mu(\epsilon)$ as an endpoint. Hence there are locally two solutions (i.e. $\delta > 0$ and $\delta < 0$) on one side of

$\lambda = \mu(\epsilon)$ and none on the other side; this justifies calling $\lambda = \mu(\epsilon)$, $\tau = \tau(\epsilon)$ a branching point. If $\tau_1(0) \neq 0$ there are two values of ϵ, with opposite sign, for which $\tau = \tau(\epsilon)$. Thus, by the above argument, there is another branching point in this case, and we find that $\nu_\delta(\epsilon,0)$ has a sign change as ϵ changes sign (the denominator in (12a) changes sign). It follows that these two families of solutions which exist for the same τ-value "turn away" from each other and there is a λ-interval about λ_0 for which no "small" solutions exist.

Of course the cases in which $\lambda_1(0) = \mu_1(0) = 0$ are easily treated by our techniques but space does not permit us to show the results here; they are conceptually similar. Also solutions can be shown to exist for (small) τ-values of sign opposite from that in (10c). However these families of solutions do not in general contain branching points.

Plate Buckling Under Combined Loading. In appropriate dimensionless variables the von Karman equations for radially symmetric deformations of a thin circular plate subjected to a uniform edge thrust, λ, and a radially symmetric normal load, $\tau q(x)$, can be reduced to the form [7]:

$$(13) \qquad G(u,v,\lambda,\tau) \equiv \begin{pmatrix} Lu + \lambda xu + uv + \tau q \\ Lv - u^2 \end{pmatrix} = 0\,; \quad B(u,v) \equiv \begin{pmatrix} u(0) & u(1) \\ v(0) & v(1) \end{pmatrix} = 0.$$

Here $L\phi \equiv x\left(x^{-1}(x\phi)_x\right)_x$ and τ is simply a scale factor which we have introduced to relate our treatment to the general theory. For $\tau = 0$ (13) has the trivial solution $u = v = 0$ for all λ. The possible bifurcation points are obtained by considering the eigenvalue problem:

$$(14) \quad a) \qquad G_{u,v}(0,0,\lambda,0) \equiv \begin{pmatrix} L+\lambda x & \\ & L \end{pmatrix}\begin{pmatrix} \phi \\ \psi \end{pmatrix} = 0\,; \quad B(\phi,\psi) = 0.$$

We easily find that the possible eigenvectors $(\phi_0, \psi_0)^T$ and eigenvalues, λ_0 are:

$$(14) \quad b) \qquad \begin{pmatrix} \phi_0 \\ \psi_0 \end{pmatrix} \equiv \begin{pmatrix} J_1(\sqrt{\lambda_0}\,x) \\ 0 \end{pmatrix}, \quad \text{where } J_1(\sqrt{\lambda_0}) = 0\,.$$

To find the bifurcating solutions $\left(u_0(\epsilon,x),\ v_0(\epsilon,x),\ \lambda(\epsilon)\right)$ we use the iteration scheme (4b), applied to (13), which becomes:

$$(15) \quad a) \qquad \begin{aligned} (L+\lambda_0 x)u_0^{\nu+1} &= (\lambda_0 - \lambda^{\nu+1})xu_0^{\nu} - u_0^{\nu} v_0^{\nu}\,; \quad \langle \phi_0(x),\, u_0^{\nu+1} - \epsilon\,\phi_0(x)\rangle = 0 \\ Lv_0^{\nu+1} &= (u_0^{\nu})^2\,; \end{aligned}$$

and (4c), the orthogonality condition becomes:

(15) b) $$\lambda^{\nu+1} = \lambda_0 - \frac{\langle \phi_0(x), u_0^{\nu}(x)\, v_0^{\nu}(x)\rangle}{\langle \phi_0(x), x u_0^{\nu}(x)\rangle} .$$

We now easily find, using $u_0{}^0(x) = \epsilon\phi_0(x)$, $v_0{}^0(x) = 0$ and carrying out one iteration that: $\lambda^1(\epsilon) = \lambda_0$, $u_0{}^1(x) = \epsilon J_1(\sqrt{\lambda_0}x)$, $v_0 = -\epsilon^2\int_0^1 g(x,t)J_1{}^2(\sqrt{\lambda_0}t)dt$ and

(16) $$\lambda^2(\epsilon) = \lambda_0 + \epsilon^2 \frac{\langle J_1(\sqrt{\lambda_0}x), J_1(\sqrt{\lambda_0}x)\int_0^1 g(x,t)J_1{}^2(\sqrt{\lambda_0}t)dt\rangle}{\langle J_1(\sqrt{\lambda_0}x)\, ,\, xJ_1(\sqrt{\lambda_0}\, , x)\rangle} = \lambda(\epsilon) + O(\epsilon^4)$$

Here $g(x,t)$ is the (positive) Green's function for $L\psi$ subject to $\psi(0) = \psi(1) = 0$. Higher order iterates are simply obtained using (15). These bifurcation results have previously been obtained and justified using an extension of the Poincaré continuation method in [5]. The above procedure however is much simpler and more general.

To find the branching solutions for (13) with $\tau \neq 0$ we simply apply the scheme (8). This yields the systems [with $\phi_0 \equiv J_1(\sqrt{\lambda_0}x)$] :

(17)

a) $$\begin{cases} (L+\lambda_0 x)u^{\nu+1} = (\lambda_0-\mu^{\nu+1})xu^{\nu} - u^{\nu}v^{\nu} - \tau^{\nu+1}q(x)\, , & B(u^{\nu+1}, v^{\nu+1}) = 0\, ; \\ Lv^{\nu+1} = (u^{\nu})^2 & , \quad \langle\phi_0, u^{\nu+1}\rangle = \epsilon\langle\phi_0, \phi_0\rangle\, ; \end{cases}$$

b) $$\begin{cases} (L+\lambda_0 x)\phi^{\nu+1} = (\lambda_0-\mu^{\nu+1})x\phi^{\nu} - v^{\nu}\phi^{\nu} - u^{\nu}\psi^{\nu} & , \quad B(\phi^{\nu+1}, \psi^{\nu+1}) = 0\, ; \\ L\psi^{\nu+1} = 2u^{\nu}\phi^{\nu} & , \quad \langle\phi_0, \phi^{\nu+1}\rangle = \langle\phi_0, \phi_0\rangle\, ; \end{cases}$$

and the orthogonality conditions give:

(17) c) $$\mu^{\nu+1} = \lambda_0 - \frac{\langle\phi_0(x), v^{\nu}\phi^{\nu} + u^{\nu}\psi^{\nu}\rangle}{\langle\phi_0, x\phi^{\nu}\rangle}\, , \qquad \tau^{\nu+1} = \frac{(\lambda_0-\mu^{\nu+1})\langle\phi_0, xu^{\nu}\rangle - \langle\phi_0, u^{\nu}v^{\nu}\rangle}{\langle\phi_0, q(x)\rangle} .$$

With: $u^0 = \epsilon\phi_0(x)$, $v^0 = 0$, $\phi^0 = \phi_0$, $\psi^0 = 0$ we get that $\mu^1 = \lambda_0$, $\tau^1 = 0$ and:

(18) a) $$u^1 = \epsilon J_1(\sqrt{\lambda_0}x)\, , \quad v^1 = -\epsilon^2\int_0^1 g(x,t)J_1{}^2(\sqrt{\lambda_0}t)dt\, ,$$
$$\phi^1 = J_1(\sqrt{\lambda_0}x)\, , \quad \psi^1 = -2\epsilon\int_0^1 g(x,t)\, J_1{}^2(\sqrt{\lambda_0}t)dt\, ,$$

Thus it is found that

$$\mu^2(\epsilon) = \lambda_0 + \epsilon^2 \frac{3\langle J_1(\sqrt{\lambda_0}x), J_1(\sqrt{\lambda_0}x)\int_0^1 g(x,t)J_1^2(\sqrt{\lambda_0}t)dt\rangle}{\langle J_1(\sqrt{\lambda_0}x), xJ_1(\sqrt{\lambda_0}x)\rangle} = \mu(\epsilon) + O(\epsilon^4)$$

(18) b)

$$\tau^2(\epsilon) = -\epsilon^3 \frac{2\langle J_1(\sqrt{\lambda_0}x), J_1(\sqrt{\lambda_0}x)\int_0^1 g(x,t)J_1^2(\sqrt{\lambda_0}t)dt\rangle}{\langle J_1(\sqrt{\lambda_0}x), q(x)\rangle} = \tau(\epsilon) + O(\epsilon^4)$$

If ϵ is eliminated between $\mu(\epsilon)$ and $\tau(\epsilon)$ we get, from (18b), that:

$$(19) \qquad \mu(\epsilon) = \lambda_0 + \tau^{2/3}(\epsilon)\left[\frac{3}{\sqrt[3]{4}} \frac{\langle J_1, q\rangle^{2/3} \langle J_1, J_1\int_0^1 gJ_1^2\,dt\rangle^{1/3}}{\langle J_1, xJ_1\rangle}\right] + O(\epsilon^3) .$$

This is the basic relation between "buckling load," μ, and imperfection magnitude, τ. Such relations can now be obtained for large classes of imperfection sensitive structures by the simple but rigorous procedure presented here. Similar results for columns and spherical caps will be presented elsewhere [1]. The iterations are easily continued and we could also easily determine the family of solutions through each branching solution.

It should be observed that in the present example what corresponds to $\langle\phi_0^*, G_{uu}(0,\lambda_0,0)\phi_0^2\rangle$ vanishes. Specifically now we have that $G_{(u,v),(u,v)}(0,\lambda_0,0)\begin{pmatrix}\phi\\ \psi\end{pmatrix}\begin{pmatrix}\phi\\ \psi\end{pmatrix} = \begin{pmatrix}2\phi\psi\\ -2\phi^2\end{pmatrix}$ and so $\langle\begin{pmatrix}\phi\\ \psi\end{pmatrix}, G_{(u,v),(u,v)}\begin{pmatrix}\phi\\ \psi\end{pmatrix}\begin{pmatrix}\phi\\ \psi\end{pmatrix}\rangle = \int_0^1 [2\phi^2\psi - 2\phi^2\psi]dx = 0.$ This accounts for the absence of linear terms in ϵ. One result, following from (18b), is that branching solutions exist for $\tau(\epsilon)$ of both signs (simply take $\epsilon > 0$ and $\epsilon < 0$). Physically this is clear as it is simply equivalent to changing the direction of the applied normal load.

References

[1] J. P. Keener, Imperfection sensitivity of buckling for columns and spherical caps (in preparation).

[2] ____________, Some modified bifurcation problems with application to imperfection sensitivity in buckling, Ph.D. thesis, Caltech, Pasadena, California, (1972).

[3] J. P. Keener and H. B. Keller, Perturbed bifurcation theory (in preparation).

[4] ____________, Positive solutions of nonlinear eigenvalue problems (in preparation).

[5] H. B. Keller, J. B. Keller and E. L. Reiss, Buckled states of circular plates, Q. Appl. Math 20 (1962) 55-65.

[6] H. B. Keller and W. Langford, Iterations, perturbations and multiplicities in nonlinear bifurcation problems, Arch. Rat. Mech. Anal (in press).

[7] H. B. Keller and E. L. Reiss, Iterative solutions for the nonlinear bending of circular plates, Comm. Pure Appl. Math, 11 (1958) 273-292.

[8] D. Sather, Branching of solutions of an equation in Hilbert space, Arch. Rat. Mech. and Anal. 36 (1970) 47-64.

[9] D. Westreich, Bifurcation theory in a Banach space, Ph.D. thesis, Yeshiva Univ., New York City (1971).

Oscillation Properties of Nonselfadjoint Differential Equations of Even Order

K. Kreith

Consider the nonselfadjoint differential equation

$$\ell u \equiv \sum_{k=0}^{n} (-1)^k \left(p_k(x)u^{(k)}\right)^{(k)} + \sum_{k=0}^{n-1} (-1)^k \left(q_k(x)u^{(k+1)}\right)^{(k)} = 0, \tag{1}$$

where the coefficients of (1) are real, $p_k(x)$ and $q_k(x)$ are of class C^k, and $p_n(x) > 0$ on $[\alpha,\infty)$. The smallest $\beta > \alpha$ such that

$$u(\alpha) = u'(\alpha) = \cdots = u^{(n-1)}(\alpha) = 0 = u(\beta) = u'(\beta) = \cdots = u^{(n-1)}(\beta) \tag{2}$$

is satisfied nontrivially by a solution of (1) is denoted by $\eta_1(\alpha)$ and called the <u>first conjugate point of</u> α (with respect to (1)).

In case (1) is selfadjoint (i.e., $q_k(x) \equiv 0$ for $k=0,1,\ldots,n-1$), the standard techniques used to establish Sturmian comparison and oscillation theorems for second order Sturm-Liouville equations lend themselves to appropriate generalizations. Such generalizations involve expressing a selfadjoint differential equation

$$Lv \equiv \sum_{k=0}^{n} (-1)^k \left(P_k(x)v^{(k)}\right)^{(k)} = 0 \tag{3}$$

as a vector system with matrix coefficients, and then relating the location of conjugate points of α (with respect to (3)) to the singularities of a certain "conjoined" matrix solution of this system [1], [2].

Attempts to generalize the Sturmian theory to nonselfadjoint equations of the form (1) invariably encounter the problem of nonexistence of conjoined matrix solutions of the corresponding vector systems. This difficulty makes it possible to establish <u>non</u> oscillation theorems by established techniques [3], [4] but appears to block the development of oscillation criteria in the general nonselfadjoint case.

In this note we develop a class of non oscillation criteria for (1) which contain the results of [3] as a special case. Letting $\mu_1(\alpha)$ denote the first conjugate point of α (with respect to (3)), the following theorem establishes conditions under which $\eta_1(\alpha) \geq \mu_1(\alpha)$.

Theorem 1. If for $\alpha \leq x \leq \eta_1(\alpha)$

(i) $p_k(x) \geq P_k(x) + \frac{1}{2}q_k'(x)$, $k=0,1,\cdots,n-1$.

(ii) $p_n(x) \geq P_n(x) > 0$,

then $\eta_1(\alpha) \geq \mu_1(\alpha)$.

Proof. Exploiting an idea used by W. Allegretto [5] in the case of second order elliptic equations, we shall show that the selfadjoint equation $\frac{1}{2}(\ell+\ell^*)u = 0$ "oscillates faster" than (1). Since

$$\ell^* u = \sum_{k=0}^{n} (-1)^k \left(p_k(x)u^{(k)}\right)^{(k)} - \sum_{k=0}^{n-1} (-1)^k \left(q_k(x)u^{(k+1)}\right)^{(k)}$$

$$- \sum_{k=0}^{n-1} (-1)^k \left(q_k'(x)u^{(k)}\right)^{(k)},$$

we have

$$\frac{1}{2}(\ell+\ell^*)u = (-1)^n \left(p_n(x)u^{(n)}\right)^{(n)} + \sum_{k=0}^{n-1} (-1)^k \left(\left(p_k(x) - \frac{1}{2}q_k'(x)\right)u^{(k)}\right)^{(k)}.$$

If $u(x)$ is a nontrivial solution of (1) which realizes the first conjugate point $\eta_1(\alpha)$, then integration by parts shows that

$$\mathcal{S}[u] \equiv \int_{\alpha}^{\eta_1(\alpha)} \left[p_n(u^{(n)})^2 + \sum_{k=0}^{n-1}\left(p_k - \frac{1}{2}q_k'\right)\left(u^{(k)}\right)^2\right]$$

$$= \int_{\alpha}^{\eta_1(\alpha)} u\ell u \, dx = 0.$$

From the hypotheses (i) and (ii) we have

$$(4) \qquad \int_{\alpha}^{\eta_1(\alpha)} \left[(p_n - P_n)(u^{(n)})^2 + \sum_{k=0}^{n-1}\left(p_k - \frac{1}{2}q_k' - P_k\right)\left(u^{(k)}\right)^2\right]dx \geq 0,$$

so that

$$\mathcal{E}[u] \equiv \int_{\alpha}^{\eta_1(\alpha)} \sum_{k=0}^{n} P_k(u^{(k)})^2 \, dx \leq 0.$$

According to a criterion of W. T. Reid [6; Theorem 5.2], the fact that $\mathcal{E}[u] \leq 0$

for some nontrivial "admissible" $u(x)$ assures that $\mu_1(\alpha) \leq \eta_1(\alpha)$, and this completes the proof.

The non oscillation criteria of Theorem 1 are different from those stated in [3]. To establish a connection we generalize a device of Picard. Let $h_k(x)$ be continuously differentiable functions for $k=0,1,\cdots,n-1$ so that

$$\frac{d}{dx}\left[h_k(u^{(k)})^2\right] = h_k'(u^{(k)})^2 + 2h_k u^{(k)}u^{(k+1)}.$$

If $u(x)$ has $(n-1)$st order zeros at $x = \alpha$ and $x = \beta$, then

$$0 = \int_\alpha^\beta \frac{d}{dx}\left[h_k(u^{(k)})^2\right]dx = \int_\alpha^\beta \left[h_k'(u^{(k)})^2 + 2h_k u^{(k)}u^{(k+1)}\right]dx.$$

Thus condition (4) could be replaced by

$$(4') \qquad \int_\alpha^{\eta_1(\alpha)} \Big[(p_n - P_n)(u^{(n)})^2 + \sum_{k=0}^{n-1}(p_k - \tfrac{1}{2}q_k' - P_k + h_k')(u^{(k)})^2 + 2\sum_{k=0}^{n-1} h_k u^{(k)}u^{(k+1)}\Big]dx \geq 0,$$

leading to the following generalization of Theorem 1.

<u>Theorem 2</u>. <u>If there exist continuously differentiable functions</u> $h_k(x)$ <u>such that</u> (4') <u>is satisfied for the solution which realizes the first conjugate point of</u> α <u>with respect to</u> (1), <u>then</u> $\mu_1(\alpha) \leq \eta_1(\alpha)$.

The choice $h_k(x) \equiv 0$ for $k=0,1,\cdots,n-1$ leads to Theorem 1. The choice $h_k(x) \equiv \frac{1}{2}q_k(x)$ for $k=0,1,\cdots,n-1$ leads essentially to the criteria of [3].

In case $n = 1$ it can be shown that the selfadjoint equation

$$\left[\tfrac{1}{2}(\ell+\ell^*) + \frac{q_0^2}{4p_1}\right] \equiv -(p_1u')' + \left(p_0 - \tfrac{1}{2}q_0' + \frac{q_0^2}{4p_1}\right)u = 0$$

"oscillates slower" than $\ell u = 0$. This fact makes it possible to establish oscillation criteria for the nonselfadjoint equation $\ell u = 0$ when $n = 1$. It would be of considerable interest to find a selfadjoint differential operator $\tilde{\ell}$ of order $\leq 2n - 2$ such that $[\frac{1}{2}(\ell+\ell^*) + \tilde{\ell}]u = 0$ "oscillates slower" than $\ell u = 0$ in the case $n > 1$. Such a result would lead to oscillation criteria for (1) which do not appear to be accessible by matrix techniques.

References

1. W. T. Reid, Oscillation criteria for selfadjoint differential systems, Trans. Amer. Math. Soc. 101(1961), 91-106.

2. K. Kreith, A comparison theorem for conjugate points of general selfadjoint differential equations, Proc. Amer. Math. Soc. 25(1970), 656-661.

3. K. Kreith, Disconjugacy criteria for nonselfadjoint differential equations of even order, Can. J. Math. 23(1971), 644-652.

4. W. T. Reid, A disconjugacy criterion for higher order linear vector differential equations, Pacific J. Math., to appear.

5. W. Allegretto, Eigenvalue comparison and oscillation criteria for elliptic operators, J. London Math. Soc. 3(1971), 571-575.

6. W. T. Reid, Riccati matrix differential equations, Pacific J. Math. 13(1963), 665-685.

On the Non-Autonomous van der Pol Equation with Large Parameter

N.G. Lloyd

Consider the equation

$$\ddot{x} + k(x^2 - 1)\dot{x} + x = b\lambda k \cos\varphi , \qquad (1)$$

where $\varphi = \lambda t+\mu$, b,λ,k,μ are independent of $x,\dot{x},t$ and b,λ,μ are independent of k. We are interested in (1) for large values of k. In [2], [3], [4], Littlewood and Cartwright have looked at this problem in detail when $b < \frac{2}{3}$. They suggested (see [5]) that when $b > \frac{2}{3}$, there is a unique periodic solution of (1); this we shall prove :

Theorem. If $b > \frac{2}{3}$, there exists $k_0(b,\lambda,\mu)$ such that for $k \geqslant k_0$, equation (1) has a unique periodic solution. This solution has period $2\pi\lambda^{-1}$ and all solutions converge to it.

We shall write (1) in the form

$$k^{-1}(X + \dot{x}) = -F(x) + b \sin\varphi + C,$$

where $F(x) = \int (x^2-1)\,dx$, $X(t) = \int_0^t x(s)\,ds$ and C is constant on a solution. The letter A is used as a generic symbol for a constant - its value is not the same at each occurrence. Most of the results are valid only for sufficiently large values of k; the necessary adjustment will be made tacitly.

It is known ([1]) that every solution of (1) satisfies

$$|x| < A, \quad |\dot{x}| < A(k+1) . \qquad (2)$$

eventually, and that there exists at least one solution of period $2\pi\lambda^{-1}$. From [2], [3] , we quote

Lemma A. Suppose $\delta > 0$. At all points Q on a trajectory Γ which have been immediately preceded by a section of Γ lying in $x \geqslant 1+\delta$ and lasting a time $k^{-1}\log^2 k$, we have $x^{(n)} = O(1)$ as $k \to \infty$ for all n. (Γ is said to be 'settled' at such points Q.)

Let $\bar{x}$ be the unique real root of $F(x) = b$ and let

$$\mathcal{B} = \{x;\ -\bar{x} + k^{-\frac{1}{2}} \leqslant x \leqslant \bar{x} - k^{-\frac{1}{2}}\}.$$

It is easily shown that (for large enough k) no trajectory that satisfies (2) can lie entirely within $\mathcal{B}$ for any interval congruent modulo π to $\frac{1}{2}\pi \leqslant \varphi \leqslant \frac{3}{2}\pi$; also no solution remains for all time in $x>0$ or in $x<0$. Hence every trajectory has a maximum in $x > \bar{x} - k^{-\frac{1}{2}}$ or a minimum in $x < -\bar{x} + k^{-\frac{1}{2}}$; we shall suppose the first alternative - the second is just a reflection in $x=0$ - and denote the set of such solutions by $\mathcal{S}$.

We first trace out solutions by a careful consideration of their extrema; it turns out that all solutions converge to a 'strip' defined by

$$\dot{x} = 0,\ |x - \bar{x}| < k^{-\frac{1}{3}},\quad |\varphi - \tfrac{1}{2}\pi| < k^{-\frac{1}{5}}\ (\mathrm{mod.}\ 2\pi). \qquad (3)$$

Then we consider the variational equations relative to a solution in (3) and find that every such solution is asymptotically stable; (this is where our approach differs from that in [2], [3], [4]). It is then an easy matter to obtain our theorem.

It is immediate that at a 'settled' maximum (minimum), $\varphi \equiv \frac{1}{2}\pi + O(k^{-1})$ ($\varphi \equiv \frac{3}{2}\pi + O(k^{-1})$), modulo 2π. This, however, is not usually known to be so. We have the following sequence of results (each has an obvious analogue replacing 'maximum' with 'minimum' etc.; φ is always modulo 2π):

Lemma B. (1) A maximum occurs only when $\frac{1}{2}\pi - Ak^{-1} < \varphi < \frac{3}{2}\pi + Ak^{-1}$.

(2) If Γ remains in $x > 1 + k^{-\frac{1}{2}}$ or $x < -1-k^{-\frac{1}{2}}$ between a minimum and the next maximum, those two points cannot both have $\varphi = \frac{1}{2}\pi m + O(k^{-1})$ any m.

(3) A maximum can occur in $x > 2$ only when $\varphi = \frac{1}{2}\pi + O(k^{-\frac{1}{4}})$.

(4) A minimum can occur in $x \geqslant 1 + k^{-\frac{1}{2}}$ only when $\varphi = \frac{3}{2}\pi + O(k^{-\frac{1}{4}})$.

(5) Every translate of the interval $\frac{1}{2}\pi \leqslant \varphi \leqslant \frac{5}{2}\pi$ by 2π contains a minimum.

(6) If P_0 is a maximum in $x \geq 1+\delta$, then $\dot{x} = O(1)$ between P_0 and the first subsequent crossing of $x = 1+\delta$. In particular, Γ settles within o(1) of P_0 (until $\dot{x} = 1+\delta$ for the first time after P_0).

We are looking at a Γ that has a maximum in $x > \bar{x} - k^{-\frac{1}{2}}$ (> 2); let this maximum be P and let the next minimum be Q. We have, by Lemma B, that $\varphi_P = \frac{1}{2}\pi + O(k^{-\frac{1}{4}})$ and $\varphi_Q < \frac{5}{2}\pi$.

<u>Lemma C</u>. (1) <u>After P</u>, x <u>decreases at least until</u> $\varphi = \frac{3}{2}\pi - Ak^{-1}$.
(2) <u>There is a maximum</u> (R, <u>say</u>) <u>with</u> $\varphi = \frac{5}{2}\pi + O(k^{-\frac{1}{4}})$.

The behaviour of solutions depends on the value of x_P:

<u>Lemma D.</u> (1) <u>If</u> $F(x_P) \geq 2b - \frac{2}{3} - k^{-\frac{1}{3}}$, <u>then</u> $\int_P^R x\,dt > A$.
(2) <u>If</u> $2b - \frac{2}{3} - k^{-\frac{1}{3}} \geq F(x_P) \geq b + \frac{1}{2}k^{-\frac{1}{3}}$, <u>then</u> $\int_P^R x\,dt > Ak^{-\frac{1}{3}}$.

By careful applications of Lemmas B,C and D, it is possible to show that all solutions eventually have points in the strip (3); the shape of solutions in the strip is also derived in some detail. The most important points are that there is a maximum (R) with $\varphi = \frac{5}{2}\pi + O(k^{-\frac{1}{4}})$ and just one minimum Q between P and R. We suppose, of course, that φ is measured so that $\varphi_P = \frac{1}{2}\pi + O(k^{-\frac{1}{4}})$; then $\varphi_Q = \frac{3}{2}\pi + O(k^{-\frac{1}{4}})$.

Suppose u(t) is a solution in (3); the variational equations of (1) relative to u(t) are

$$\dot{\xi} = -kp(t)\xi + \eta$$
$$\dot{\eta} = -\xi,$$

where $p(t) = (u(t))^2 - 1$. Define $\zeta^2 = \xi^2 + \eta^2$; we show that over the interval $\varphi_P < \varphi < \varphi_P + 2\pi$, ζ^2 decreases, from which it follows that every solution in (3) is asymptotically stable. We first need some more estimates. After P, let Γ cross $x = 1$ first at V and $x = -1$ at W; K is to be the point on PV with $x = 1 - k^{-\frac{1}{3}}$.

Lemma E. (1) At $x = 1 + k^{-\frac{1}{3}}$ in PV, $\dot{x}$ lies between constant multiples of $k^{\frac{1}{3}}$.

(2) In I' : $1+k^{-\frac{1}{3}} \geqslant x \geqslant 1$, $|\dot{x}|$ lies between multiples of $k^{\frac{1}{3}}$; I' is traversed in a time $O(k^{-\frac{2}{3}})$; $b \sin\varphi_V \doteq -\frac{2}{3} + O(k^{-\frac{1}{3}})$.

(3) $\int_K^W (1-x^2)\, dt = \frac{2}{3}k^{-1}\log k + O(k^{-1})$; $\varphi_K - \varphi_V$ and $\varphi_W - \varphi_K$ are $O(k^{-\frac{2}{3}})$

Next we trace the changes of ξ and η along our Γ; we may suppose that $\xi^2 + \eta^2 = 1$ at P and $\xi > 0$. Clearly r^2 decreases over PV; it may be shown that $\xi = O(k^{-\frac{2}{3}})$ from a point a time $O(k^{-\frac{1}{3}})$ after P until $x = 1+k^{-\frac{1}{3}}$. Thereafter the sequence of results is : $\xi = O(k^{-\frac{1}{3}})$ until $x=1$; $\xi = O(k^{-\frac{2}{3}})$ in $1 \geqslant x \geqslant 1-k^{-\frac{1}{3}}$; $\xi = O(1)$ in $1-k^{-\frac{1}{3}} \geqslant x \geqslant -1$. We can show that $\xi = O(k^{-\frac{2}{3}})$ within a time $O(k^{-\frac{1}{3}})$ of $x = -1-\delta$ and we know that Γ reaches $x = -2$ within $O(k^{-\frac{1}{4}})$ of V. So $\xi = O(k^{-\frac{2}{3}})$ within a time $o(1)$ later than V - and certainly before $\varphi = \min(\varphi_P + \pi, \varphi_Q)$.

We may similarly follow the changes in η. This splits into three cases, depending on the value of η at P. The result of these estimates is that r^2 decreases over the interval $\varphi_P \leqslant \varphi \leqslant \varphi_P + 2\pi$.

The whole proof is concluded by using the asymptotic stability of solutions in our strip and a standard index number argument.

References

[1] M.L.Cartwright, 'Forced oscillations in nonlinear systems' (Contributions to the theory of nonlinear oscillations, Vol.1, Annals of Mathematics studies No.20, Princeton, 1950).

[2] M.L.Cartwright and J.E.Littlewood, 'On nonlinear differential equations of the second order, I '. J.London Math. Soc. 20 (1945), 180 - 189.

[3] J.E.Littlewood, 'On nonlinear differential equations of the second order, III '. Acta Math. 97 (1957), 267 - 308.

[4] J.E.Littlewood, 'On nonlinear differential equations of the second order, IV '. Acta Math. 98 (1957), 1 - 110.

[5] J.E.Littlewood, Some problems in real and complex analysis (Heath, 1968).

Global Attraction and Asymptotic Equilibrium for Nonlinear Ordinary Equations

D.L. Lovelady

Let Y be a Banach space with norm $|\ |$, and let R^+ be the set of all nonnegative real numbers. Let A be a (strongly) continuous function from $R^+ \times Y$ to Y, and let α be a continuous real-valued function on R^+. This author and R.H. Martin, Jr. [4] have shown that if

(1) $$|x - y - c[A(t,x) - A(t,y)]| \geq [1 - c\alpha(t)]|x - y|$$
whenever (t,x,y) is in $R^+ \times Y \times Y$ and c is a positive number

it is then the case that if z is in Y there is exactly one continuously differentiable function u_z from R^+ to Y such that

$$u_z'(t) = A(t,u_z(t)) \ ; \ \ u_z(0) = z$$

whenever t is a positive number.

A member z of Y is called a *point of asymptotic equilibrium* if and only if

$$\lim_{t\to\infty} u_z(t) = z$$

and is called a *point of global attraction* if and only if

$$\lim_{t\to\infty} u_x(t) = z$$

whenever x is in Y. We now give sufficient conditions for a point z to be either a point of asymptotic equilibrium or a point of global attraction.

<u>THEOREM</u>: *Let A and α be as above, suppose that (1) is true, and let z be in Y. If*

$$\lim_{t\to\infty} \int_0^t |A(s,z)| \exp[\int_s^t \alpha(r)dr]ds = 0,$$

then z is a point of asymptotic equilibrium. If, in addition,

$$\lim_{t\to\infty} \exp[\int_0^t \alpha(s)ds] = 0,$$

then z is a point of global attraction.

Let m_- be that real-valued function on $Y\times Y$ given by

$$m_-[x,y] = \lim_{\delta\to 0-} (1/\delta)(|x + \delta y| - |x|).$$

Now (1) is equivalent to requiring that

$$m_-[x - y, A(t,x) - A(t,y)] \le \alpha(t)|x - y|$$

whenever (t,x,y) is in $R^+ \times Y \times Y$ (compare [1, p.3]). It is known that if u is a differentiable function from R^+ to Y, and P is given on R^+ by $P(t) = |u(t)|$, then P is left-differentiable on $(0,\infty)$ and $P'_-(t) = m_-[u(t),u'(t)]$ whenever t is in $(0,\infty)$ [1, p.3]. Also, $m_-[x,y + z] \le m_-[x,y] + |z|$ whenever (x,y,z) is in $Y \times Y \times Y$ (see [3, *Lemma 6*]).

Suppose that (1) holds and let (x,z) be in $Y \times Y$. Let Q be given on R^+ by $Q(t) = |u_x(t) - z|$. Now $Q(0) = |x - z|$ and, if t is a positive number,

$$\begin{aligned} Q'_-(t) &= m_-[u_x(t) - z, u'_x(t)] \\ &= m_-[u_x(t) - z, A(t,u_x(t)) - A(t,z) + A(t,z)] \\ &\le m_-[u_x(t) - z, A(t,u_x(t)) - A(t,z)] + |A(t,z)| \\ &\le \alpha(t)Q(t) + |A(t,z)|. \end{aligned}$$

Hence [2, *Theorem 1.4.1*, p.15],

$$Q(t) \le |x - z| \exp\left[\int_0^t \alpha(s)ds\right] + \int_0^t |A(s,z)| \exp\left[\int_s^t \alpha(r)dr\right]ds$$

whenever t is in R^+. Our theorem now follows easily (for the first conclusion put $x = z$ in the above computation).

In light of the theorem it is interesting to note that if p and q are in $(1,\infty)$, if $p + q = pq$, if

$$(2) \qquad \int_0^\infty |A(s,z)|^p ds \ \text{ is finite,}$$

and if there is a number K such that

$$\int_0^t (\exp[\int_s^t \alpha(r)dr])^q ds \leq K^q$$

whenever t is in R^+, then

$$\lim_{t\to\infty} \int_0^t |A(s,z)| \exp[\int_s^t \alpha(r)dr]ds = 0$$

and

$$\lim_{t\to\infty} \exp[\int_0^t \alpha(s)ds] = 0.$$

This follows directly from [*1, Lemma 1, p.68*], from the Hölder inequality, and from the fact that (2) implies

$$\lim_{t\to\infty} [\int_t^\infty |A(s,z)|^p ds]^{1/p} = 0.$$

REFERENCES

[1] W.A. Coppel, *Stability and asymptotic behavior of differential equations*, D.C. Heath & Co., Boston, 1965.

[2] V. Lakshmikantham and S. Leela, *Differential and integral inequalities, vol. 1*, Academic Press, New York, 1969.

[3] D.L. Lovelady, *A functional differential equation in a Banach space*, Funkcialaj Ekvacioj, 14(1971),111-122.

[4] __________ and R.H. Martin, Jr., *A global existence theorem for a nonautonomous differential equation in a Banach space* (to appear).

The Functional-Differential Equation $y'(x) = ay(\lambda x) + by(x)$ and Generalisations

J. B. McLeod

1. Introduction

In a recent paper [1] Kato and McLeod have discussed the functional-differential equation

$$y'(x) = ay(\lambda x) + by(x) \qquad (0 \leq x < \infty), \qquad (1.1)$$

where a is a possibly complex constant, b a real constant, and λ a non-negative constant. We may always assume that $a \neq 0$, $\lambda \neq 0,1$, since $a = 0$ or $\lambda = 0$ or $\lambda = 1$ reduces the question to a rather trivial ordinary differential equation. Our object here is to indicate the type of result obtained in [1], and to look briefly at some extensions which have been or might be discussed.

By a solution of (1.1) we mean a complex-valued continuous function $y(x)$ defined in some subinterval of $0 \leq x < \infty$ and satisfying (1.1). If the original interval of definition includes an interval of the form $[\lambda x_0, x_0]$ $(\lambda < 1)$ or $[x_0, \lambda x_0]$ $(\lambda > 1)$ for some fixed $x_0 > 0$, then the solution can be extended uniquely to the right (increasing values of x) or to the left (decreasing values of x) by using (1.1), and <u>by a solution we shall always mean a solution extended as far as it can be in either direction</u>.

We shall be interested in behaviour as $x \to \infty$. If $\lambda < 1$, it is easy to see (as is pointed out in [1]) that any solution becomes increasingly smooth as x increases, and thus every solution continues to exist as $x \to \infty$. It is thus reasonable to ask for the asymptotic behaviour of <u>all</u> solutions. But if $\lambda > 1$, the solutions become less smooth, and in fact a solution can exist for all $x \geq 0$ only if it is C^∞. We can in this case expect to consider the asymptotic behaviour of only some solutions.

Looking intuitively at the kind of asymptotic behaviour that arises, we might hope most simply for an asymptotic behaviour that is algebraic. If y is to behave like x^k, then since y' is presumably negligible compared with y, we determine k from the equation

$$0 = a(\lambda x)^k + bx^k ,$$

so that $$k = \log(-b/a) / \operatorname{Log} \lambda , \tag{1.2}$$

Log denoting the principal branch of log, so that $-\pi < \operatorname{im}(\operatorname{Log} z) \leq \pi$. The equation (1.2) does not determine k uniquely, but it does determine re k (= κ, say), and so $|x^k|$, for

$$\kappa = (\operatorname{Log} |b/a|) / \operatorname{Log} \lambda . \tag{1.3}$$

We will let k_0 be any particular solution of (1.2).

2. Some theorems on asymptotic behaviour

One typical theorem from [1] is the following.

__Theorem 1.__ __Let__ $\lambda < 1$, $b < 0$.

(i) __No solution__ $y \neq 0$ __of__ (1.1) __is__ $o(x^\kappa)$ __as__ $x \to \infty$.

(ii) __Every solution of (1.1) has the asymptotic form__

$$x^{k_0} g(\operatorname{Log} x) + o(x^\kappa) , \tag{2.1}$$

__where__ $g(s)$ __is a__ C^∞ __function, periodic of period__ $\operatorname{Log} \lambda$ __and satisfying__

$$|g^{(n)}(s)| \leq M K^n \lambda^{-\frac{1}{2}n^2} \quad (n = 0,1,2,\ldots) \tag{2.2}$$

__for some positive constants__ M __and__ K __and all__ s.

(iii) __For each__ $g(s)$ __with the properties described above, there is one (and by (i) only one) solution of (1.1) with the asymptotic behaviour (2.1)__.

There are similar theorems for $\lambda < 1$, $b > 0$, and for $\lambda < 1$, $b = 0$. We will merely remark here that, for $\lambda < 1$, $b > 0$, there is (as is intuitively very reasonable) at least one solution behaving like e^{bx} as $x \to \infty$. If $y_0(x)$ is one such solution, then the asymptotic behaviour of any solution is, apart from the possible addition of a constant multiple of $y_0(x)$, exactly the same as in Theorem 1. If $\lambda < 1$, $b = 0$, the behaviour is altogether more complicated, in that any solution is asymptotic to

$$x^l (\operatorname{Log} x)^m \exp\left\{-\tfrac{1}{2}(\operatorname{Log} x - \operatorname{Log}\operatorname{Log} x)^2 / \operatorname{Log}\lambda\right\} g(\operatorname{Log} x - \operatorname{Log}\operatorname{Log} x),$$

where $l = \frac{1}{2} - (\operatorname{Log}\lambda)^{-1} - \operatorname{Log}(-ac)/\operatorname{Log}\lambda$, $m = -1 + \operatorname{Log}(-ac)/\operatorname{Log}\lambda$, and g is again a C^{∞} function, periodic of period Log λ, and satisfying inequalities similar to (2.2) which we will not write down here. The discussion of this case, which is much more delicate than that of Theorem 1, was first carried out by de Bruijn [2].

Turning now to the case $\lambda > 1$, we need, in order to state any of the theorems fully, the idea of an asymptotic distribution. We say that a distribution $g(s)$ is <u>asymptotically zero</u>, in symbols $g(s) \overset{d}{\sim} 0$ as $s \to \infty$, if the sequence $\{g(s + h_n)\}$ converges to zero for any increasing sequence h_n of real numbers with $h_n \to \infty$, i.e. if for any test function $\phi(s)$,

$$\int_{-\infty}^{\infty} g(s + h_n)\, \phi(s)\, ds \to 0 \quad \text{as} \quad n \to \infty .$$

(We use the symbolic notation of the integral for the functional $\langle g, \phi \rangle$.) Further, we write $g_1(s) \overset{d}{\sim} g_2(s)$ if $g_1(s) - g_2(s) \overset{d}{\sim} 0$.

If we think of a distribution intuitively as a generalised derivative, so that integration of it sufficiently often yields an ordinary function, we can see why it is important to introduce the concept here. For given a certain asymptotic behaviour for a solution, we can think of obtaining the solution by extending to the left, and since $\lambda > 1$, this has the smoothing effect of integration, necessarily repeated an infinite number of times to reach any finite value of x. Thus a distributional asymptotic behaviour will yield a solution that is an ordinary function. The following theorem is proved in [1].

<u>Theorem 2</u>. <u>Let</u> $\lambda > 1$, $b > 0$. <u>Let</u> $s = \text{Log}\, x$.

(i) <u>No solution</u> $y \neq 0$ <u>of (1.1) is</u> $o(x^{K})$. <u>More generally, no solution</u> $y(e^s) \neq 0$ <u>satisfies</u>

$$e^{-k_0 s}\, y(e^s) \overset{d}{\sim} 0 \quad \underline{\text{as}} \quad s \to \infty .$$

(ii) <u>Let</u> $y(x)$ <u>be a solution of (1.1) with</u>

$$y^{(n)}(x) \leq M K^n \lambda^{\frac{1}{2}n^2} \qquad (x_0 \leq x \leq \lambda x_0;\ n = 0,1,2,\ldots) \qquad (2.3)$$

<u>for some positive constants</u> x_0, M <u>and</u> K. <u>Then there is a periodic</u>

distribution $g(s)$ *with period* $\text{Log}\,\lambda$ *such that*

$$e^{-k_0 s}\, y(e^s) \overset{d}{\sim} g(s) \quad \textit{as} \quad s \to \infty \,. \tag{2.4}$$

(iii) *For any periodic distribution* $g(s)$ *of period* $\text{Log}\,\lambda$, *there is one (and by (i) only one) solution of (1.1) satisfying (2.4). For each* $x_0 > 0$, *the solution satisfies (2.3) for some positive constants* M *and* K *which depend on* x_0.

The rather beautiful duality between (2.2) and (2.3) can only arise by the inclusion of distributions. There are similar theorems for $\lambda > 1$, $b < 0$ and $\lambda > 1$, $b = 0$, but we will not go into them here.

3. The case b complex

This case has been considered by Kato, in a paper shortly to appear. Theorems 1 and 2 continue to hold, except that the conditions $b < 0$, $b > 0$ respectively become replaced by $\text{re}\, b < 0$, $\text{re}\, b > 0$. The intriguing case is the case $\text{re}\, b = 0$. If $\lambda > 1$, $\text{re}\, b = 0$, then Kato obtains some results on the lines of Theorem 2, but less complete; if $\lambda < 1$, $\text{re}\, b = 0$, then Kato has no results at all, and this leaves an intriguing open problem.

4. The case of varying coefficients

We are interested here in the difference-differential equation

$$y'(s) = a(s)\, y(s-1) + b(s)\, y(s)\,, \tag{4.1}$$

to which (1.1) may be reduced by making the substitution $s = \text{Log}\, x$. If $\phi(s)$ is any particular solution of (4.1), and if we set $y(s) = \phi(s)\, z(s)$, then it is easy to verify that z satisfies an equation of the form

$$z'(s) = f(s)\, \{y(s-1) - y(s)\}, \tag{4.2}$$

and it is in this form that we consider it.

The case $f(s) > 0$ has been considered by de Bruijn [3]. Under conditions which certainly include $f(s) = s^{\alpha} (\alpha > 1)$ or $f(s) = \exp(s^{\beta})$ $(\beta > 0)$, he shows that any solution must tend as $s \to \infty$ to a periodic function of period 1, and that no solution not identically zero tends to zero. This is the analogue of Theorem 1. De Bruijn has no inequalities corresponding to (2.2), but this is not perhaps to be expected unless $f(s)$ is specified more precisely. Work is also currently in progress to discuss the case $f(s) < 0$, which should correspond to the case $\lambda < 1$, $b > 0$ for (1.1), and it does indeed seem to be indicated that there is a 'large' solution (corresponding to e^{bx}) modulo which the asymptotic behaviour is as described by de Bruijn.

The case $s \to -\infty$ would lead to an analogue of Theorem 2, but does not so far seem to have been investigated.

A rather intriguing open question arises with the apparently very simple equation

$$y'(s) = s\left\{y(s-1) - y(s)\right\} .$$

This is not quite covered by de Bruijn's results as so far stated, although he proves separately that any solution is asymptotic to $g(s - \operatorname{Log} s)$, g periodic of period 1. But whether solutions not identically zero can tend to zero is not known, and may be difficult. (It is worth remark that the case $b = 0$ of (1.1), when reduced to the form (4.2), yields a function such that $f(s) \sim s$ as $s \to \infty$.)

5. The matrix case

For brevity, we confine our remarks to the case which should most closely correspond to Theorem 1. Consider

$$y'(x) = Ay(\lambda x) + By(x) \tag{5.1}$$

where y is now an n-dimensional vector and A and B are $n \times n$ matrices. We take $\lambda < 1$ and suppose B hermitian with negative eigenvalues. Then we would expect algebraic asymptotic behaviour, and if $y(x) \sim x^k y_0$ for some constant vector y_0, then $(A\lambda^k + B)\, y_0 = 0$, so that y_0 is an eigenvector of $B^{-1}A$ corresponding to the eigenvalue $-\lambda^{-k}$. General experience with matrices then suggests that there may be difficulties if $B^{-1}A$ does not have distinct eigenvalues, or more precisely if it is not diagonalisable, and in fact the theorem stated below is not true if the diagonalisable requirement on $B^{-1}A$ is dropped. (A counter-example may be constructed by setting

$$A = \begin{pmatrix} 1 & 1 \\ 0 & 1 \end{pmatrix}, \qquad B = \begin{pmatrix} b & 0 \\ 0 & b \end{pmatrix}, \quad b < 0 \quad .)$$

Theorem 3. *Suppose* $\lambda < 1$, B *hermitian with negative eigenvalues,* $B^{-1}A$ *diagonalisable.* *If* $-\lambda^{-k_1}$, $-\lambda^{-k_n}$ *are eigenvalues of* $B^{-1}A$ *such that* k_1 *has the largest possible real part, and* k_n *the smallest possible, then, as* $x \to \infty$,

(i) *no solution* $y \neq 0$ *of (5.1) is* $o(x^{k_n})$,

(ii) *every solution of (5.1) is* $O(x^{k_1})$.

Part (i) of this theorem has been proved by J. Dyson, the argument being a modification of that for the scalar case. Part (ii), on which depends the subsequent development of the matrix problem, remains at the time of writing obstinately unproved.

References

[1] T. Kato and J.B. McLeod, 'The functional-differential equation $y'(x) = ay(\lambda x) + by(x)$', *Bull. Amer. Math. Soc.* 77 (1971), 891-937.

[2] N.G. de Bruijn, 'The difference-differential equation $F'(x) = e^{\alpha x+\beta}F(x-1)$', I,II, *Nederl. Akad. Wetensch. Proc. Ser.* A56 = *Indag. Math*. 15 (1953), 449-464.

[3] ______________, 'The asymptotically periodic behaviour of the solutions of some linear functional equations', *Amer. J. Math.* 71 (1949), 313-330.

Multiple Solutions of a Singularly Perturbed Boundary Value Problem Arising in Chemical Reactor Theory

J. Chen* and R.E. O'Malley, Jr.**

1. Summary:

Boundary value problems of the form

$$(1) \quad \begin{cases} \varepsilon u'' + u' = g(t,u), & 0 \leqslant t \leqslant 1 \\ u'(0) - au(0) = A \\ u'(1) + bu(1) = B \end{cases}$$

arise in the study of adiabatic tubular chemical flow reactors with axial diffusion (cf., e.g., Raymond and Amundson {7} and Burghardt and Zaleski {1}). We are interested in obtaining asymptotic solutions $u(t,\varepsilon)$ of (1) as the positive parameter ε tends to zero. This corresponds to the Peclet number becoming large. Progress toward solving the problem has been made by Cohen {2} , Keller {3}, and Parter {6} . We have succeeded in obtaining asymptotic solutions to this and certain other problems of the form

$$\begin{cases} \varepsilon u'' + f(t,u,\varepsilon)\, u' = g(t,u,\varepsilon) \\ m(u(0),\, u'(0),\varepsilon) = 0 \\ n(u(1),\, u'(1),\varepsilon) = 0 \end{cases} ,$$

but shall consider only the physical problem (1) in this short note. We shall assume that $g(t,u)$ is infinitely differentiable in both arguments. Then we obtain: /

* Department of Mathematics, University Heights, New York University.
This author's work was supported in part by Air Force Office of Scientific Research under grant number AFOSR-67-1062D.

** University of Edinburgh.
This author's work was supported in part by the Science Research Council of the United Kingdom while he was on leave from New York University.

Theorem

Suppose the equation

(2) $g(1,\alpha) + b\alpha = B$

has a root α_0 *of finite multiplicity* m *such that the corresponding terminal value problem*

$$(3)\quad \begin{aligned} u' &= g(t,u) \\ u(1) &= \alpha_0 \end{aligned}$$

has a solution $U_o(t)$ *throughout* $0 \leq t \leq 1$. *If* $m > 1$, *suppose further that*

$$(4)\quad \begin{cases} U_o''(1) \neq 0 & \text{if } m \text{ is odd} \\ \text{and} & \\ \dfrac{\partial^m g(1,\alpha_0)}{\partial u^m} U_o''(1) > 0 & \text{if } m \text{ is even.} \end{cases}$$

Then, the boundary value problem (1) *has a solution for* ε *sufficiently small of the form*

(5) $u(t,\varepsilon) = U(t,\varepsilon^{1/m}) + \varepsilon\, v(\frac{t}{\varepsilon}, \varepsilon^{1/m})$

where

(6) $U(t, \varepsilon^{1/m}) \sim \sum_{j=o}^{\infty} U_j(t)\varepsilon^{j/m}$

and

(7) $v(\tau, \varepsilon^{1/m}) \sim \sum_{j=o}^{\infty} v_j(\tau)\varepsilon^{j/m}$

Here, $\varepsilon^{1/m} > 0$, $v_o(\tau) = (U_o'(0) - a\,U_o(0) - A)\, e^{-\tau}$, *and all the* $v_j(\tau)$*'s* $\to 0$ *as* $\tau \to \infty$ *Further, there are two such expansions when* m *is even.*

2. The Outer Expansion

Familiarity with elementary singular perturbation problems (cf. e.g. Wasow [8]) leads us to suspect nonuniform convergence of derivatives of the solution $u(t,\varepsilon)$ at $t = 0$ with convergence elsewhere to the corresponding derivatives of the solution of the reduced problem /

$$(8) \qquad \begin{cases} U_o' = g(t,U_o) \\ U_o'(1) + bU_o(1) = B \end{cases}$$

Further, away from $t = 0$ one expects the solution to be asymptotically equal to an outer expansion $U^o(t,\varepsilon)$ which satisfies

$$(9) \qquad \begin{cases} \varepsilon U^{o}{}'' + U^{o}{}' = g(t, U^o) \\ U^{o}{}'(1,\varepsilon) + bU^o(1,\varepsilon) = B \end{cases}$$

and converges to a solution $U_o(t)$ as $\varepsilon \to 0$ throughout $0 \leq t \leq 1$. Note that $U_o(t)$ will be completely determined by its terminal value $U_o(1) = \alpha_o$ which satisfies the nonlinear equation (2).

Suppose, first, that α_o is a root of (2) of multiplicity $m=1$ and let $U^o(t,\varepsilon)$ have an asymptotic power series expansion $U(t,\varepsilon)$ in ε as in (6). Since U_o satisfies (8), higher order coefficients U_j must satisfy linear problems of the form

$$(10) \qquad \begin{cases} U_j'(t) = g_u(t,U_o)U_j + \tilde{\tilde{G}}_{j-1}(t) \\ U_j'(1) + bU_j(1) = \tilde{\tilde{B}}_{j-1} \end{cases}$$

where $\tilde{G}_{j-1}$ and $\tilde{B}_{j-1}$ are known successively. (These equations are obtained by simply equating coefficients of ε^j in (9)). Note that the terminal value $U_j(1)$ is uniquely determined successively since

$$(g_u(1, \alpha_o) + b)\, U_j(1) = \tilde{\tilde{B}}_{j-1} - \tilde{\tilde{G}}_{j-1}(1)$$

and $g_u(1,\alpha_o) + b \neq 0$. Thus, the outer expansion can be completely obtained when $m = 1$. For $m > 1$, however, $g_u(1,\alpha_o) + b = 0$, so this expansion scheme fails.

For $m > 1$, we shall seek an outer expansion of the form (6). Substituting into (9), we successively equate coefficients of $\varepsilon^{j/m}$. Thus, U_1 must satisfy the linear problem

$$\begin{cases} U_1' = g_u(t,U_o)U_1 \\ U_1'(1) + b\, U_1(1) = 0\ . \end{cases}$$

Note that the terminal condition can be rewritten as $(g_u(1,\alpha_o)+b)U_1(1)=0$.

It/

It is therefore satisfied whatever $U_\ell(1)$. In general, the U_j for $j \geq 1$ must satisfy linear equations of the form

(11) $$U_j' = g_u(t,U_o)U_j + G^{**}_{j-1}$$

where $G^{*}_{j-1}(t)$ is determined by lower order coefficients. The terminal conditions become

$$U_j'(1) + b\,U_j(1) = 0 .$$

Using the differential equations (11), however, these conditions are automatically satisfied for $j < m$ since

$$\frac{\partial^\ell g(1,\alpha_o)}{\partial u^\ell} = 0 \quad \text{for} \quad j = 0, 1, \ldots, m-1 .$$ For $j = m$, however the boundary condition becomes

$$\frac{1}{m!}\,\frac{\partial^m g(1,\alpha_o)}{\partial u^m}\,(U_1(1))^m = U_o''(1) ,$$

while for $j > m$, we have

$$\frac{\partial^m g(1,\alpha_o)}{\partial u^m}\,(U_1(1))^{m-1}\,U_{j-m+1}(1) = B^{*}_{j-m}$$

where B^{**}_{j-m} is known in terms of the U_ℓ's with $\ell \leq j-m$.

Using the hypothesis (4), then,

(12) $$U_1(1) = \left\{ m!\,U_o''(1) \Big/ \frac{\partial^m g(1,\alpha_o)}{\partial u^m} \right\}^{1/m}$$

and

(13) $$U_\ell(1) = \tilde{B}_{\ell-1} \Big/ \left\{ (U_1(1))^{m-1}\,\frac{\partial^m g(1,\alpha_o)}{\partial u^m} \right\}$$

for each $\ell > 1$. Knowing the terminal values $U_\ell(1)$ uniquely specifies the $U_\ell(t)$'s as solutions of the linear equations (11). Thus, we obtain two real-valued outer expansions when m is even and one real-valued outer expansion when m is odd. Note further that the expansion procedure fails when hypothesis (4) is not satisfied.

3. The Boundary Layer Correction

The asymptotic solution to (1) cannot, of course, be given by the outer expansions $U(t,\varepsilon^{1/m})$ since they do not, in general, satisfy the/

the initial condition. To remedy this, we shall seek asymptotic solutions $u(t,\varepsilon)$ of the form

(14) $$u(t,\varepsilon) = U(t,\varepsilon^{1/m}) + \varepsilon\, v(\tau,\varepsilon^{1/m})$$

where the boundary layer correction $v(\tau,\varepsilon^{1/m})$ has an expansion of the form (7) where the coefficients v_j and their derivatives $v_{j\tau}$ tend to zero as the boundary layer coordinate

(15) $$\tau = {}^{t}/_{\varepsilon}$$

tends to infinity. Away from $t = 0$, then, the solutions will converge to the outer expansion and the boundary layer correction accounts for the nonuniform convergence at $t = 0$.

Since $u(t,\varepsilon)$ satisfies (1), v must satisfy the nonlinear initial value problem

(16) $$\begin{cases} v_{\tau\tau} + v_{\tau} = g(\varepsilon\tau,\ U(\varepsilon\tau,\varepsilon^{1/m}) + \varepsilon v(\tau,\varepsilon^{1/m})\ ) \\ \qquad\qquad -g(\varepsilon\tau,\ U(\varepsilon\tau,\ \varepsilon^{1/m})\ ) \\ v_{\tau}(0,\varepsilon^{1/m}) = A - U'(0,\varepsilon^{1/m}) + a\, U(0,\varepsilon^{1/m}) + \varepsilon\, a v(0,\varepsilon^{1/m}) . \end{cases}$$

The coefficients $v_j(\tau)$ will satisfy the linear initial value problems obtained by equating coefficients of ε^j in (16). Thus, v_o satisfies

$$\begin{cases} v_{o\tau\tau} + v_{o\tau} = 0 \\ v_{o\tau}(0) = \beta_{-1} \equiv A - U_o'(0) + a\, U_o(0) . \end{cases}$$

Integrating and taking limits as $\tau \to \infty$, we obtain

(17) $$v_o(\tau) = -\beta_{-1}\, e^{-\tau} .$$

Likewise, higher order coefficients $v_j(\tau)$ must satisfy

$$\begin{cases} v_{j\tau\tau} + v_{j\tau} = F_{j-1}(\tau) \\ v_{j\tau}(0) = \beta_{j-1} \end{cases}$$

where F_{j-1} and β_{j-1} are known successively. Thus,

(18) $$v_j(\tau) = \beta_{j-1}\, e^{-\tau} - \int_{\tau}^{\infty}\int_{0}^{r} e^{-(r-s)}\, F_{j-1}(s)\, ds\, dr .$$

By/

By induction, F_{j-1} is exponentially decaying as $\tau \to \infty$ and this implies the exponential decay of the $v_j(\tau)$'s. Thus we can formally obtain the unique expansion for the boundary layer correction.

That the formally obtained expansion represents an asymptotic solution of the problem (1) follows by applying known results concerning asymptotic solutions of initial value problems(cf. O'Malley (45)).

References

(1) A. Burghardt and T. Zaleski, "Longitudinal Dispersion at Small and Large Peclet Numbers in Chemical Flow Reactors," Chem. Eng. Sci. 23(1968), 575-591.

(2) D.S. Cohen, "Multiple Solutions of Singular Perturbation Problems", SIAM J. Math. Anal., to appear

(3) H.B. Keller, "Existence Theory of Multiple Solutions of a Singular Perturbation Problem," to appear.

(4) R.E. O'Malley, Jr., "Singular Perturbation of a Boundary Value Problem for a System of Nonlinear Differential Equations," J. Diff. Eq. 8(1970), 431-447.

(5) R.E. O'Malley, Jr., "On Initial Value Problems for a Nonlinear System of Differential Equations with Two Small Parameters," Arch. Rational Mech. Anal. 40(1971), 209-222.

(6) S.V. Parter, "Remarks on the Existence Theory for Multiple Solutions of a Singular Perturbation Problem," to appear.

(7) L.R. Raymond and N.R. Amundson, "Some Observations on Tubular Reactor Stability," Can. J. Chem. Eng. 42(1964), 173-177.

(8) W. Wasow, Asymptotic Expansions for Ordinary Differential Equations, Interscience, New York, 1965.

A Liapounov Function for an Automomous Second-Order Ordinary Differential Equation

C.J.S. Petrie

Summary: Two results on the stability of a singular point of the differential equation

$$x'' = f(x)(1+x'^2) + g(x)\sqrt{(1+x'^2)} + h(x,x') \qquad (1)$$

are proved, in the form of conditions on the functions f, g, h. The results are in the style of many results for Lienard's equation, and the proofs rest on the choice of a suitable Liapounov function. The arguments leading to this choice may be extended to any equation "close" to one for which a first integral may be obtained.

We shall assume throughout that f, g and h are continuous functions of their arguments which satisfy sufficient conditions for the uniqueness of solutions of (1) in some region. We write (1) as a pair of first order equations,

$$\begin{aligned} x' &= y \\ y' &= f(x)(1+y^2) + g(x)\sqrt{(1+y^2)} + h(x,y) \end{aligned} \qquad (2)$$

and observe that, for $y \neq 0$, this has the same trajectories as

$$y(1+y^2)^{-\frac{1}{2}} \frac{dy}{dx} = f(x)(1+y^2)^{\frac{1}{2}} + g(x) + h(x,y)(1+y^2)^{-\frac{1}{2}} \qquad (3)$$

which is a linear equation in $(1+y^2)^{\frac{1}{2}}$, apart from the last term. Using the integrating factor

$$I(x) = \exp[-\int f(x)dx] \qquad (4)$$

and writing

$$V(x,y) = (1+y^2)^{\frac{1}{2}}I(x) - I(s) - \int_s^x g(u)I(u)du \qquad (5)$$

we obtain formally

$$V(x,y) = \int_s^x h(u,y(u))(1+y(u)^2)^{-\frac{1}{2}}I(u)\,du \qquad (6)$$

for the solution of (3) satisfying the initial condition $y(s) = 0$. If the curves $V(x,y) = \text{constant}$ are closed then we may deduce from the sign of the right-hand side of (6) whether the trajectories of (3), and hence of (2), cross these curves from outside to inside as t increases, or vice versa. (See Pearson & Petrie (1970), appendix (vi).)

This approach may be extended to any second order autonomous equation whose trajectories are given implicitly by an equation of the form

$$dV(x,y(x))/dx = F(x,y(x)) \qquad (7)$$

(corresponding to (3) above). This equation is "close" to one that we can integrate, in the sense that only the term $F(x,y)$ prevents us, and is analogous to the archetypal situation for Liapounov's direct method, with an "energy" function, V, and a "dissipation" term, F. Use of Liapounov's theorems will enable us to demonstrate stability of a singular point if V is positive definite and its time derivative along a trajectory,

$$V' = dV(x(t),y(t))/dt = yF(x,y) , \qquad (8)$$

is negative definite.

In the particular case considered here V has the derivative

$$V'(x,y) = yh(x,y)(1+y^2)^{-\frac{1}{2}}I(x) \qquad (9)$$

and we may now proceed to the proof of

__Result 1__: The solution $x = s$, $y = 0$ of equations (2) is asymptotically stable if

(i) $yh(x,y) < 0$ for all (x,y) for which $y \neq 0$;

(ii) $(x-s)[f(x) + g(x)] < 0$ for all $x \neq s$.

__Result 2__: If, in addition to the conditions of result 1, we have

(iii) $xf(x) < k < 0$ for some k and for $|x|$ large;

then the solution $(s,0)$ is globally asymptotically stable.

__Proof__: We use the standard theorems (see Hahn (1967), theorems 26.2 and 26.3) with an implicit change of origin to $(s,0)$. "Let V be a positive definite function with a negative semi-definite derivative. Suppose it is known that apart from the singular point no positive half-trajectory lies entirely in the region $V' = 0$. Then the singular point is asymptotically stable."

"If, in addition, V is radially unbounded then the singular point is globally asymptotically stable."

We observe that $I(x) > 0$ for all x, so that condition (i) ensures that V' is negative semi-definite. Continuity of h, with (i), implies that $h(x,0) = 0$, so that the singular points of (2) are all of the form $(z,0)$ where z satisfies the equation

$$f(z) + g(z) = 0 .$$

Then condition (ii) and the continuity of f and g imply that $z = s$ is the only solution of this equation. It remains, for result 1, to show that V is positive definite. Clearly $V(s,0) = 0$ and $V(x,y) > V(x,0)$ for all x. Now

$$V_x(x,y) = -I(x)[f(x)(1+y^2)^{\frac{1}{2}} + g(x)]$$

and condition (ii) ensures that $(x-s)V_x(x,y) > 0$ for all x, so that $V(x,0) > V(s,0)$ and hence

$$V(x,y) > 0 \quad \text{for all } (x,y) \text{ except } (s,0).$$

Thus we deduce from the first of the quoted theorems that V is a Liapounov function for (2) and that $(s,0)$ is an asymptotically stable singular point for (2).

Result 2 follows immediately from the second quoted theorem since condition (iii) ensures that $V_x(x,y)$ is bounded away from zero and hence that $V(x,y)$ is radially unbounded.

Application: We use result 2 to show that the region of asymptotic stability of the singular point in the half-plane $r > 0$ of the equation

$$2r^2(1+r^2)r'' = r(1-3r^2)(1+r'^2) + nr^2\sqrt{(1+r'^2)} - mr' \qquad (10)$$

is the whole of that half-plane. The equation is an extension of one discussed in Pearson & Petrie (1970) and gives the shape of a tubular bubble acted on by an internal pressure, an axial force, surface tension and viscous forces. ($r(t)$ is the dimensionless bubble radius expressed as a function of the dimensionless axial distance, t .)

We write $x = \log r$, $y = r'$ giving a one-one mapping between the half-plane $r > 0$ and the whole xy-plane. The equation corresponding to (3) is

$$\frac{y}{\sqrt{(1+y^2)}}\frac{dy}{dx} = \frac{(1-3e^{2x})\sqrt{(1+y^2)}}{2(1+e^{2x})} + \frac{ne^x}{2(1+e^{2x})} - \frac{my}{2e^x(1+e^{2x})\sqrt{(1+y^2)}}$$

and conditions (i) and (iii) readily seen to be satisfied. The singular point is at (s,0), where

$$s = \log[(n + \sqrt{(n^2+12)})/6]$$

and

$$f(x) + g(x) = (1 + ne^x - 3e^{2x})/2(1 + e^{2x})$$

which is positive for $x > s$ and negative for $x < s$, giving us condition (ii). This completes the demonstration of the stated result.

An application for which we obtain a region of asymptotic stability rather than a proof of global stability is the similar equation (arising from the same physical problem)

$$2r^2(r^2-1)r'' = -r(1+3r^2)(1+r'^2) + nr^2\sqrt{(1+r'^2)} - mr' \quad . \qquad (11)$$

Here there is a singular point in $r > 1$ if $n > 4$, namely (s,0) where

$$s = [n + \sqrt{(n^2-12)}]/6 \quad .$$

Equation (5) now gives us the Liapounov function

$$V(x,y) = (x^2-1)\sqrt{[(1+y^2)/x]} - n\sqrt{x} - (s^2-1)/\sqrt{s} + n\sqrt{s}$$

with derivative (for $x \neq 1$)

$$V' = -\tfrac{1}{2}my^2/\sqrt{[(1+y^2)x^5]} \quad .$$

The region of asymptotic stability we obtain is bounded by the largest closed curve $V(x,y) = \text{constant}$ enclosing the singular point $(s,0)$, namely the curve

$$V(x,y) = -n + (1+ns-s^2)/\sqrt{s} \quad .$$

This meets $x = 1$ at $y^2 = n^2/16 - 1$, and we remark that the region of asymptotic stability so obtained is not best possible. As $x \to 1+$ the region of asymptotic stability is bounded by positive y values tending to the positive root of

$$4(1+y^2) - n\sqrt{(1+y^2)} - my = 0$$

since there is a separatrix which tends to this point, and our region of asymptotic stability is bounded by a curve tending to the smaller value, $\sqrt{(n^2/16 - 1)}$.

References

W. Hahn (1967) "Stability of Motion" , Springer, Berlin.

J.R.A.Pearson & C.J.S.Petrie (1970) "The flow of a tubular film. Part 2: Interpretation of the model and discussion of solutions." , J.Fluid Mech., 42, 609-625.

Boundary Values of Holomorphic Functions and the Iterated Wave Equation

J.W. de Roever

1. Introduction

In this lecture we construct the fundamental solution D(t,x) of the following hyperbolic differential equation

$$(1)\quad \left(\frac{1}{c_1^2}\frac{\partial^2}{\partial t^2} - \Delta\right)\left(\frac{1}{c_2^2}\frac{\partial^2}{\partial t^2} - \Delta\right) D = 0 \quad , \quad t > 0$$

with the initial values

$$(2)\quad \lim_{t\downarrow 0} \frac{\partial^k D}{\partial t^k}(t,x) = \begin{cases} 0 & \text{for } k = 0,\ 1,\ 2 \\ \delta(x) & \text{for } k = 3 \end{cases}$$

Here Δ denotes $\frac{\partial^2}{\partial x_1^2} + \dots + \frac{\partial^2}{\partial x_n^2}$ and x the space variables $(x_1,\dots,x_n) \in \mathbb{R}^n$.

We abbriviate $\frac{\partial}{\partial t}$ by ∂ and $\lim_{t\downarrow 0} \partial^k D(t,x)$ by $\partial^k D(0,x)$.

It will turn out that D can be expressed as a linear combination of primitives with respect to t of the fundamental solutions $D_j^{\,1}(t,x)$ of the single wave equations

$$(3)\quad \left(\frac{1}{c_j^2}\partial^2 - \Delta\right) D_j^{\,1} = 0 \quad , \quad t > 0$$

with the initial values

$$(4)\quad D_j^{\,1}(0,x) = 0 \quad \text{and} \quad \partial D_j^{\,1}(0,x) = \delta(x) \qquad (j=1,2).$$

$D_j^{\,1}$ is a distribution in $D'(\mathbb{R}^n)$ or $S'(\mathbb{R}^n)$ depending on a parameter t. Therefore we first consider a way of calculating primitives of distributions. Since some primitives we need can be written as convolutions, we discuss a method of performing convolutions on distributions too. We use a representation of distributions $f \in D'$ or S' as differences of boundary values of analytic functions. The method discussed in this lecture can be extended to iterated wave equations consisting of more than two wave operators.

2. Integration of distributions

As is well known the restriction of an arbitrary distribution in D' or S' to a compact domain can be expressed as differences of boundary values of holomorphic functions (see lit.[1], pag.153 and 155). In particular a solution f of the wave equation can be expressed as the difference of the boundary values of only two holomorphic functions. See for this lit.[3] pag.268 and 270; there it is shown that in any compact domain we have

$$f(t,x) = \lim_{\varepsilon \downarrow 0} \{f^{+}(t+i\varepsilon,x) - f^{-}(t-i\varepsilon,x)\}$$

with f^{+} and f^{-} functions holomorphic in respectively $\mathbb{R}^{n+1} + iC$ and $\mathbb{R}^{n+1} - iC$, where C is the cone $\{(y_0,y_1,\dots,y_n) \mid y_0 > 0,\ c_j^2\, y_0^2 - y_1^2 - \dots - y_n^2 > 0\}$. Here we have to consider f (and in particular D_j^{1}) as distributions in $D'(\mathbb{R}^{n+1})$ or $S'(\mathbb{R}^{n+1})$ acting on testfunctions of the variables (t,x).

Convolutions of distributions may be computed by performing them on the pertinent holomorphic functions, being classical integrals and by taking the boundary values after the integrations have been performed.

Let $f(x) = \lim_{\varepsilon \downarrow 0} \{f^{+}(x+i\varepsilon) - f^{-}(x-i\varepsilon)\}$ be a distribution in D' or S'.

Then if g is a regular distribution such that $f * g$ exists, we can calculate for each testfunction ϕ $\langle f*g,\phi\rangle$ as follows (notation $f_\varepsilon^{\pm}(x) = f^{\pm}(x \pm i\varepsilon)$) $\langle f*g,\phi\rangle =$
$= \langle f(x), \langle g(\xi),\phi(x+\xi)\rangle\rangle = \lim_{\varepsilon \downarrow 0} \langle f^{+}(x+i\varepsilon) - f^{-}(x-i\varepsilon), \langle g(\xi), \phi(x+\xi)\rangle\rangle =$
$= \lim_{\varepsilon \downarrow 0} \langle\{f_\varepsilon^{+} - f_\varepsilon^{-}\} * g, \phi\rangle$. Hence

$$(5) \qquad f * g = \lim_{\varepsilon \downarrow 0} (f_\varepsilon^{+} * g - f_\varepsilon^{-} * g) .$$

It may happen that in spite of the existence of $f * g$ the convolution $f_\varepsilon^{\pm} * g$ does not exist; e.g. for n = 1 $f(x) = \delta(x)$, $g(x) = 1$ and $f_\varepsilon^{\pm} = -\frac{1}{2\pi i}\,\frac{1}{x \pm i\varepsilon}$.

In such a case we have to take

$$f * g = \lim_{n\to\infty} \lim_{\varepsilon \downarrow 0} \{(e_n f_\varepsilon^{+}) * g - (e_n f_\varepsilon^{-}) * g\}$$

with $e_n(x) \in C^\infty$, $e_n(x) \equiv 1$ for $|x| \le n$ and $e_n(x) \equiv 0$ for $|x| \ge 2n$.

Also a distribution F with $\frac{\partial}{\partial x_i} F = f$ can be found by classical integration; let $F^{\pm}(z) = \int^{x} f^{\pm}(\xi+iy)\,d\xi$ be primitives of $f^{\pm}$ respectively, then

$$(6) \qquad F(x) = \lim_{\varepsilon \downarrow 0} \{F^{+}(x+i\varepsilon) - F^{-}(x-i\varepsilon) .$$

3. The fundamental solution D(t,x)

3.1 The reduction of D to D_j^1

In order to construct the fundamental solution D we introduce the distributions $D_j^2(t,x)$ (j=1,2) satisfying the differential equations

$$(7) \qquad \left(\frac{1}{c_j^2}\partial^2 - \Delta\right) D_j^2 = 0 \quad , \quad t > 0$$

$$(8) \qquad \frac{1}{c_j^2}\partial^2 D_j^2 = D_j^1$$

and the conditions

$$(9) \qquad \partial^k D_1^2(0,x) = \partial^k D_2^2(0,x) \quad \text{for } k = 0,1$$

Then it is clear that the fundamental solution D can be written as

$$(10) \qquad D(t,x) = B_1 D_1^2(t,x) + B_2 D_2^2(t,x)$$

with $B_1 = \dfrac{1}{c_1^2 - c_2^2}$ and $B_2 = \dfrac{1}{c_2^2 - c_1^2}$

Hence the problem for constructing and calculating D consists in solving the system (7), (8), (9).

The elementary solutions $D_j^1(t,x)$ are given by (see lit.[3],pag.289 or[2],pag.234)

$$D_j^1(t,x) = \frac{1}{2c_j}\sigma(t)\,\theta(c_j^2t^2 - x^2) \quad , \quad n = 1$$

$$(11) \qquad D_j^1(t,x) = \frac{1}{2c_j}\pi^{-m}\sigma(t)\,\delta^{(m-1)}(c_j^2t^2 - |x|^2), \quad n = 2m+1 \ , \quad m \geq 1$$

$$D_j^1(t,x) = \frac{1}{2c_j}(-1)^{m-1}\pi^{-m-\frac{1}{2}}\Gamma(m-\tfrac{1}{2})\,\sigma(t)\,\theta(c_j^2t^2-|x|^2).\left[c_j^2t^2-|x|^2\right]^{-m+\frac{1}{2}}, \ n=2m$$

with $\sigma(t) = +1$ or -1 for respectively $t > 0$ and $t < 0$, $|x| = x_1^2 + x_2^2 + \ldots + x_n^2$.
θ denotes the Heaviside function and $\delta^{(m-1)}(c_j^2t^2-|x|^2)$ is the (m-1)-derivative of the distribution $\delta(c_j^2t^2-|x|^2)$, which is concentrated on the cone $c_j^2t^2 - |x|^2 = 0$.
For the precise definition of $\delta^{(k)}(c_j^2t^2-|x|^2)$ (see lit.[3],pag.284 or [2] pag.228).
The representation of D_j^1 by holomorphic functions is (see lit.[3], pag.284)

$$D_j^1(t,x) = \lim_{\varepsilon \downarrow 0}\,[D_j^+(t+i\varepsilon,x) - D_j^-(t-i\varepsilon,x)] \quad \text{with}$$

$$(12)\qquad D_j^{\pm}(t\pm i\varepsilon,x) = \frac{1}{4\pi^{(n+1)/2} i c_j}\,\Gamma\left(\frac{n-1}{2}\right)\frac{1}{\{|x|^2-c_j^{\,2}(t\pm i\varepsilon)^2\}^{(n-1)/2}}$$

It seems that a solution of the system (7), (8) is readily obtained by

$$(13)\qquad \overline{D}_j^{\,2} = c_j^{\,2} t_+ \underset{t}{*} D_j^{\,1}$$

where the convolution is taken with respect to t and $t_+ = t\theta(t)$.

However the convolution (13) exists only for $n \geq 4$.

3.2. The case $n \geq 4$

If $n \geq 4$, we may write according to (5)

$$(14)\qquad \overline{D}_j^{\,2}(t,x) = \lim_{\varepsilon\downarrow 0} c_j^{\,2}\, t_+ \underset{t}{*} \{D_j^{\pm}(t+i\varepsilon,x) - D_j^{-}(t-i\varepsilon,x)\}\ .$$

These convolutions are now classical integrals and they can easily be computed.
In order to satisfy condition (9) we take for $D_j^{\,2}(t,x)$ the distribution $D_j^{\,2}(t,x) = \overline{D}_j^{\,2}(t,x) - \overline{D}_j^{\,2}(0,x)$. Condition (9) is now clearly satisfied for $k = 0$.
Using (12) and (14) we see that $D_j^{\,2}(t,x)$ is of the form $\frac{1}{c_j} F(c_j t,x)$ and hence $\partial D_j^{\,2}(0,x)$ is independent of c_j.
Finally the fundamental solution D is given by (10).

3.3. The case $n = 1, 2, 3$

Because the convolution (13) does not exist for $n = 1, 2, 3$ we have to find another way of solving the system (7), (8), (9).
It is not difficult to show that for any solution $\overline{D}_j^{\,2}$ of equation (8) the relations

$$(7)\qquad \left(\frac{1}{c_j^{\,2}}\partial^2-\Delta\right)\overline{D}_j^{\,2} = 0 \quad \text{and}$$

$$(15)\qquad \Delta\partial^k\,\overline{D}_j^{\,2}(0,x) = \partial^k D_j^{\,1}(0,x)\quad (k=0,1)\qquad \text{are equivalent.}$$

Let $\overline{D}_j^{\,2}$ be an arbitrary second order primitive with respect to t of $D_j^{\,1}$, which can be obtained according to (6), then we take for $D_j^{\,2}$ the expression

$$(16)\qquad D_j^{\,2}(t,x) = \overline{D}_j^{\,2}(t,x) - \overline{D}_j^{\,2}(0,x) - t.\partial\,\overline{D}_j^{\,2}(0,x) + t.f_j(x)$$

with $f_j(x)$ a solution of the equation $\Delta f = \delta(x)$.
According to (15) and (4) the distribution $D_j^{\,2}$ satisfies besides equation (8) also equation (7). In order to satisfy (9) we take $f_1(x) = f_2(x)$ and hence (16) yields the distribution $D_j^{\,2}$ we looked for.
For $n = 1, 2$ the distributions $D_j^{\,1}$ are integrable functions (see (11)) and we find

in these cases $D_j^2(t,x) = c_j^2 \int_0^t \{ \int_0^\tau D_j^1(\sigma,x)\, d\sigma\} d\tau + t\, f(x)$

with $f(x) = x\theta(x)$ for $n = 1$ and $f(x) = \frac{1}{2\pi} \ln|x|$ for $n = 2$.

The case $n = 3$ is the most interesting case.

For $\overline{D}_j^2$ we take now according to (6) the distribution

$$\overline{D}_j^2(t,x) = \lim_{\varepsilon \downarrow 0} c_j^2 \int^t [\int^\tau \{D_j^+(\sigma+i\varepsilon,x) - D_j^-(\sigma-i\varepsilon,x)\}\, d\sigma]\, d\tau$$

Applying (12) we obtain after a simple calculation

$$c_j^2 \int^t [\int^\tau D_j^{\pm}(\sigma \pm i\varepsilon,x)\, d\sigma]\, d\tau = \frac{1}{8\pi^2\, i\, c_j} \ln(-c_j^2 z^2 + |x|^2) + \frac{c_j z}{|x|} \ln \frac{c_j z + |x|}{c_j z - |x|}$$

with $z = t \pm i\varepsilon$ and hence

$$(17) \quad \overline{D}_j^2(t,x) = \frac{-1}{4\pi} \{ \frac{\sigma(t)}{c_j} \theta(c_j^2 t^2 - |x|^2) + \frac{t}{|x|} \theta(-c_j^2 t^2 + |x|^2 \} .$$

It is easily seen that $\overline{D}_j^2$ has the initial values $\overline{D}_j^2(0,x) = 0$ and $\partial\, \overline{D}_j^2(0,x) = \frac{-1}{4\pi} \frac{1}{|x|}$,

so (17) also gives us the distribution $D_j^2(t,x)$, since $f_j(x) = \frac{-1}{4\pi} \frac{1}{|x|}$.

Finally we obtain the fundamental solution D from (10) and after a rearrangement of the terms we get

$$D(t,x) = \frac{1}{4\pi} \frac{1}{c_1^2 - c_2^2} \Big[\sigma(t) \{\frac{1}{c_1} - \frac{1}{c_2}\} \theta(c_2^2 t^2 - |x|^2) + \{\frac{t}{|x|} - \frac{\sigma(t)}{c_1}\} \{\theta(c_1^2 t^2 - |x|^2) - \theta(c_2^2 t^2 - |x|^2)\} \Big]$$

if we take $c_1 > c_2$. The first term has its support in the inner cone $\{(t,x) \mid c_2^2 t^2 \geq |x|^2\}$, while the second term has its support between this cone and the outer cone $\{(t,x) \mid c_1^2 t^2 \geq |x|^2\}$.

Literature

[1] H. Bremermann : Distributions, complex variables and Fouriertransforms.

[2] J.M. Gelfland and G.E. Shilov : Generalized functions, Volume I.
(translated from Russian)

[3] V.S. Vladimirov : Les fonctions de plusieurs variables complexes.
(translated from Russian).

Détermination, à l'Aide d'Index, des Solutions Périodiques de l'Equation de Duffing $x''+2x^3 = \lambda \cos t$

B.V. Schmitt

On développe ici un exemple de recherche numérique de solutions périodiques d'un système différentiel à coefficients T-périodiques, d'ordre deux, du type

$$x' = f(t, x, \lambda) \qquad (f)$$

par la méthode des index exposée en détail dans [1]. L'équation en question est l'équation de Duffing sans dissipation

$$x'' + 2x^3 = \lambda \cos t \qquad (d_\lambda)$$

dont l'étude a été amorcée dans [1].

1. Index en un point, de l'équation (f)

Soit l'équation (f), du deuxième ordre, à coefficients T-périodiques, dont on désire rechercher les solutions périodiques harmoniques, ou sous-harmoniques d'ordre $1/k$ (donc de période minimale kT).

Soit $x(t, t_o, x^o, \lambda)$ la solution, supposée unique, continue, définie pour tout t, issue de x^o au temps t_o ; soit $T^k_{t_o}$ l'application $E \times \Lambda \rightarrow E$ définie par

$$T^k_{t_o}(x^o, \lambda) = x(t_o + kT, t_o, x^o, \lambda) .$$

La courbe fermée $T^k_{t_o}(x^o, \lambda)$, $t_o \in [0, T]$, est dite courbe des kT-différences de (f) au point (x^o, λ) ; l'ensemble $N_k(f) \subset E \times \Lambda$ $(x^o \in E, \lambda \in \Lambda)$ des points tels que $T^k_{t_o}(x_o,\lambda) \ni x^o$ est dit ensemble des <u>points</u> k-<u>exceptionnels</u> de (f). C'est l'ensemble des points de $E \times \Lambda$ desquels est issue au moins une solution kT-périodique. <u>L'index</u> $i_k(x^o,\lambda)$ défini pour $(x^o,\lambda) \notin N_k(f)$, est l'ordre de $T^k_{t_o}(x^o,\lambda)$ relativement à x^o. La recherche des points exceptionnels est facilitée par le théorème suivant :

THEOREME. - <u>L'index</u> $i_k(x^o,\lambda)$ <u>est constant dans toute composante connexe de</u> $E \times \Lambda - N_k(f)$.

Le type elliptique ou hyperbolique des points exceptionnels ainsi obtenus peut alors être déterminé à l'aide du théorème de Seifert [1].

2. Détermination numérique de la courbe des kT - différences

Ce travail numérique a été effecuté à l'aide d'un calculateur IBM 1800 associé à un ensemble d'unités de dialogue ; on a ainsi pu visualiser sur écran les courbes $T^k_{t_o}(x^o)$, intervenir sur machine à écrire à clavier pour modifier au besoin les échelles et les paramètres, enfin dessiner sur papier, à l'échelle désirée, les courbes obtenues, pour stocker les informations. Ce matériel a été décrit en détail dans [2].

3. Les résultats

Ils ont été consignés dans la figure 1 (solutions harmoniques issues de $x \in [0;3]$ avec $\lambda \in [0;15]$) et dans la figure 2 (solutions sous-harmoniques d'ordre $\frac{1}{2}$ issues du point $a_{1/2}$). On lit ces figures de la manière suivante : pour trouver les abscisses x ces points k-exceptionnels (k=1 ou 2) de l'équation (d_λ) (correspondant à des solutions $2k\pi$-périodiques issues du point x avec une vitesse $x' = 0$) on trace la droite d'ordonnée λ, qui coupe les séparatrices aux points exceptionnels d'abscisse x cherchés. En faisant varier λ on peut suivre l'évolution de ces points exceptionnels. Les points exceptionnels situés sur des lignes continues sont des points elliptiques, ceux situés sur des lignes en pointillés, sont des points hyperboliques (donc instables).

a) Solutions harmoniques (figure 1)

Les solutions $\frac{2\pi}{\ell}$- périodiques, $\ell \in \mathbb{N}$, donc 2π- périodiques de l'équation autonome $x'' + 2x^3 = 0$, sont issues des points a_ℓ d'abscisse $\ell\alpha$ où α est voisin de 0,834 . L'évolution des points exceptionnels issus des points a_ℓ $(\ell = 1, 2, 3)$ est donnée figure 1 . Les séparatrices doublées au départ des points $a_\ell(\ell = 2, 3)$ sont doubles, c'est-à-dire qu'il existe deux solutions périodiques distinctes issues de ces points, à des instants différents ; quelsques-uns des index calculés ont été placés à titre d'exemple dans les régions connexes correspondantes.

figure 1

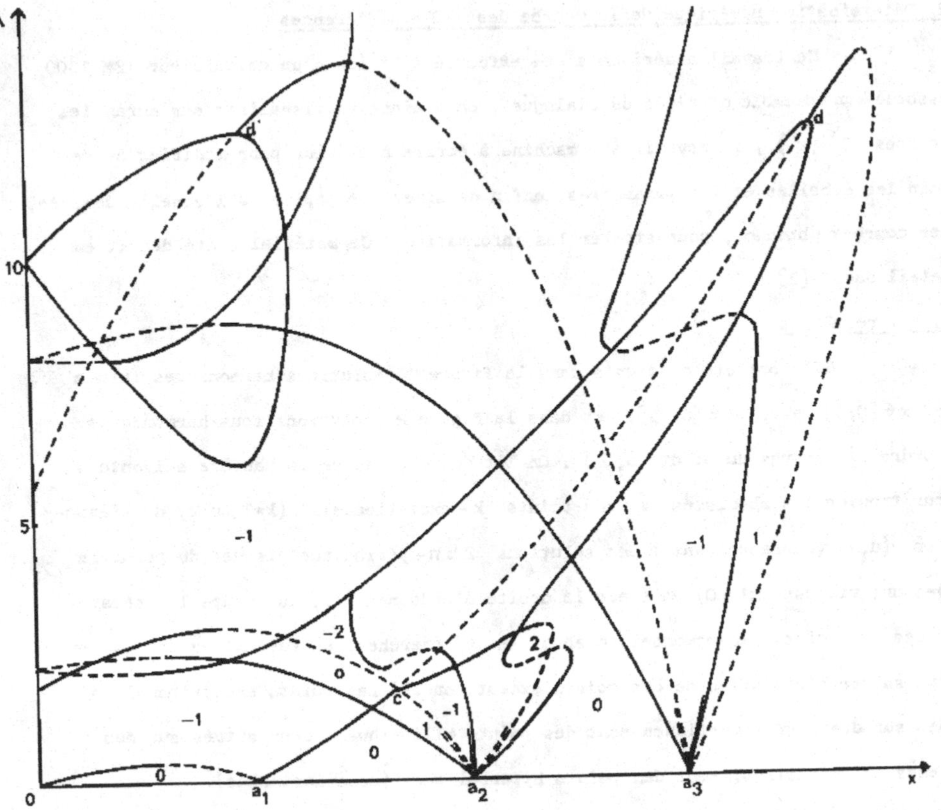

Remarques

Les séparatrices issues des points O (respectivement a_1 et a_3) correspondent à des solutions périodiques harmoniques, au nombre de 1 (respectivement 2 et 2) ; elles ne contiennent que des harmoniques impaires. Par contre les solutions périodiques (au nombre de 4) issues de a_2 contiennent des harmoniques paires. Les points c et d sont des bifurcations ; de ces points sont issues une famille de solutions harmoniques impaires (a_1 c d a_3) et deux familles de solutions contenant des harmoniques paires (c a_2 , famille double, et d a_2) ; le passage continu de c (ou d) à a_2 réalise un passage continu d'une solution n'ayant que des harmoniques impaires (en c et d) à une solution n'ayant que des harmoniques paires (en a_2) .

b. Solutions sous-harmoniques d'ordre $\frac{1}{2}$:(figure 2)

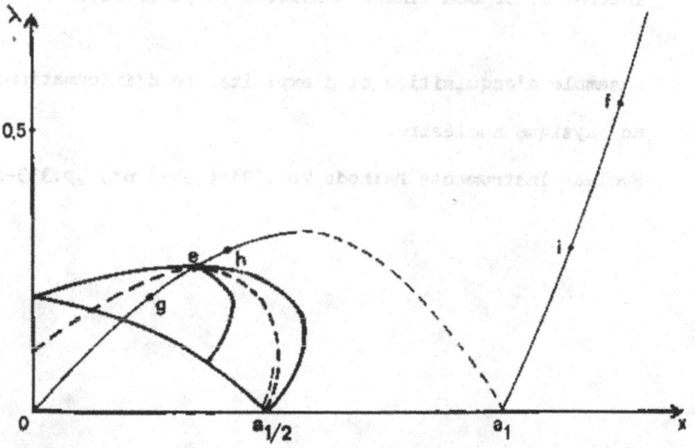

Les seules familles représentées (au nombre de 4) sont celles issues du point $a_{1/2}$, d'abscisse $\alpha/2$. Elles comprennent des harmoniques d'ordre $\frac{1}{2}$, paires. Ces quatre familles peuvent être obrenues par bifurcation au point e($\lambda \simeq 0{,}26$; $x \simeq 0{,}29$) de la famille de solutions harmoniques joignant O à a_1 (représentée en traits fins). De même, le point f est un point de bifurcation de 2 famillesde solutions sous-harmoniques d'ordre 1/2 , allant de f à $a_{3/2}$. Les points g , h et i sont des points de bifurcation de solutions harmoniques d'ordre 1/3 , allant de g à $a_{1/3}$, de h à $a_{2/3}$ et de i à $a_{4/3}$.

Je tiens à remercier M. WERY et son équipe de Laboratoire des Basses Energies (Centre de Recherches Nucléaires de Strasbourg) pour la possibilité qu'ils m'offrent d'utiliser l'unité de dialogue associée à ordinateur qu'ils ont mise au moint, et en particulier B. Bueb, qui a bien voulu l'adapter à mon problème mathématique.

BIBLIOGRAPHIE

[1] SCHMITT B.V. — Sur l'existence et la localisation de solutions périodiques de systèmes différentiels à coefficients périodiques. Application à l'équation de Duffing. Intern. J. of non linear Mechanics (A paraître).

[2] WERY M., C. RING, D. BUEB et P. WITTMER — Ensemble d'acquisition et d'exploitation d'informations en physique nucléaire. Nuclear Instruments Methods Vol. 91 (1971) n°3 pp.333-339.

Periodic Solutions of Nonlinear Differential Systems

K. Schmitt

1. In this paper we present some applications of topological degree theory (the Brouwer degree) to the study of periodic solutions of both "small" and "large" periods of nonlinear differential systems. The results presented in sections 2 and 3 of the paper extend in various directions a result of Seifert [11] and results of the author [9]. In the last part of the paper, we illustrate the applicability of our method by proving a theorem for a class of differential equations which has been studied extensively by various authors, in particular in [1-4] and [6-8]. For the sake of brevity and clarity we shall not establish the theorem in its greatest generality but consider only a very special case.

2. Let I denote the compact interval $[0,T]$, $R = (-\infty,\infty)$ and let $f : I \times R^{nm} \to R^m$ be continuous. We consider the differential system

$$(1) \qquad x^{(n)} + f(t,x,x',\ldots,x^{(n-1)}) = 0 \ ,$$

and give conditions which ensure that (1) have solutions satisfying the periodic boundary conditions

$$(2) \qquad x^{(i)}(0) = x^{(i)}(\omega) \ , \ i = 0,\ldots,n-1 \ .$$

Such a solution will henceforth be called ω-periodic. Further, we say that (1) has periodic solutions of small period in case there exists $\omega_o > 0$ such that for every ω, $0 < \omega \leq \omega_o$, (1) has a solution of period ω. This notion of periodicity does not conflict with the usual terminology since such a solution may be extended to be a periodic solution in the usual sense whenever its period is equal to the period of f. The main result of the paper is the following theorem.

Theorem 1. Let there exist a nonempty bounded open set $A \subset R^m$ such that $f(0,x,0,\ldots,0) \neq 0$ for $x \in \partial A$ (the boundary of A) and let $\deg(f(0,x,0,\ldots,0),A,0) \neq 0$. Then (1) has periodic solutions of small periods.

The proof of Theorem 1 is somewhat lengthy and will not be given in its entirety here. (see [10] for a complete proof). We will only discuss that part of the proof which is important in obtaining estimates on the magnitude of possible periods of solutions.

Let $N > 0$, $k > 1$, be given and let $B \subset R^m$ be a bounded open set such that $\bar{A} \subset B$. Let Ω and Σ be defined as ($\|\cdot\|$ is the norm induced by the usual inner product in R^m)

$$\Omega = \{(x,x',\ldots,x^{(n-1)}) : x \in A, \|x^{(i)}\| < N, i = 1,\ldots,n-1\}$$

$$\Sigma = \{(x,x',\ldots,x^{(n-1)}) : x \in B, \|x^{(i)}\| < kN, i = 1,\ldots,n-1\}$$

Lemma. There exists $\omega_o > 0$, $0 < \omega_o \leq T$ such that any solution trajectory $(x(t), x'(t),\ldots,x^{(n-1)}(t))$ which starts in $\partial\Omega$ stays in Σ and cannot return to its initial value for $t \in (0,\omega_o]$.

The number ω_o, whose existence follows from the lemma, depends on the arbitrarily chosen constants N and k and the region B, thus in applying Theorem 1 and its lemma, one would vary N, k, and B in order to maximize ω_o; this will be illustrated later.

3. In this section, we consider some applications of Theorem 1 to obtain the existence of solutions of small periods.

Corollary 1. Let $f : I \times R^{nm} \to R^m$ be continuous and let there exist a constant $r > 0$ such that

(3) $$x \cdot f(0,x,0,\ldots,0) > 0 \quad , \quad \|x\| = r$$

or

(4) $$x \cdot f(0,x,0,\ldots,0) < 0 \quad , \quad \|x\| = r \ .$$

Then (1) has periodic solutions of small periods.

Proof. Let $A = \{x \in R^m : \|x\| < r\}$. Then either (3) or (4) implies that $\deg(f(0,x,0,\ldots,0),A,0)$ is defined. Further, either condition implies that the vector fields $f(0,x,0,\ldots,0)$ and $f(0,-x,0,\ldots,0)$ do not have the same directions for $x \in \partial A$ which in turn means that $f(0,x,0,\ldots,0)$ is homotopic to an odd vector field and hence $\deg(f(0,x,0,\ldots,0),A,0) \neq 0$.

Corollary 2. Consider the differential equation

(5) $$x^{(n)} + h(t,x,x',\ldots,x^{(n-1)}) + g(x) = p(t) \ ,$$

where $h : I \times R^n \to R$, $g : R \to R$, $p : I \to R$ are continuous and are such that $h(0,x,0,\ldots,0) = 0$, $xg(x) > 0$ or $xg(x) < 0$, $x \neq 0$, and $|g(x)| \to \infty$ as $|x| \to \infty$. Then (5) has periodic solutions of small period.

Proof. The conditions on g imply the existence of a constant $r > 0$ so that either

$$x(g(x) - p(0)) > 0 \quad , \quad |x| = r$$

or

$$x(g(x) - p(0)) < 0 \quad , \quad |x| = r \ ;$$

one hence may apply the previous corollary.

Remark. Taking $h \equiv 0$, $n = 2$, one obtains an extension of the result of Seifert [11]. Also if $n = 2$ and $h = k(x,x')$ one obtains the existence of periodic solutions of small periods of the forced generalized Liènard equation under extremely weak assumptions on k and g . Other variations on the theme of Corollary 1 yield results similar to the results in [9]. It is to be emphasized here that no local Lipschitz conditions nor uniqueness assumptions for initial value problems need to be made here, which were needed in previous work, see eg. [9], [11].

4. Let us consider now the system

$$x'' + f(t,x) = 0 , \tag{6}$$

where $f : I \times R^m \to R^m$ is continuous and satisfies the condition

$$x \cdot f(t,x) < 0 , \; 0 \leq t \leq T , \; \|x\| = r \tag{7}$$

where r is a positive constant. Under these conditions, (6) will have a T-periodic solution. Note that if we let $A = \{x : \|x\| < r\}$ then by (7) $\deg(f(0,x),A,0) \neq 0$, hence we only need to show that the constant ω_o of the lemma may be chosen to be T . To show this we make the following observations.

Lemma 1. If $x(t)$ is a solution of (6) such that $\|x(t_1)\| \leq r$, where $0 \leq t_1 < t_2 \leq T$, then $\|x(t)\| \leq r$, $t_1 \leq t \leq t_2$.

Lemma 2. There exists $t_o \in (0,T]$ and a constant $N = N(t_o)$ such that every solution $x(t)$ of (7) satisfying $\|x(0)\| \leq r$ and $\|x'(0)\| \geq N$ is such that $\|x(t_o)\| > r$ and $x(t) \neq x(0)$, $0 < t \leq t_o$.

The proof of Lemma 1 follows easily from (7), whereas Lemma 2 follows from the fact that equation (6) has the property that a uniform bound on solutions implies a uniform bound on their derivatives.

Thus we choose $\Omega = \{(x,x') : \|x\| < r , \|x'\| < N\}$.

To show that $\omega_o = T$ it suffices to show that if $(x(0),x'(0)) \in \partial\Omega$ is the initial point of the trajectory $(x(t),x'(t))$, then as long as $0 < t \leq T$, $(x(t),x'(t)) \neq (x(0),x'(0))$. In case $\|x'(0)\| = N$, this easily follows by applying Lemma 2 and then Lemma 1. In case $\|x(0)\| = r$ and $\|x'(0)\| < N$ we argue indirectly, use Lemma 1 and then (7) to obtain a contradiction.

Remark. Using a more detailed analysis, one may obtain similar results for more general equations. We refer the interested reader to [1], [4], [6], [8] where equations similar to (6) are studied.

REFERENCES

1. Bebernes, J. W. and K. Schmitt, "Periodic boundary value problems for systems of second order differential equations," (submitted for publication).

2. Hartman, P., "On a boundary value problem for systems of ordinary nonlinear, second order differential equations," Trans. Amer. Math. Soc. 96(1960), pp. 493-509.

3. Knobloch, H. W., "Eine neue Methode zur Approximation periodischer Lösungen nichtlinearer Differential-gleichungen zweiter Ordnung," Math. Z. 82(1963), pp. 177-197.

4. Knobloch, H. W., "On the existence of periodic solutions of second order vector differential equations," J. Differential Equations 9(1971), pp. 67-85.

5. Krasnosel'skii, M. A., Translation Along Trajectories of Differential Equations, Transl. of Russian Monograph Ser. Vol. 19, Amer. Math. Soc., Providence, Rhode Island, 1968.

6. Mawhin, J., "Existence of periodic solutions for higher order differential systems that are not of class D," J. Differential Equations 8(1970), pp. 523-530.

7. Schmitt, K., "Periodic solutions of nonlinear second order differential equations," Math. Z. 98(1967), pp. 200-207.

8. Schmitt, K., "Periodic solutions of systems of second order differential equations," J. Differential Equations 11(1972), pp. 180-192

9. Schmitt, K., "A note on periodic solutions of second order ordinary differential equations," SIAM J. Appl. Math. 21(1971), pp. 491-494.

10. Schmitt, K., "Periodic solutions of small period of systems of nth order differential equations," (submitted for publication).

11. Seifert, G., "A note on periodic solutions of second order differential equations without damping, " Proc. Amer. Math. Soc. 10(1959), pp. 396-398.

Bifurcation Theory for Equations with Nonlocal Potentials

C.A. Stuart

Let B denote a real Hilbert space and let $\mathbb{R}$ denote the real numbers. Set $E = B \times \mathbb{R}$. A connected subset C of E is called a continuum if every closed bounded subset of C is compact.

Recently, P. H. Rabinowitz [1] has investigated the existence and behaviour of continua of solutions of equations of the form, $u = \lambda G(u)$, where $G : B \to B$ is a completely continuous operator. Although this theory yields a detailed bifurcation theory for nonlinear eigenvalue problems associated with ordinary and partial differential operators on bounded domains, it is generally not applicable to the corresponding problems on unbounded domains, since the relevant inverse operator is not completely continuous.

Here we consider the set of solutions of the equation,

$$u = \lambda G(u) \equiv \lambda\{ A(u) + H(u) \} \quad \text{for} \quad (u,\lambda) \in E , \tag{1}$$

where we assume that the following conditions are satisfied.

H1) Suppose that $A : B \to B$ is a bounded self-adjoint operator with $r_e(A) < 1$, where $r_e(A)$ denotes the radius of the essential spectrum of A.

H2) Suppose that $H : B \to B$ is a completely continuous operator such that

$$\|H(u)\| / \|u\| \to 0 \text{ as } \|u\| \to 0.$$

A point $(0,\lambda) \in E$ is called a trivial solution of (1). Let S denote the set of non-trivial solutions of (1) and let $S_1 = S \cup \{(0, \lambda) : \lambda \text{ is a characteristic value of } A\}$.

In [2] , the abstract theory of P. H. Rabinowitz was extended to the case where $G : B \to B$ is a k-set contraction for some $k < 1$. It was also observed ([2]) that a self-adjoint adjoint operator is an r_e-set contraction. Hence Theorem 1. 4 of [2] has, as a special case, the following result about the problem (1), under the hypotheses H1) and H2).

Theorem 1 Suppose that μ is a characteristic value of odd multiplicity of A, and that $|\mu| r_e(A) < 1$. Let C_μ denote the component of S_1 which contains $(0,\mu)$.

Then C_μ satisfies at least one of the following conditions.

(i) C_μ is an unbounded subset of E.

(ii) C_μ meets $(0,\hat{\mu}) \in S_1$, where $\hat{\mu} \neq \mu$.

(iii) $\inf\{ |\lambda \pm a| : (u,\lambda) \in C_\mu \} = 0$, where $a = [r_e(A)]^{-1}$.

We shall apply Theorem 1 to the following problem. To be brief, we consider only the one-dimensional case; the corresponding results for higher dimensions can easily be constructed. Consider the problem,

$$\begin{aligned} -(p(x)u'(x))' + q(x)u(x) - \int_0^\infty k(x,y)f(y,u(y))u(y)dy &= \lambda u(x) \\ \text{for } x \in (0,\infty), & \\ u(0) = 0, & \end{aligned} \tag{2}$$

where $'$ denotes differentiation with respect to x. Associated with (2), we have the following linear problem,

$$\begin{aligned} Lu(x) \equiv -(p(x)u'(x))' + q(x)u(x) &= \lambda u(x) \\ \text{for } x \in (0,\infty), & \\ u(0) = 0. & \end{aligned} \tag{3}$$

All functions in the present discussion are real-valued.

For $p \geq 1$, L^p will denote the real function space $L^p(0,\infty)$, with norm $\| \ \|_p$.

$H_0^2 = \{u \in L^2 : u', u'' \in L^2 \text{ and } u(0) = 0\}$

is a closed subspace of the space $W_2^2(0,\infty)$, with norm $\| \ \|$. The norm of a bounded linear operator from L^2 into L^2 will be denoted by $\| \ \|$.

We make the following assumptions about the functions p, q, f and k.

A1) Suppose that $p : [0,\infty) \to \mathbb{R}$ is continuous, and is continuously differentiable in $(0,\infty)$, with p' bounded. Suppose also that p is strictly positive, with $0 < P_1 \leq p(x) \leq P_2 < \infty$ for all $x \in [0,\infty)$.

A2) Suppose that $q : [0,\infty) \to \mathbb{R}$ is continuous, with $-\infty < Q_1 \leq q(x) \leq Q_2 < \infty$ for all $x \in [0,\infty)$.

A3) Suppose that $f : [0,\infty) \times \mathbb{R} \to \mathbb{R}$ is continuous, with $|f(x, \pm 1)| \leq C_1$ for all $x \in [0,\infty)$. Suppose also that there exists $\alpha > \frac{1}{2}$ such that

$$f(x, \pm t) = t^{\alpha} f(x, \pm 1) \text{ for all } t \geq 0 \text{ and } x \in [0, \infty).$$

A4) Suppose that $k : [0,\infty) \times [0,\infty) \to \mathbb{R}$ is measurable and that $\int_0^\infty\int_0^\infty |k(x,y)|^2 dxdy < \infty$.

<u>Remark</u> The hypotheses A1) and A2) are unnecessarily severe. It is only necessary to ensure that the following result is valid.

<u>Lemma 1</u> (a) The differential operator L has a unique self-adjoint extension $\tilde{L}$ in L^2, with $\mathfrak{D}(\tilde{L}) = H_0^2$ and $\sigma_e(\tilde{L}) \subset [Q,\infty)$ where $Q \equiv \lim_{x\to\infty} \inf q(x)$ and $\sigma_e(\tilde{L})$ denotes the essential spectrum of $\tilde{L}$.

(b) There exists a positive constant C_2 such that, for all $\xi \geq C_2$, the operator $(\tilde{L} + \xi)^{-1}$ is defined on all of L^2, and is a linear homeomorphism of L^2 onto H_0^2. ($(\tilde{L} + \xi)u = \tilde{L}u + \xi u$ for all $u \in H_0^2$.)

<u>Proof</u> (a) See [3].

(b) This follows from Theorem 24(v) and lemmas 23 and 25 of [4].

We shall also need the following lemma of Sobolev type, [4].

<u>Lemma 2</u> For $s \in [2,\infty)$, we have

$$H_0^2 \subset L^s \text{ and } \|u\|_s \leq d_s \,|||u||| \text{ for all } u \in H_0^2,$$

where d_s is a positive constant depending only on s.

In order to discuss the nonlinear term in (2), it is convenient to introduce the following notation.

Let $F(u)(x) = f(x,u(x))$,

$$J(u)(x) = f(x,u(x))u(x),$$

$$K(u)(x) = \int_0^\infty k(x,y)u(y)dy,$$

$$T(u)(x) = KJ(u)(x), \text{ for all } x \in [0,\infty).$$

<u>Lemma 3</u> The operator $T : H_0^2 \to L^2$ is completely continuous and $\| Tu \| / |||u||| \to 0$ as $|||u||| \to 0$.

<u>Proof</u> Since A3) implies that

$$|f(x,t)t| \leq C_1 |t|^{\alpha+1} \text{ for all } (x,t) \in [0,\infty) \times \mathbb{R},$$

it follows from Theorem 19. 1 of [5] that the operator $J : L^{2(\alpha+1)} \to L^2$ is continuous and bounded. Hence $T : L^{2(\alpha+1)} \to L^2$ is completely continuous, since $K : L^2 \to L^2$ is a compact linear operator. Appealing to Lemma 2, this shows that $T : H_0^2 \to L^2$ is completely continuous.

Using Theorem 19. 1 of [5] and the inequality,

$$|f(x,t)| \leq C_1 |t|^{\alpha} \text{ for all } (x,t) \in [0,\infty) \times \mathbb{R},$$

it follows that $F : L^{4\alpha} \to L^4$ is continuous and bounded. But

$$\| Tu \|_2 \leq \| K \| \ \| u \|_4 \ \| F(u) \|_4 \text{ for all } u \in H_0^2,$$

and so $0 \leq \| Tu \|_2 / |||u||| \leq d_4 \| K \| \ \| F(u) \|_4 \to 0$

as $|||u||| \to 0$, since $F(0) = 0$ and $F : H_0^2 \to L^4$ is continuous.

This establishes the lemma.

A point $(0,\lambda) \in H_0^2 \times \mathbb{R}$ is called a trivial solution of (2). Let $\mathcal{S}$ denote the set of non-trivial solutions of (2) in $H_0^2 \times \mathbb{R}$. Let $\mathcal{S}_1 = \mathcal{S} \cup \{(0,\lambda) : \lambda$ is an eigenvalue of $\tilde{L}\}$.

<u>Theorem 2</u> Suppose that μ is an eigenvalue of $\tilde{L}$ contained in the interval $(-\infty, Q)$. Let $\mathcal{C}_\mu$ denote the component of $\mathcal{S}_1$ which contains $(0,\mu)$.

Then $\mathcal{C}_\mu$ satisfies at least one of the following conditions.

(i) $\mathcal{C}_\mu$ is an unbounded subset of $H_0^2 \times \mathbb{R}$.

(ii) $\mathcal{C}_\mu$ meets $(0,\hat{\mu}) \in \mathcal{S}_1$, where $\hat{\mu} \neq \mu$.

(iii) $\inf\{Q - \lambda : (u,\lambda) \in \mathcal{C}_\mu\} = 0$.

<u>Proof</u> Choose $\xi > C_3 \equiv \max\{|\mu|, C_2, 1-Q\}$. Then consider the equation

$$u = \lambda G(u) \equiv \lambda\{A(u) + H(u)\} \text{ for } (u,\lambda) \in L^2 \times \mathbb{R}, \tag{4}$$

where $A = (\tilde{L} + \xi)^{-1}$ and $H = TA$.

From Lemmas 1 and 3, it follows that G satisfies the hypotheses of Theorem 1, with $B = L^2$ and $r_e(A) \leq (Q+\xi)^{-1} < 1$. Furthermore, $\mu+\xi$ is a characteristic value of multiplicity one of A, and

$$|\mu + \xi| r_e(A) = (\xi + \mu) r_e(A) < (\xi + Q)(Q + \xi)^{-1} = 1.$$

Let $C_{\mu+\xi}$ denote the continuum of solutions of (4) in $L^2 \times \mathbb{R}$, emanating from $(0, \mu+\xi)$, discussed in Theorem 1. Since $\mu+\xi$ is positive, and zero is not a characteristic value of A, it follows that $C_{\mu+\xi} \subset L^2 \times (0,\infty)$. Let

$$\mathscr{C}_\mu(\xi) = \{(\lambda^{1/\alpha} Au, \lambda - \xi) : (u,\lambda) \in C_{\mu+\xi}\}.$$

Using Lemma 1 (b), one checks that $\mathscr{C}_\mu = \bigcup_{\xi > C_3} \mathscr{C}_\mu(\xi)$

and thus that $\mathscr{C}_\mu$ has the properties stated.

<u>Related problems</u> 1. From the above proof, it is clear that a similar treatment of the problem,

$$Lu(x) = \lambda\{u(x) + \int_0^\infty k(x,y)f(y,u(y))u(y)dy\} \text{ for } x \in (0,\infty),$$

$$u(0) = 0,$$

is possible.

2. Replacing Lemma 3 by the corresponding result for the operator S, defined by $S(u)(x) = u(x).KF(u)(x)$ for $x \in [0,\infty)$, problems of the form,

$$Lu(x) - u(x)\int k(x,y)f(y,u(y))dx = \lambda u(x) \text{ for } x \in (0,\infty),$$

$$u(0) = 0,$$

can be studied by the above method. Equations of this form are studied in [6] , where the results are extended to include the Hartree equation.

3. In [2], similar methods are used to discuss the problems,

$$Lu(x) - f(x,u(x)) = \lambda u(x) \quad \text{for} \quad x \in (0,\infty),$$

$$u(0) = 0,$$

and $$Lu(x) = \lambda\{u(x) + f(x,u(x))\} \quad \text{for } x \in (0,\infty)$$

$$u(0) = 0.$$

References

[1.] Rabinowitz, P. H.,
Some global results for nonlinear eigenvalue problems, J. Functional Analysis, 7, 487-513, (1971).

[2.] Stuart, C. A.,
Some bifurcation theory for k-set contractions, (to appear).

[3.] Dunford, N., and J. T. Schwartz,
Linear Operators, Part II, Interscience, (1963).

[4.] Browder, F. E.,
On the spectral theory of elliptic differential operators. I, Math. Annalen, 142, 22-130, (1961).

[5.] Vainberg, M. M.,
Variational Methods for the Study of Nonlinear Operators, Holden-Day, (1964).

[6.] Stuart, C. A.,
Bifurcation theory for the Hartree equation, (to appear).

Multi-Parameter Eigenvalue Problems and k-linear Operators

B.D. Sleeman

§1. Introduction

Let H_r, $(r = 1, 2, \ldots, k)$ be separable Hilbert spaces and let $u_r, v_r \in H_r$. In H_r we define the scalar product by

$$(u_r, v_r)_r = \int_{a_r}^{b_r} u_r(x_r)\,\overline{v}_r(x_r)\,\rho_r(x_r)dx_r, \qquad (1)$$

where in general u_r, v_r are complex valued functions defined on H_r and ρ_r is a real valued and continuous function such that

$$\rho_r(x_r) > 0 \;\forall\; x_r \in [a_r, b_r].$$

In H_r, $(r = 1, 2, \ldots, k)$ we consider ordinary differential operators A_r in $D(A_r)$ defined by

$$A_r u_r = \frac{1}{\rho_r}\{-(p_r u_r')' + q_r(x_r)u_r\}, \qquad u_r \in D(A_r) \qquad (2)$$

with p_r, p_r', q_r real valued and continuous for all $x_r \in [a_r, b_r]$, and

$$D(A_r) = \{u_r \mid u_r \in C^2[a_r, b_r], U_r(u_r(a_r)) = V_r(u_r(b_r)) = 0\}. \qquad (3)$$

The quantities U_r, V_r introduced in (3) are used to denote homogeneous boundary conditions of Sturm-Liouville or periodic type.

By the multi-parameter eigenvalue problem for ordinary differential equations we mean the study of a system of k ordinary linear formally self-adjoint differential equations in k-parameters $\lambda_s (s = 2, \ldots, k, k \geq 2)$ for which solutions are sought in the k-domains $D(A_r)$, $(r = 1, \ldots, k)$. In its most general form the problem may be conveniently stated as

$$A_r y_r = \sum_{s=1}^{k} \lambda_s\, p_{rs}(x_r) y_r, \quad y_r \in D(A_r), \quad (r = 1, \ldots k) \qquad (4)$$

where $P(x_1, x_2, \ldots x_k) = \det\{p_{rs}(x_r)\}_{r,s=1}^{k} > 0 \ \forall \ \mathbf{x} \in I^k$,

(the cartesian product of the k-intervals $[a_r, b_r]$, $(r = 1, \ldots, k)$) and

$p_{rs}(x_r) \in C[a_r, b_r]$.

The problem posed by (4) is only now beginning to attract its due share of attention. There are several approaches to the problem and these have been discussed by Atkinson (1968) and Sleeman (1971a) where references to some of the most recent contributions may be found. Among the methods of attack which have proved fruitful we cite (i) the work of Faierman (1969) on completeness and expansion theorems in which the system (4) is replaced by its discrete analogue and (ii) the reduction of (4) to a system of one-parameter eigenvalue problems for partial differential operators, c.f. Sleeman (1971a). The latter approach has been exploited by the author and Arscott and his co-workers in a number of interesting particular cases.

In this paper we introduce a further approach which leads to some interesting problems in the theory of non-linear integral equations and touches on the important topic of bifurcation theory.

§2. Formulation in Tensor products of Hilbert spaces

We now form a tensor product of each Hilbert space $H_r (r = 1, 2, \ldots, k)$ and denote this product by

$$H = \prod_{s=1}^{k} \otimes H_s. \qquad (6)$$

For $u, v \in H$ we define

$$H = \left\{ u \mid \int_{I^k} |u|^2 \, P \, \rho_1 \rho_2 \ldots \rho_k dx < \infty \right\}, \qquad (7)$$

with scalar product

$$(u, v) = \int_{I^k} u\bar{v} \, P \, \rho_1 \rho_2 \ldots \rho_k dx . \qquad (8)$$

The method of construction of H is similar to that employed by Murray and Von Neuman (1936). It is to be noted that if $f_r(x_r)$ $(r = 1, \ldots, k)$ such that $f_r : H_r \to H_r$ is a set of endomorphisms of the H_r, then the f_r are not operators on the tensor product. However they will induce endomorphisms of H which may be denoted by

$$\tilde{f}_r = I_1 \otimes I_2 \otimes \ldots \otimes I_{r-1} \otimes f_r \otimes \ldots \otimes I_k. \tag{9}$$

where $I_r (r = 1, \ldots, k)$ is the identity operator in H_r.

In Sleeman (1971a) it was shown that the k eigenvalue problems defined by (2) and (3) in the k Hilbert spaces $H_r (r = 1, 2, 3, \ldots, k)$ may be replaced by the following k eigenvalue problems for partial differential operators defined in H, i.e.

$$L_r Y_r = \lambda_r Y_r, \quad \forall \quad Y_r \in D, \; r = 1, 2, \ldots, k, \tag{10}$$

where

(i) $$L_r = \frac{1}{P(x_1, x_2, \ldots x_k)} \sum_{s=1}^{k} p^*_{sr} \tilde{A}_s \tag{11}$$

(ii) $$\tilde{A}_s = I_1 \otimes \ldots \otimes I_{s-1} \otimes A_s \otimes \ldots \otimes I_k \tag{12}$$

(iii) p^*_{sr} is the cofactor of p_{rs} in the $k \times k$ non-singular matrix $(p_{rs})^k_{r,s=1}$

and (iv) $D = \prod_{p=1}^{k} \otimes D(A_r)$ and so is the same for each member of (10).

We remark that the eigenfunctions satisfying the problems (10) are not necessarily decomposable into tensor product form. Such decomposable tensors must be sought in a dense subspace of D and a prime aim of this paper is to achieve such a decomposition.

§3. A partial differential equation

From equation (10) we may formally write

$$\lambda_r = Y_r^{-1} L_r Y_r \quad (r = 1, \ldots, k) \;\; Y_r \in D. \tag{13}$$

That is equation (13) gives an expression relating the eigenvalue λ_r to the eigenfunction Y_r.

Consider the l-th member of the system (4) and replace the independent variable x_l by ξ. Thus we have

$$A_l y_l(\xi) = \sum_{s=1}^{k} \lambda_s p_{ls}(\xi)\, y_l(\xi), \quad y_l(\xi) \in D(A_l) \tag{14}$$

and on substituting for the spectral parameters λ_r ($r = 1, \ldots k$) from (13) we obtain

$$A_l y_l(\xi) = \sum_{s=1}^{k} Y_s^{-1} L_s Y_s\, p_{ls}(\xi)\, y_l(\xi), \quad y_l(\xi) \in D(A_l). \tag{15}$$

Observe that (13) is satisfied by the product

$$Y_r = \prod_{m=1}^{k} y_m(x_m) \;\forall\; r$$

and using this fact in (15) and writing

$$K(x_1, x_2, \ldots, x_l, \ldots, x_k; \xi) = y_l(\xi) \prod_{m=1}^{k} y_m(x_m) \tag{16}$$

we obtain the partial differential equation

$$A_l K = \sum_{s=1}^{k} p_{ls}(\xi)\, L_s K\,. \tag{17}$$

Solutions of equation (17) for $K \in D \cup D(A_l)$ are central to the main result of this paper which is given in the following section.

§4. k-linear operators

Theorem 1 For the problem (4), the operator I_l which is such that

$$I_l : D \to D(A_l), \quad D \subset H, \quad D(A_l) \subset H_l$$

is the k-linear operator

$$I_l\{y_1(x_1) y_2(x_2) \ldots y_k(x_k)\} = \int_{I^k} K(x_1, x_2, \ldots, x_k; \xi) P(x_1, x_2, \ldots, x_k) \times \rho_1 \rho_2 \ldots \rho_k\, y_1(x_1) \ldots y_k(x_k) dx. \tag{18}$$

That is

$$y_1(\xi) = \mu_1 \, I_1\{y_1(x_1)y_2(x_2) \ldots y_k(x_k)\} \tag{19}$$

where the eigenvalue μ_1 depends on the normalisation of the eigenfunctions $y_r(x_r)$, $r = 1, 2, \ldots, k$.

The proof of this result is straight forward and proceeds by substituting (19) into the 1-th member of (4) integrating twice with respect to each variable $x_1, x_2, \ldots, x_k$ and making use of the boundary condition (3).

An important subclass of the problem defined by (4) is the case in which we have a single differential equation containing k spectral parameters and solutions are required to satisfy a k + 1 point boundary condition. In this case we have the problem

$$Ay = \sum_{s=1}^{k} \lambda_s \, p_{rs}(x)y, \; y \in D(A) \tag{20}$$

where A has the same form as (2) and

$$D(A) = \{y \mid y \in C^2[a, b], \; U_r(y(\eta_r)) = V_r(y(\eta_{r+1})) = 0\} \; r = 1, 2, \ldots, k, \tag{21}$$

where

$$a = \eta_1 < \eta_2 < \ldots < \eta_{k+1} = b.$$

This problem may be easily cast into the form (4) by linear transformations, see Sleeman (1971b) for the case k = 2. For this situation the theorem corresponding to theorem 1 is

<u>Theorem 2</u> For the problem (20) the operator I which is such that $I : D \to D(A)$, is the homogeneous operator

$$I = \int_{I^k} K(x_1, x_2, \ldots, x_k;\xi) \, P(x_1, x_2 \ldots x_k) \, \rho(x_1) \ldots \rho(x_k) \, y(x_1) \ldots y(x_k) dx \, . \tag{22}$$

That is

$$y(\xi) = \int_{I^k} K(x_1, \ldots, x_k;\xi) \, P(x_1, \ldots, x_k) \, \rho(x_1) \ldots \rho(x_k) \, y(x_1) \ldots y(x_k) dx \, . \tag{23}$$

In this theorem $x_r \in [\eta_r, \eta_{r+1}]$, $r = 1, \ldots, k$, $\xi \in [a, b]$ and I^k is the cartesian product of the k intervals $[\eta_r, \eta_{r+1}]$, $r = 1, 2, \ldots, k$.

The proof of theorem 2 proceeds in the same manner as in theorem 1 and for the case k=2 a proof may be found in Arscott (1964a).

§5. Examples

In this section we give some illustrative examples which are known in the case k = 2.

1. Lame' polynomials (Sleeman (1968)).

$$E_n^m(\gamma) = \lambda \int_{-2K}^{2K} \int_{K-2iK'}^{K+2iK'} \frac{1}{R(P, P_0)} E_n^m(\alpha)\, E_n^m(\beta)(\mathrm{sn}^2\alpha - \mathrm{sn}^2\beta)\, d\alpha\, d\beta, \tag{24}$$

where $\gamma \in (K + iK', \gamma_0)$ and $R(P, P_0)$ denotes the distance, in ellipsoidal coordinates, of $P(\alpha, \beta, \gamma)$ from $P_0(\alpha_0, \beta_0, \gamma_0)$.

2. Ellipsoidal wave functions (Sleeman (1968))

$$\mathrm{el}_n^m(\gamma) = \lambda \int_{-2K}^{2K} \int_{K-2iK'}^{K+2iK'} \frac{\exp(i\chi R)}{R}\, \mathrm{el}_n^m(\alpha)\, \mathrm{el}_n^m(\beta)(\mathrm{sn}^2\alpha - \mathrm{sn}^2\beta)\, d\alpha\, d\beta \tag{25}$$

Other examples are to be found for example in Sleeman (1969), Arscott (1964b).

§6. Remarks

As mentioned in theorems 1 and 2 the operators I_1 and I are k-linear operators (see Hille (1948, ch. 4), Krasnosel'skii (1964)). If the kernels K are symmetric in each of their arguments then I_1 and I are completely continuous symmetric operators. This fact may be of use in establishing completeness and expansion properties of the eigenfunction for the problems (4) and (20), thus providing an alternative approach to that of Faierman (1969).

The integral equations (19) (23) are not entirely new. They seem to have first been discussed by Schmidt (1908) who with Lyapunov (1906) spearheaded the now important subject of bifurcation theory. It would be of interest therefore to pursue this connection between the multi-parameter eigenvalue problem and bifurcation theory.

REFERENCES

1. Arscott F M (1964a). Two-parameter eigenvalue problems in differential equations, Proc. Lond. Math Soc. 14, 459-470.
2. Arscott F M (1964b). Periodic differential equations, (Pergamon).
3. Atkinson F V (1968). Multi-parameter spectral theory, Bull. Amer. Math. Soc. 74, 1-27.
4. Faierman M (1969). The completeness and expansion theorems associated with the multi-parameter eigenvalue problem in ordinary differential equations, J. Diff. Equations, 5,197-213.
5. Hille E (1948). Functional analysis and semi-groups, Amer. Math. Soc. Publication.
6. Krasnosel'skii M A (1964). Topological Methods in the Theory of Nonlinear Integral Equations, (Pergamon).
7. Lyapunov A M (1906). Sur les figures d'equilibre peu differentes des ellipsoides d'une masse liquide homogene douee d'un mouvement de rotation, Zap. Akad. Nauk St Petersburg, 1.
8. Murray F J and Von Neumann J (1936). On rings of operators, Ann. of Math. 37, 116-229.
9. Schmidt E (1908). Zur theorie der linear und nicht linearen integralgleichungen, Math. Ann. 65, 370-399.
10. Sleeman B D (1968). Integral equations and relations for Lame' functions and ellipsoidal wave functions, Proc. Camb. Phil. Soc. 64, 113-126.
11. Sleeman B D (1969). Non-linear integral equations for Heun functions, Proc. Edin. Math. Soc. 16, 281-289.
12. Sleeman B D (1971a). Multi-parameter eigenvalue problems in ordinary differential equations, Bull. Inst. Poli. Jassy 17 (21) 51-60.
13. Sleeman B D (1971b). The two-parameter Sturm-Liouville problem for ordinary differential equations, Proc. Roy. Soc. Edin. 69, 139-148.

An Epidemic Model with Two Populations

P. Waltman

We consider two populations of N_1 and N_2 individuals and the spread of an infection which is transmitted only by members of one population to members of the other. For example, this might be malaria spread from human to mosquito to human or a venereal disease spread between opposite sexes. Such problems, particularly with an intermediate host to spread the disease, have been studied in particular in Bailey [1, p. 30]. We use here a threshold type model posed first by Cooke [2] and investigated by Hoppensteadt and Waltman [3], [4], .

The first population is divided into four disjoint classes. Let

$S(t)$ denote the number of susceptible but unexposed individuals at time t,

$E(t)$ denote the number of exposed but not infective individuals at time t,

$I(t)$ denote the number of infective individuals at time t,

$R(t)$ denote the number of individuals removed from the population (recovered and immune, dead, isolated, etc.).

The second population is divided in the same way and we denote the classes by $\tilde{S}$, $\tilde{E}$, $\tilde{I}$, and $\tilde{R}$.

The infection spreads according to the following rules:

(i) an individual who becomes infective at time t recovers and becomes immune at time $t+\sigma$ $(t+\tilde{\sigma})$;

(ii) the rate of change (decrease) in the susceptible population is proportional to the number of susceptibles in the population and the number of infectives in the other population;

(iii) an individual in the first population who is first exposed at time τ becomes infective at time t if

$$\int_{\tau}^{t}(\rho_1(x)+\rho_2(x)\tilde{I}(x))dx = m;$$

similarly, for the other population, if

$$\int_{\tilde{\tau}}^{t}(\tilde{\rho}_1(x)+\tilde{\rho}_2(x)I(x))dx = \tilde{m};$$

(iv) the number of individuals in each population is constant.

A discussion of the model, of the parameters, and the previously studied special cases that may be obtained, can be found in [2], [3]. Using essentially the same derivation as in [3] leads to the following system of (functional) differential equations in the unknowns τ, $\tilde{\tau}$, I, $\tilde{I}$, S, $\tilde{S}$. (The behavior of the functions $E(t)$, $\tilde{E}(t)$, $R(t)$, $\tilde{R}(t)$ can be obtained from these.)

$$(1)\quad \begin{cases} \int_{\tau(t)}^{t}(\rho_1(x)+\rho_2(x)\tilde{I}(x))dx = m \\ S'(t) = -r(t)\,S(t)\,\tilde{I}(t) \\ I(t) = S(\tau(t-\sigma)) - S(\tau(t)) \\ \int_{\tilde{\tau}}^{t}(\tilde{\rho}_1(x)+\tilde{\rho}_2(x)I(x))dx = \tilde{m} \\ \tilde{S}'(t) = -\tilde{r}(t)\,\tilde{S}(t)\,I(t) \\ \tilde{I}(t) = \tilde{S}(\tilde{\tau}(t-\tilde{\sigma}) - \tilde{S}(\tilde{\tau}(t)). \end{cases}$$

Two initial functions $I_0(t)$, $\tilde{I}_0(t)$ describe the initial infective populations on $[-\sigma,0]$ and $[-\tilde{\sigma},0]$. We assume $I_0(t)=0$, $t\le-\sigma$, $\tilde{I}_0(t)=0$, $t\le-\tilde{\sigma}$. Initial conditions for $S(t)$, $\tilde{S}(t)$

are $S(0) = N_1 - I_0(0)$, $\tilde{S}(0) = N_2 - \tilde{I}_0(0)$. Thus $R(0) = \tilde{R}(0) = E(0) = \tilde{E}(0) = 0$. In order for the infection to spread it is necessary (but not sufficient) that either there is a $t_0 < \sigma$ or a $\tilde{t}_0 < \tilde{\sigma}$ such that

$$\int_0^{t_0} [\rho_1(x) + \rho_2(x) I_0(x)]dx = m$$

$$\int_0^{\tilde{t}_0} [\tilde{\rho}_1(x) + \tilde{\rho}_2(x) \tilde{I}_0(x)]dx = \tilde{m}.$$

We investigate existence and uniqueness of solutions of the system (1). We can also include the case when immunity is only temporary at the expense of greater complication in the model.

THEOREM: Let $\rho_1, \rho_2, \tilde{\rho}_1, \tilde{\rho}_2, r, \tilde{r}$ be continuous nonnegative functions on R with $\rho_1 > 0$, $\tilde{\rho}_1 > 0$. Let S_0, $\tilde{S}_0$ be given constants and let I_0, $\tilde{I}_0$ be continuous monotone increasing functions on $[-\sigma, 0]$, $[-\tilde{\sigma}, 0]$ with $I_0(-\sigma) = \tilde{I}_0(-\tilde{\sigma}) = 0$. Then there exist unique continuous nonnegative functions $\tau(t)$, $\tilde{\tau}(t)$, $I(t)$, $\tilde{I}(t)$, $S(t)$, $\tilde{S}(t)$ satisfying (1).

The proof proceeds via a contraction mapping argument similar to that used in [4] but taking into account the interaction of the population.

REFERENCES

[1] N. T. J. Bailey: The Mathematical Theory of Epidemics, Hafner, New York, 1957

[2] K. L. Cooke: Functional-differential equations: Some models and perturbation problems, in Differential Equations and Dynamical Systems (J. K. Hale and J. P. LaSalle, eds.), Academic Press, 1967.

[3] F. Hoppensteadt and P. Waltman: A Problem in the Theory of Epidemics, Math. Biosciences 9(1970), 71-91.

[4] F. Hoppensteadt and P. Waltman: A problem in the Theory of Epidemics II, <u>Math</u>. <u>Biosciences</u> 12(1971), 133-145.

Existence Theorems for Operators of Monotone Type

J.R.L. Webb

One of the features of the development of nonlinear Functional Analysis over the last few years has been the search for interesting classes of operators, which actually arise from concrete problems in Partial Differential Equations, whose structure can be determined sufficiently well to enable existence or regularity of solutions of Partial Differential Equations to be asserted.

The monotone operators constitute such a class. They arose fairly naturally from the study of nonlinear boundary value problems of elliptic type and have been extensively studied. The basic theory has now been extended to deal with more general classes of operators such as the pseudomonotone ones.

A useful tool in nonlinear problems is the recently developed theories of generalized topological degree and these can be applied to monotone operators. This degree theory enabled Browder [1] to announce a powerful existence theorem for pseudomonotone operators which is generalised below. This generalisation could have been carried through using the same methods as did Browder but here an elementary argument is used which avoids the use of the degree theory. The results below were obtained by the author in conjunction with Bruce Calvert in the case when the underlying Banach space was separable [3]; the nonseparable case is treated here.

Throughout X will stand for a reflexive real Banach space and (f,x) will denote the value of $f \in X^*$ at $x \in X$. We shall assume that X is endowed with an equivalent norm with respect to which X and X^* are locally uniformly convex; a recent result of Trojanskii assures that this can be done. We recall that X is called locally uniformly convex if for each x_0, $\|x_0\| = 1$, and each $\varepsilon > 0$, there exists $\delta > 0$ such that $\|x - x_0\| < \varepsilon$ whenever $\|x\| \leq 1$ and $\|x + x_0\| \geq 2 - 2\delta$. The operators under study here satisfy the following criteria: Let $T: X \longrightarrow X^*$ be a (not necessarily linear) mapping. T is called quasimonotone if it satisfies.

(Q): Whenever $\{x_j\}$ is a sequence weakly convergent to an element x,

$$\limsup\,(Tx_j - Tx,\ x_j - x) \geq 0.$$

Obviously a monotone map, that is, one for which $(Tx - Ty, x - y) \geq 0$ for all x,y in X, is quasimontone, as is a compact mapping and the sum of a monotone and a compact map. We also need the following notion for a map T, due to Browder (e.g.[2]):

$(S)_+$: Whenever $x_j \longrightarrow x$ weakly and $\limsup (Tx_j - Tx, x_j - x) \leq 0$, $x_j \longrightarrow x$ (in the norm topology).

As an example of a map satisfying $(S)_+$ we cite the duality map $J: X \longrightarrow X^*$ determined by the requirements $\|Jx\| = \|x\|$ and $(Jx,x) = \|x\|^2$. Moreover, J is demicontinuous, that is, $x_n \longrightarrow x$ implies $Jx_n \longrightarrow Jx$ weakly. The following result characterizes the quasimonotone operators:

__Proposition.__ Let $T: X \longrightarrow X^*$ be bounded and demicontinuous. Then T is quasimonotone if and only if $T + \varepsilon J$ satisfies $(S)_+$ for all $\varepsilon > 0$.

__Proof.__ Let T be quasimonotone. Then

$$\limsup((T + \varepsilon J)(x_j) - (T + \varepsilon J)(x), x_j - x) \leq 0 \text{ implies that}$$
$$\limsup \varepsilon(J(x_j) - J(x), x_j - x) \leq 0. \text{ However, } J \text{ satisfies } (S)_+ \text{ so}$$

that $x_j \longrightarrow x$ as required.

Conversely, if T is not quasimonotone, there is a sequence $\{x_j\}$, $x_j \longrightarrow x$ weakly and $\limsup(Tx_j - Tx, x_j - x) = -\delta$, $\delta > 0$. Note that this shows that x_j does not converge to x by the demicontinuity of T. The sequence $\{x_j\}$ is bounded say $\|x_j\| \leq M$ so that for $\varepsilon \leq \frac{\delta}{4M}$ we have

$$\limsup((T + \varepsilon J)x_j - (T + \varepsilon J)x, x_j - x) \leq 0$$

but $x_j \not\longrightarrow x$ so that $T + \varepsilon J$ does not satisfy $(S)_+$. This Proposition shows that every pseudomonotone operator is quasimonotone; the converse is false. In order to prove the Theorem we need a special case. This is

__Lemma.__ Let $T: X \longrightarrow X^*$ be demicontinuous and satisfy $(S)_+$ and suppose that T is odd for $\|x\| = r$, that is, $T(-x) = -T(x)$. Then there exists x_0, $\|x_0\| \leq r$ such that $Tx_0 = 0$.

__Proof.__ Let F be a finite dimensional subspace of X, $j_F: F \longrightarrow X$ the injection map and $j_F^*: X^* \longrightarrow F^*$ its adjoint. Let

$T_F = j_F^* T j_F : F \longrightarrow F^*$. Obviously for $x \in F$, $\|x\| = r$,

$T_F(-x) = -T_F(x)$, so, by the finite dimensional Borsuk theorem for continuous maps, there exists $x_F \in F$, $\|x_F\| \leq r$, such that $T_F x_F = 0$. For a finite dimensional subspace F_0 let $W_{F_0} = \bigcup_{F_0 \subset F} \{x_F\}$. The family of all such sets has the finite intersection property so the same holds for their weak closures. Hence there exists a point x_0 belonging to the intersection of the weak closures of the W_{F_0}. Let $x_0 \in F_1$. As W_{F_1} is bounded it is weakly relatively compact, so there is a sequence $x_n = x_{F_n} \in W_{F_1}$ converging weakly to x_0. Now,

$$\begin{aligned}(Tx_n - Tx_0, x_n - x_0) &= (j_{F_n}^* Tx_n - Tx_0, x_n - x_0) \text{ because } x_n - x_0 \in F_n \\ &= (T_{F_n} x_n - Tx_0, x_n - x_0) \\ &= -(Tx_0, x_n - x_0) \longrightarrow 0 \text{ as } n \longrightarrow \infty\end{aligned}$$

Hence, by $(S)_+$, $x_n \longrightarrow x_0$ and by demicontinuity of T, $Tx_n \longrightarrow Tx_0$ weakly. However, $Tx_n \longrightarrow 0$ weakly. For let $v \in X$ and let F_1 contain v and x_0. Then $(Tx_n, v) = (T_{F_n} x_n, v) = 0$. This completes the proof.

The above result is invariant under homotopy as is shown by the next Theorem.

<u>Theorem 1.</u> Let B be a ball in X, let $T: B \times [0,1] \longrightarrow X^*$ and let $T_t = T(\cdot, t)$ be bounded, demicontinuous mappings satisfying $(S)_+$. Suppose $t \longmapsto T_t$ is continuous uniformly for x in B and that $T_t(x) \neq 0$ for all $t \in [0,1]$ and all $x \in \partial B$. Then, if T_0 is odd on ∂B, there exists $x_1 \in B$ such that $T_1(x_1) = 0$.

<u>Proof.</u> Without loss of generality we can assume B to be the closed unit ball in X. Let $A: B \longrightarrow X^*$ be defined by

$$Ax = \begin{cases} T\left(\frac{x}{\|x\|}, 2(1-\|x\|)\right) & \text{if } \frac{1}{2} \leq \|x\| \leq 1 \\ T(2x, 1) & \text{if } \|x\| \leq \frac{1}{2} \end{cases}$$

One may easily prove that A is demicontinuous. In fact, it satisfies $(S)_+$ for a subsequence. To see this, suppose that $x_n \longrightarrow x$ weakly and $\limsup(Ax_n - Ax, x_n - x) \leq 0$ and let $\{x_j\}$ be the subsequence for which $\|x_j\| > \frac{1}{2}$. If there are only finitely many such x_j the assertion is obvious, otherwise we can assume that $\|x_j\| \longrightarrow \alpha \in [\frac{1}{2},1]$. Then, $\limsup (T(\frac{x_j}{\|x_j\|}, 2(1-\alpha)), x_j - x) \leq 0$ by the continuity $t \longmapsto T_t$. Furthermore, $\lim(T(\frac{x_j}{\|x_j\|}, 2(1-\alpha)), \frac{x_j}{\|x_j\|} - \frac{x_j}{\alpha}) = 0$ as each T_t is bounded. Hence,

$\limsup(T(\frac{x_j}{\|x_j\|}, 2(1-\alpha)), \frac{x_j}{\|x_j\|} - \frac{x}{\alpha}) \leq 0$, and by $(S)_+$ applied to $T_{2(1-\alpha)}$, $\frac{x_j}{\|x_j\|} \longrightarrow \frac{x}{\alpha}$, that is $x_j \longrightarrow x$ and $\|x\| = \alpha$.

As A coincides with T_0 on ∂B, A has a zero in B by the Lemma. This implies that T_1 has a zero since, to the contrary, A could have no zero.

The promised existence theorem is the following.

Theorem 2

Let $T_t: X \longrightarrow X^*$, $0 \leq t \leq 1$, be bounded, demicontinuous, quasimonotone mappings and let $t \longmapsto T_t$ be continuous uniformly on bounded subsets of X. Suppose there is a continuous real valued function ϕ such that for any $f \in X^*$ and $t \in [0,1]$, $T_t x = f$ implies the a priori bound $\|x\| \leq \phi(\|f\|)$ holds. Suppose T_0 is odd for $\|x\| \geq r$ and that T_1 maps each closed ball into a closed set. Then T_1 is surjective.

Proof Let f be any element of X^* and let $r_1 \geq r$ be so large that $\|T_t(x)\| \geq 1 + \|f\|$ for $\|x\| = r_1$; this is possible by the a priori bound assumption. For $0 < \varepsilon < \frac{1}{r_1}$ set $F_t(x) = T_t(x) + \varepsilon J(x) - tf$. $F_t(x) \neq 0$ for $\|x\| = r_1$ since $\|F_t(x)\| \geq \|T_t(x)\| - t\|f\| - \varepsilon\|Jx\| > 0$. An application of Theorem 1 yields a zero of F_1 so there exists x_ε, $\|x_\varepsilon\| \leq r_1$, such that $T_1(x_\varepsilon) + \varepsilon J(x_\varepsilon) = f$. As $\varepsilon \longrightarrow 0$, $\|\varepsilon J(x_\varepsilon)\| \leq \varepsilon r_1 \longrightarrow 0$ so $T_1(x_\varepsilon) \longrightarrow f$ and f belongs to the closure of $T_1(B_{r_1})$ which is closed by assumption: thus $f \in T_1(B_{r_1})$ and the Theorem is proved.

In [3] an example is given to show that this Theorem may be applied to situations in which the operator is not pseudomontone. Other applications may be given to the same kind of boundary value problems that are considered in [1].

References

1. F. E. Browder: "Nonlinear Elliptic boundary value problems and the generalized topological degree", Bull. Amer. Math. Soc. 76, 995-1005 (1970).

2. F. E. Browder: "Nonlinear operators and nonlinear equations of evolution in Banach Spaces", Proc. Symposium Pure Math. Vol. 18 Part 2, Amer. Math.Soc.

3. B. Calvert and J. R. L. Webb: "An existence theorem for quasimonotone operators", to appear in Atti dell 'Accademia dei Lincei, Rend.

Elliptic Partial Differential Equations may be Related by a Change of Independent Variables

D.H. Wood

The equation $U_{xx} + U_{yy} + e^{2x} U = 0$ (subscripts denote partial differentiation) can be transformed into an equation with constant coefficients, $U_{\xi\xi} + U_{\eta\eta} + U = 0$, by changing the independent variables to $\xi = \Xi(x, y) = e^x \cos y$ and $\eta = N(x, y) = e^x \sin y$. We will discuss the possibility of changing the independent variables in the equation

$$U_{xx} + U_{yy} + C^{-2}(x, y)\, U = 0 , \tag{1}$$

where C is a positive function, to obtain a more tractable equation, usually Eq. (1) with $C(x, y)$ replaced by a function of a single variable.

Under the most general transformation of Eq. (1) induced by twice continuously differentiable functions Ξ and N of two variables, the discriminant of the new equation will equal the discriminant of Eq. (1) times the square of the Jacobian $J(x, y) = \Xi_x N_y - \Xi_y N_x$. Since we will assume J is not zero — we make J positive by interchanging Ξ and N, if necessary — the new equation

$$(\nabla\Xi)^2 U_{\xi\xi} + \nabla\Xi\cdot\nabla N\, U_{\xi\eta} + (\nabla N)^2 U_{\xi\xi} + \Delta\Xi\, U_\xi + \Delta N U_\eta + C^{-2} U = 0 \tag{2}$$

(∇F is the gradient of F and ΔF is the Laplacian of F) is also elliptic and, without any loss of generality, in canonical form. That is, $\nabla\Xi\cdot\nabla N = 0$ and both $(\nabla\Xi)^2$ and $(\nabla N)^2$ are the same function, temporarily denoted by G. Twice the square root of the discriminant of Eq. (2), $2G = 2(\Xi_x N_y - \Xi_y N_x)$, minus $(\nabla\Xi)^2 = G$, minus $(\nabla N)^2 = G$, yields the Cauchy-Riemann equations, $(\Xi_x - N_y)^2 + (\Xi_y + N_x)^2 = 0$. Since $\Delta\Xi = \Delta N = 0$, Eq. (2) is the same as Eq. (1) with the function C replaced by the function Ψ defined by

$$\Psi^2(\Xi(x, y),\ N(x, y)) \equiv J(x, y)\, C^2(x, y) \ . \tag{3}$$

That is, $\Psi(\xi,\eta)$ expresses the function $J\,C^2$ of x and y as a function of the new independent variables ξ and η. From Eq. (3), C is at least as differentiable as Ψ, and conversely (using the inverse transformation).

An equivalence relation is defined on the set of functions of two variables with continuous fourth partial derivatives by saying that Ψ and C are equivalent when some conjugate harmonic functions Ξ and N satisfy Eq. (3).

For equivalent Ψ and C, Eq. (3) implies $\Delta \ln \Psi = \Delta \ln C$ because $\ln J$ is the real part of the analytic function $\ln f'(z)$, where $f(z) = f(x+iy) = \Xi(x,y) + iN(x,y)$. Multiplying the equation $\Delta \ln C = \Delta \ln \Psi = J\left[(\ln\Psi)_{\xi\xi} + (\ln\Psi)_{\eta\eta}\right]$ by Eq. (3) and canceling J from both sides, we obtain

$$\Phi(\Xi(x,y),\ N(x,y)) \equiv F(x,y), \tag{4}$$

where $\Phi(\xi,\eta) = \Psi^2(\xi,\eta)\left[(\ln\Psi)_{\xi\xi} + (\ln\Psi)_{\eta\eta}\right]$ and $F(x,y) = C^2(x,y)\left[(\ln C)_{xx} + (\ln C)_{yy}\right]$, as a necessary consequence of the equivalence of Ψ and C. We use this to show that distinct functions of one variable usually belong to distinct equivalence classes.

<u>THEOREM 1.</u> Aside from the exceptional case of F(x, y) identically constant and disregarding linear transformations, (i) C(x, y) is equivalent to at most one function of a single variable $\Psi(\xi)$; (ii) C(x, y) is equivalent to some $\Psi(\xi)$ only if $(\Delta F)/(\nabla F)^2$ is a function of F, denoted H(F); and (iii) $\Xi = G(F(x,y))$ and the transformation are determined by a non-constant solution of $G'' + HG' = 0$, when C(x, y) is equivalent to some $\Psi(\xi)$.

<u>PROOF.</u> If $\Psi(\xi)$ and C(x) are equivalent, Eq. (4) becomes $\Phi(\Xi) = F(x)$. Aside from the exceptional case, Φ is somewhere locally invertible, implying that Ξ is a linear function because it is a harmonic function of one variable. We also prove (ii) by expressing $\Xi = \Phi^{-1}(F(x,y))$. Since Ξ is harmonic, $(\Phi^{-1})''(\nabla F)^2 + (\Phi^{-1})'\,\Delta F = 0$, which implies $(\Delta F)/(\nabla F)^2$ is a function H(F). Knowing H, we find Ξ to prove (iii). Since $\Xi = G(F(x,y))$ for some solution of $G'' + HG' = 0$, we have

$\Xi = a\int^{F} \exp(-\int^{t} H(s)\,ds)\,dt + b$, $a \neq 0$, but we disregard the a and b as linear transformations of the variable ξ. The transformation is determined by finding a harmonic function conjugate to Ξ.

REMARK. A given equivalence class contains a function of ξ alone if and only if it contains a function of $r = \sqrt{x^2 + y^2}$ alone. The transformation $\xi + i\eta = f(z) = f(x + iy) = \ln(x + iy)$ connects a function $C^2(r)$ to $\Psi^2(\xi,\eta) = J(x,y)\,C^2(r) = |f'(z)|^2 \times C^2(r) = r^{-2}\,C^2(r)$, which is a function of ξ alone because $r = e^{\Xi}$. Note that some linear changes of the variable ξ correspond to arbitrary powers of r.

When F is not constant, the only transformation that can take $C(x, y)$ into $\Psi(\xi)$ has been found. One must verify that Ψ defined by the transformation and Eq. (3) really is a function of a single variable. When F is a nonzero constant, the next theorem identifies the only possible functions $\Psi(\xi)$, but unfortunately, not the transformation. When F is zero, Theorem 3 yields both $\Psi \equiv 1$ and the transformation.

THEOREM 2. If $C(x, y)$ is a member of the equivalence class of $\Psi(\xi,\eta) = \xi$ or $\Psi(\xi,\eta) = \cosh\xi$, $F(x, y)$ must be identical to a negative constant or a positive constant, respectively. Disregarding linear transformations, the only members of these equivalence classes depending on a single variable are sinh, the identity function, and sin when $F < 0$ and cosh when $F > 0$.

PROOF. Linear transformations can normalize the constants mentioned above to unit magnitude, but cannot change their sign. Since $\Psi = \xi$ gives $\Phi = -1$ and $\Psi = \cosh\xi$ gives $\Phi = 1$, their normalized equivalent functions will have $F = -1$ and $F = 1$, respectively. Members of either equivalence class that depend on one variable must satisfy $C\,C'' - (C')^2 = k$ for some constant k. Differentiating this equation gives $C\,C''' - C''\,C' = 0$, or $(C''/C)' = 0$. Only nonzero multiples of cosh, sin, the identify function, and sinh of linear arguments satisfy $(C''/C)' = 0$ and make $F < 0$, except cosh which makes $F > 0$. The candidates sinh and sin are in the equivalence class of $\Psi = \xi$ because the transformations $\xi + i\eta = \tanh(x + iy)$ for $C = \sinh 2x$

and $\xi + i\eta = \tan(x + iy)$ for $C = \sin 2x$ satisfy Eq. (3) with $\Psi = 2\xi$.

THEOREM 3. $C(x, y)$ is equivalent to $\Psi(\xi, \eta) \equiv 1$ if and only if $F(x, y) \equiv 0$. A transformation is $\xi + i\eta = f(z) = f(x + iy) = \int^z \exp\left[-\ln C(x, y) - iH(x, y)\right] dz$ where $H(x, y)$ is a harmonic function conjugate to the (necessarily harmonic) function $\ln C(x, y)$. The most general function $C(x)$ that makes $F = 0$ is $a e^{bx}$.

PROOF. $F = 0$ implies that $\ln C(x, y)$ is harmonic and determines an analytic function $g(z) = g(x + iy) = \ln C(x, y) + i H(x, y)$. The transformation defined by $\xi + i\eta = f(z) = f(x + iy) = \int^z \exp\left[-g(z)\right] dz$ connects $C(x, y)$ and $\Psi^2(\xi, \eta) = J(x, y)\, C^2(x, y) = \left|f'(z)\right|^2 C^2(x, y) \equiv 1$. The last statement of the theorem is obvious.

The next theorem yields conditions necessary for the equivalence of two given functions. Its application requires testing the identity of functions involving several differentiations of the given functions. This task is unattractive to us but merely routine to symbol manipulation computers that can differentiate functions.

THEOREM 4. If $C(x, y)$ and $\Psi(\xi, \eta)$ are equivalent and have continuous fifth order partial derivatives, both

$$U_y + U_\xi V + U_\eta U = V_x + V_\xi U - V_\eta V \text{ and} \tag{5}$$

$$U_x + U_\xi U - U_\eta V = -V_y - V_\xi V - V_\eta U, \text{ where} \tag{6}$$

$$U = (\Phi_\xi F_x + \Phi_\eta F_y)/(\Phi_\xi^2 + \Phi_\eta^2) \quad \text{and}$$

$$V = (\Phi_\xi F_y - \Phi_\eta F_x)/(\Phi_\xi^2 + \Phi_\eta^2),$$

must be identities in the four independent variables x, y, ξ, and η.

PROOF. Differentiating Eq. (4) first with respect to x and then with respect to y and using the Cauchy-Riemann equations, we can obtain a pair of linear equations for Ξ_x and Ξ_y whose solutions are $\Xi_x = U$ and $\Xi_y = V$. Using the Cauchy-Riemann equations again, we obtain an overdetermined system of four equations, $\Xi_x = U$, $N_y = U$, $\Xi_y = V$, and $N_x = -V$ for the two unknown functions Ξ and N. Since Ξ, N, U, and V have continuous second partial derivatives, a standard theorem on overdetermined systems of first order partial differential equations [J. J. Stoker, "Differential Geometry," John Wiley and Sons, New York, 1969,

Theorem V, page 393] states that our system has a solution if and only if Eq. (5) and Eq. (6), called compatibility conditions, are satisfied identically in the variables x, y, ξ and η.

All of the above results, except Theorem 3, remain true for general elliptic partial differential equations in canonical form (identifying $C(x, y)$ with the coefficient of the unknown function) because they are necessary consequences of the existence of a transformation of variables. However, for a more general equation, these necessary conditions are less likely to also be sufficient.

Example 1 $C(x,y) = e^x$. This is the exceptional case, $F = 0$, dealt with in Theorem 3.

Example 2 $C(x,y) = x^2 - y$. In this case, $F = -2x^2 - 2y - 1$ and the ratio $\nabla^2 F/(\nabla F)^2 = 1/(-4x^2 - 1)$ is independent of y and can not be expressed as a function of F, because F depends on y. Therefore, by Theorem 1, $C(x,y)$ can not be transformed into any function of a single variable.

Example 3 $C(x,y) = (e^{-3x}/\cos y)^{\frac{1}{2}}$. Routine computations show that $F = (e^x \cos y)^{-3}$, $H(F) = -(4/3)\, F^{-1}$, and $\Xi(x,y) = e^x \cos y$. We now compute Ψ from Eq. (3): $\Psi^2 = (\nabla \Xi)^2 C^2 = e^{-x}/\cos y = 1/\Xi$. Notice that it is necessary to verify that Ψ really is a function of a single variable.

Example 4 $C(x,y) = xy$. We find that $F = -x^2 - y^2$, $H(F) = F^{-1}$, and $\Xi(x,y) = \log\left| x^2 + y^2 \right|$. This is the only possible transformation that could transform $C(x,y)$ into a function of a single variable. However, when we compute Ψ from Eq. (3), we obtain $\Psi = e^{\xi/2} \sin \eta$. Hence no member of this equivalence class is a function of a single variable.

Example 5 $C(x,y) = 1 + x^2 + y^2$. This is a function of r. Under the transformation $\xi + i\eta = \log (x + iy)$, we obtain $\Psi = \cosh \xi$. This example also illustrates Theorem 2 because $F = 4$.

Lecture Notes in Mathematics

Comprehensive leaflet on request

Vol. 74: A. Fröhlich, Formal Groups. IV, 140 pages. 1968. DM 16,-

Vol. 75: G. Lumer, Algèbres de fonctions et espaces de Hardy. VI, 80 pages. 1968. DM 16,-

Vol. 76: R. G. Swan, Algebraic K-Theory. IV, 262 pages. 1968. DM 18,-

Vol. 77: P.-A. Meyer, Processus de Markov: la frontière de Martin. IV, 123 pages. 1968. DM 16,-

Vol. 78: H. Herrlich, Topologische Reflexionen und Coreflexionen. XVI, 166 Seiten. 1968. DM 16,-

Vol. 79: A. Grothendieck, Catégories Cofibrées Additives et Complexe Cotangent Relatif. IV, 167 pages. 1968. DM 16,-

Vol. 80: Seminar on Triples and Categorical Homology Theory. Edited by B. Eckmann. IV, 398 pages. 1969. DM 20,-

Vol. 81: J.-P. Eckmann et M. Guenin, Méthodes Algébriques en Mécanique Statistique. VI, 131 pages. 1969. DM 16,-

Vol. 82: J. Wloka, Grundräume und verallgemeinerte Funktionen. VIII, 131 Seiten. 1969. DM 16,-

Vol. 83: O. Zariski, An Introduction to the Theory of Algebraic Surfaces. Second Printing. IV, 100 pages. 1972. DM 16,-

Vol. 84: H. Lüneburg, Transitive Erweiterungen endlicher Permutationsgruppen. IV, 119 Seiten. 1969. DM 16,-

Vol. 85: P. Cartier et D. Foata, Problèmes combinatoires de commutation et réarrangements. IV, 88 pages. 1969. DM 16,-

Vol. 86: Category Theory, Homology Theory and their Applications I. Edited by P. Hilton. VI, 216 pages. 1969. DM 16,-

Vol. 87: M. Tierney, Categorical Constructions in Stable Homotopy Theory. IV, 65 pages. 1969. DM 16,-

Vol. 88: Séminaire de Probabilités III. IV, 229 pages. 1969. DM 18,-

Vol. 89: Probability and Information Theory. Edited by M. Behara, K. Krickeberg and J. Wolfowitz. IV, 256 pages. 1969. DM 18,-

Vol. 90: N. P. Bhatia and O. Hajek, Local Semi-Dynamical Systems. II, 157 pages. 1969. DM 16,-

Vol. 91: N. N. Janenko, Die Zwischenschrittmethode zur Lösung mehrdimensionaler Probleme der mathematischen Physik. VIII, 194 Seiten. 1969. DM 16,80

Vol. 92: Category Theory, Homology Theory and their Applications II. Edited by P. Hilton. V, 308 pages. 1969. DM 20,-

Vol. 93: K. R. Parthasarathy, Multipliers on Locally Compact Groups. III, 54 pages. 1969. DM 16,-

Vol. 94: M. Machover and J. Hirschfeld, Lectures on Non-Standard Analysis. VI, 79 pages. 1969. DM 16,-

Vol. 95: A. S. Troelstra, Principles of Intuitionism. II, 111 pages. 1969. DM 16,-

Vol. 96: H.-B. Brinkmann und D. Puppe, Abelsche und exakte Kategorien, Korrespondenzen. V, 141 Seiten. 1969. DM 16,-

Vol. 97: S. O. Chase and M. E. Sweedler, Hopf Algebras and Galois theory. II, 133 pages. 1969. DM 16,-

Vol. 98: M. Heins, Hardy Classes on Riemann Surfaces. III, 106 pages. 1969. DM 16,-

Vol. 99: Category Theory, Homology Theory and their Applications III. Edited by P. Hilton. IV, 489 pages. 1969. DM 24,-

Vol. 100: M. Artin and B. Mazur, Etale Homotopy. II, 196 Seiten. 1969. DM 16,-

Vol. 101: G. P. Szegö et G. Treccani, Semigruppi di Trasformazioni Multivoche. VI, 177 pages. 1969. DM 16,-

Vol. 102: F. Stummel, Rand- und Eigenwertaufgaben in Sobolewschen Räumen. VIII, 386 Seiten. 1969. DM 20,-

Vol. 103: Lectures in Modern Analysis and Applications I. Edited by C. T. Taam. VII, 162 pages. 1969. DM 16,-

Vol. 104: G. H. Pimbley, Jr., Eigenfunction Branches of Nonlinear Operators and their Bifurcations. II, 128 pages. 1969. DM 16,-

Vol. 105: R. Larsen, The Multiplier Problem. VII, 284 pages. 1969. DM 18,-

Vol. 106: Reports of the Midwest Category Seminar III. Edited by S. Mac Lane. III, 247 pages. 1969. DM 16,-

Vol. 107: A. Peyerimhoff, Lectures on Summability. III, 111 pages. 1969. DM 16,-

Vol. 108: Algebraic K-Theory and its Geometric Applications. Edited by R. M. F. Moss and C. B. Thomas. IV, 86 pages. 1969. DM 16,-

Vol. 109: Conference on the Numerical Solution of Differential Equations. Edited by J. Ll. Morris. VI, 275 pages. 1969. DM 18,-

Vol. 110: The Many Facets of Graph Theory. Edited by G. Chartrand and S. F. Kapoor. VIII, 290 pages. 1969. DM 18,-

Vol. 111: K. H. Mayer, Relationen zwischen charakteristischen Zahlen. III, 99 Seiten. 1969. DM 16,-

Vol. 112: Colloquium on Methods of Optimization. Edited by N. N. Moiseev. IV, 293 pages. 1970. DM 18,-

Vol. 113: R. Wille, Kongruenzklassengeometrien. III, 99 Seiten. 1970. DM 16,-

Vol. 114: H. Jacquet and R. P. Langlands, Automorphic Forms on GL (2). VII, 548 pages. 1970. DM 24,-

Vol. 115: K. H. Roggenkamp and V. Huber-Dyson, Lattices over Orders I. XIX, 290 pages. 1970. DM 18,-

Vol. 116: Séminaire Pierre Lelong (Analyse) Année 1969. IV, 195 pages. 1970 DM 16,-

Vol. 117: Y. Meyer, Nombres de Pisot, Nombres de Salem et Analyse Harmonique. 63 pages. 1970. DM 16,-

Vol. 118: Proceedings of the 15th Scandinavian Congress, Oslo 1968. Edited by K. E. Aubert and W. Ljunggren. IV, 162 pages. 1970. DM 16,-

Vol. 119: M. Raynaud, Faisceaux amples sur les schémas en groupes et les espaces homogènes. III, 219 pages. 1970. DM 16,-

Vol. 120: D. Siefkes, Büchi's Monadic Second Order Successor Arithmetic. XII, 130 Seiten. 1970. DM 16,-

Vol. 121: H. S. Bear, Lectures on Gleason Parts. III, 47 pages. 1970. DM 16,-

Vol. 122: H. Zieschang, E. Vogt und H.-D. Coldewey, Flächen und ebene diskontinuierliche Gruppen. VIII, 203 Seiten. 1970. DM 16,-

Vol. 123: A. V. Jategaonkar, Left Principal Ideal Rings. VI, 145 pages. 1970. DM 16,-

Vol. 124: Séminare de Probabilités IV. Edited by P. A. Meyer. IV, 282 pages. 1970. DM 20,-

Vol. 125: Symposium on Automatic Demonstration. V, 310 pages 1970. DM 20,-

Vol. 126: P. Schapira, Théorie des Hyperfonctions. XI, 157 pages. 1970. DM 16,-

Vol. 127: I. Stewart, Lie Algebras. IV, 97 pages. 1970 DM 16,-

Vol. 128: M. Takesaki, Tomita's Theory of Modular Hilbert Algebras and its Applications. II, 123 pages. 1970. DM 16,-

Vol. 129: K. H. Hofmann, The Duality of Compact Semigroups and C*- Bigebras. XII, 142 pages. 1970. DM 16,-

Vol. 130: F. Lorenz, Quadratische Formen über Körpern. II, 77 Seiten. 1970. DM 16,-

Vol. 131: A Borel et al., Seminar on Algebraic Groups and Related Finite Groups. VII, 321 pages. 1970. DM 22,-

Vol. 132: Symposium on Optimization. III, 348 pages. 1970. DM 22,-

Vol. 133: F. Topsøe, Topology and Measure. XIV, 79 pages. 1970. DM 16,-

Vol. 134: L. Smith, Lectures on the Eilenberg-Moore Spectral Sequence. VII, 142 pages. 1970. DM 16,-

Vol. 135: W. Stoll, Value Distribution of Holomorphic Maps into Compact Complex Manifolds. II, 267 pages. 1970. DM 18,-

Vol. 136: M. Karoubi et al., Séminaire Heidelberg-Saarbrücken-Strasbourg sur la K-Théorie. IV, 264 pages. 1970. DM 18,-

Vol. 137: Reports of the Midwest Category Seminar IV. Edited by S. MacLane. III, 139 pages. 1970. DM 16,-

Vol. 138: D. Foata et M. Schützenberger, Théorie Géométrique des Polynômes Eulériens. V, 94 pages. 1970. DM 16,-

Vol. 139: A. Badrikian, Séminaire sur les Fonctions Aléatoires Linéaires et les Mesures Cylindriques. VII, 221 pages. 1970. DM 18,-

Vol. 140: Lectures in Modern Analysis and Applications II. Edited by C. T. Taam. VI, 119 pages. 1970. DM 16,-

Vol. 141: G. Jameson, Ordered Linear Spaces. XV, 194 pages. 1970. DM 16,-

Vol. 142: K. W. Roggenkamp, Lattices over Orders II. V, 388 pages. 1970. DM 22,-

Vol. 143: K. W. Gruenberg, Cohomological Topics in Group Theory. XIV, 275 pages. 1970. DM 20,-

Vol. 144: Seminar on Differential Equations and Dynamical Systems, II. Edited by J. A. Yorke. VIII, 268 pages. 1970. DM 20,-

Vol. 145: E. J. Dubuc, Kan Extensions in Enriched Category Theory. XVI, 173 pages. 1970. DM 16,-

Please turn over

Vol. 146: A. B. Altman and S. Kleiman, Introduction to Grothendieck Duality Theory. II, 192 pages. 1970. DM 18,–

Vol. 147: D. E. Dobbs, Cech Cohomological Dimensions for Commutative Rings. VI, 176 pages. 1970. DM 16,–

Vol. 148: R. Azencott, Espaces de Poisson des Groupes Localement Compacts. IX, 141 pages. 1970. DM 16,–

Vol. 149: R. G. Swan and E. G. Evans, K-Theory of Finite Groups and Orders. IV, 237 pages. 1970. DM 20,–

Vol. 150: Heyer, Dualität lokalkompakter Gruppen. XIII, 372 Seiten. 1970. DM 20,–

Vol. 151: M. Demazure et A. Grothendieck, Schémas en Groupes I. (SGA 3). XV, 562 pages. 1970. DM 24,–

Vol. 152: M. Demazure et A. Grothendieck, Schémas en Groupes II. (SGA 3). IX, 654 pages. 1970. DM 24,–

Vol. 153: M. Demazure et A. Grothendieck, Schémas en Groupes III. (SGA 3). VIII, 529 pages. 1970. DM 24,–

Vol. 154: A. Lascoux et M. Berger, Variétés Kähleriennes Compactes. VII, 83 pages. 1970. DM 16,–

Vol. 155: Several Complex Variables I, Maryland 1970. Edited by J. Horváth. IV, 214 pages. 1970. DM 18,–

Vol. 156: R. Hartshorne, Ample Subvarieties of Algebraic Varieties. XIV, 256 pages. 1970. DM 20,–

Vol. 157: T. tom Dieck, K. H. Kamps und D. Puppe, Homotopietheorie. VI, 265 Seiten. 1970. DM 20,–

Vol. 158: T. G. Ostrom, Finite Translation Planes. IV. 112 pages. 1970. DM 16,–

Vol. 159: R. Ansorge und R. Hass. Konvergenz von Differenzenverfahren für lineare und nichtlineare Anfangswertaufgaben. VIII, 145 Seiten. 1970. DM 16,–

Vol. 160: L. Sucheston, Constributions to Ergodic Theory and Probability. VII, 277 pages. 1970. DM 20,–

Vol. 161: J. Stasheff, H-Spaces from a Homotopy Point of View. VI, 95 pages. 1970. DM 16,–

Vol. 162: Harish-Chandra and van Dijk, Harmonic Analysis on Reductive p-adic Groups. IV, 125 pages. 1970. DM 16,–

Vol. 163: P. Deligne, Equations Différentielles à Points Singuliers Reguliers. III, 133 pages. 1970. DM 16,–

Vol. 164: J. P. Ferrier, Seminaire sur les Algebres Complètes. II, 69 pages. 1970. DM 16,–

Vol. 165: J. M. Cohen, Stable Homotopy. V, 194 pages. 1970. DM 16,–

Vol. 166: A. J. Silberger, PGL_2 over the p-adics: its Representations, Spherical Functions, and Fourier Analysis. VII, 202 pages. 1970. DM 18,–

Vol. 167: Lavrentiev, Romanov and Vasiliev, Multidimensional Inverse Problems for Differential Equations. V, 59 pages. 1970. DM 16,–

Vol. 168: F. P. Peterson, The Steenrod Algebra and its Applications: A conference to Celebrate N. E. Steenrod's Sixtieth Birthday. VII, 317 pages. 1970. DM 22,–

Vol. 169: M. Raynaud, Anneaux Locaux Henséliens. V, 129 pages. 1970. DM 16,–

Vol. 170: Lectures in Modern Analysis and Applications III. Edited by C. T. Taam. VI, 213 pages. 1970. DM 18,–

Vol. 171: Set-Valued Mappings, Selections and Topological Properties of 2^X. Edited by W. M. Fleischman. X, 110 pages. 1970. DM 16,–

Vol. 172: Y.-T. Siu and G. Trautmann, Gap-Sheaves and Extension of Coherent Analytic Subsheaves. V, 172 pages. 1971. DM 16,–

Vol. 173: J. N. Mordeson and B. Vinograde, Structure of Arbitrary Purely Inseparable Extension Fields. IV, 138 pages. 1970. DM 16,–

Vol. 174: B. Iversen, Linear Determinants with Applications to the Picard Scheme of a Family of Algebraic Curves. VI, 69 pages. 1970. DM 16,–

Vol. 175: M. Brelot, On Topologies and Boundaries in Potential Theory. VI, 176 pages. 1971. DM 18,–

Vol. 176: H. Popp, Fundamentalgruppen algebraischer Mannigfaltigkeiten. IV, 154 Seiten. 1970. DM 16,–

Vol. 177: J. Lambek, Torsion Theories, Additive Semantics and Rings of Quotients. VI, 94 pages. 1971. DM 16,–

Vol. 178: Th. Bröcker und T. tom Dieck, Kobordismentheorie. XVI, 191 Seiten. 1970. DM 18,–

Vol. 179: Seminaire Bourbaki – vol. 1968/69. Exposés 347-363. IV. 295 pages. 1971. DM 22,–

Vol. 180: Séminaire Bourbaki – vol. 1969/70. Exposés 364-381. IV, 310 pages. 1971. DM 22,–

Vol. 181: F. DeMeyer and E. Ingraham, Separable Algebras over Commutative Rings. V, 157 pages. 1971. DM 16,–

Vol. 182: L. D. Baumert. Cyclic Difference Sets. VI, 166 pages. 1971. DM 16,–

Vol. 183: Analytic Theory of Differential Equations. Edited by P. F. Hsieh and A. W. J. Stoddart. VI, 225 pages. 1971. DM 20,–

Vol 184: Symposium on Several Complex Variables, Park City, Utah, 1970. Edited by R. M. Brooks. V, 234 pages. 1971. DM 20,–

Vol. 185: Several Complex Variables II, Maryland 1970. Edited by J. Horváth. III, 287 pages. 1971. DM 24,–

Vol. 186: Recent Trends in Graph Theory. Edited by M. Capobianco/ J. B. Frechen/M. Krolik. VI, 219 pages. 1971. DM 18,–

Vol. 187: H. S. Shapiro, Topics in Approximation Theory. VIII, 275 pages. 1971. DM 22,–

Vol. 188: Symposium on Semantics of Algorithmic Languages. Edited by E. Engeler. VI, 372 pages. 1971. DM 26,–

Vol. 189: A. Weil, Dirichlet Series and Automorphic Forms. V, 164 pages. 1971. DM 16,–

Vol. 190: Martingales. A Report on a Meeting at Oberwolfach, May 17-23, 1970. Edited by H. Dinges. V, 75 pages. 1971. DM 16,–

Vol. 191: Séminaire de Probabilités V. Edited by P. A. Meyer. IV, 372 pages. 1971. DM 26,–

Vol. 192: Proceedings of Liverpool Singularities – Symposium I. Edited by C. T. C. Wall. V, 319 pages. 1971. DM 24,–

Vol. 193: Symposium on the Theory of Numerical Analysis. Edited by J. Ll. Morris. VI, 152 pages. 1971. DM 16,–

Vol. 194: M. Berger, P. Gauduchon et E. Mazet. Le Spectre d'une Variété Riemannienne. VII, 251 pages. 1971. DM 22,–

Vol. 195: Reports of the Midwest Category Seminar V. Edited by J.W. Gray and S. Mac Lane. III, 255 pages. 1971. DM 22,–

Vol. 196: H-spaces – Neuchâtel (Suisse)- Août 1970. Edited by F. Sigrist, V, 156 pages. 1971. DM 16,–

Vol. 197: Manifolds – Amsterdam 1970. Edited by N. H. Kuiper. V, 231 pages. 1971 DM 20,–

Vol. 198: M. Hervé, Analytic and Plurisubharmonic Functions in Finite and Infinite Dimensional Spaces. VI, 90 pages. 1971. DM 16,–

Vol. 199: Ch. J. Mozzochi, On the Pointwise Convergence of Fourier Series. VII, 87 pages. 1971. DM 16,–

Vol. 200: U. Neri, Singular Integrals. VII, 272 pages. 1971. DM 22,–

Vol. 201: J. H. van Lint, Coding Theory. VII, 136 pages. 1971. DM 16,–

Vol. 202: J. Benedetto, Harmonic Analysis on Totally Disconnected Sets. VIII, 261 pages. 1971. DM 22,–

Vol. 203: D. Knutson, Algebraic Spaces. VI, 261 pages. 1971. DM 22,–

Vol. 204: A. Zygmund, Intégrales Singulières. IV, 53 pages. 1971. DM 16,–

Vol. 205: Séminaire Pierre Lelong (Analyse) Année 1970. VI, 243 pages. 1971. DM 20,–

Vol. 206: Symposium on Differential Equations and Dynamical Systems. Edited by D. Chillingworth. XI, 173 pages. 1971. DM 16,–

Vol. 207: L. Bernstein, The Jacobi-Perron Algorithm – Its Theory and Application. IV, 161 pages. 1971. DM 16,–

Vol. 208: A. Grothendieck and J. P. Murre, The Tame Fundamental Group of a Formal Neighbourhood of a Divisor with Normal Crossings on a Scheme. VIII, 133 pages. 1971. DM 16,–

Vol. 209: Proceedings of Liverpool Singularities Symposium II. Edited by C. T. C. Wall. V, 280 pages. 1971. DM 22,–

Vol. 210: M. Eichler, Projective Varieties and Modular Forms. III, 118 pages. 1971. DM 16,–

Vol. 211: Théorie des Matroïdes. Edité par C. P. Bruter. III, 108 pages. 1971. DM 16,–

Vol. 212: B. Scarpellini, Proof Theory and Intuitionistic Systems. VII, 291 pages. 1971. DM 24,–

Vol. 213: H. Hogbe-Nlend, Théorie des Bornologies et Applications. V, 168 pages. 1971. DM 18,–

Vol. 214: M. Smorodinsky, Ergodic Theory, Entropy. V, 64 pages. 1971. DM 16,–